ENCOUNTERING THE NORTH

Encountering the North

Cultural Geography, International Relations
and Northern Landscapes

Edited by

FRANK MÖLLER and SAMU PEHKONEN
*Tampere Peace Research Institute,
University of Tampere, Finland*

Routledge
Taylor & Francis Group

LONDON AND NEW YORK

First published 2003 by Ashgate Publishing

Reissued 2018 by Routledge
2 Park Square, Milton Park, Abingdon, Oxon OX14 4RN
711 Third Avenue, New York, NY 10017, USA

Routledge is an imprint of the Taylor & Francis Group, an informa business

A Library of Congress record exists under LC control number: 2002043965

ISBN 13: 978-1-138-72250-7 (hbk)
ISBN 13: 978-1-138-72246-0 (pbk)
ISBN 13: 978-1-315-19350-2 (ebk)

Contents

List of Plates

List of Figures and Maps

List of Contributors

Ulrich Albrecht is Professor of Political Science and Peace Research at Free University Berlin, Germany.

Jochen Hille is Ph.D. candidate at Humboldt University Berlin, Germany.

Hanna-Mari Ikonen is Researcher in the Department of Regional Studies and Environmental Policy at the University of Tampere, Finland.

Pertti Joenniemi is Senior Research Fellow at Copenhagen Peace Research Institute, Denmark.

Ari Aukusti Lehtinen is Professor of Geography at the University of Joensuu, Finland.

Frank Möller is Research Fellow at Tampere Peace Research Institute, Finland.

Samu Pehkonen is Research Fellow at Tampere Peace Research Institute, Finland.

Juha Ridanpää is Lecturer in the Department of Geography at the University of Oulu, Finland.

Leena Suopajärvi is Lecturer in the Department of Social Sciences at the University of Lapland in Rovaniemi, Finland.

Paulo Susiluoto is Researcher at the Finnish Forest Research Institute in Kolari, Finland.

Monica Tennberg is Researcher in the Department of Social Science at the University of Lapland in Rovaniemi, Finland.

Preface

We are indebted to Tampere Peace Research Institute (TAPRI) for financial and logistic support of the project which resulted in the present volume. Furthermore, TAPRI hosted, in December 2001, an international workshop on Cultural Geography, International Relations and Northern Landscapes, at which occasion the articles commissioned for the book were first presented and discussed. We are in particular grateful to the former and current Research Directors, Tuomo Melasuo and Tarja Väyrynen, for unconditionally supporting the project.

Permission to reproduce maps and plates here is gratefully acknowledged from the following:

The Geographical Society of Finland;
Kenneth Mikko and *N66/Culture in the Barents Region*;
Uppsala University Library;
Stortingsarkivet and Teigen Fotoatelier;
Musée de l'Homme, Paris;
Department of Musical Anthropology, University of Tampere; and
Jorma Puranen.

Our special thanks go to Roland Caldbeck for editing the language of most of the chapters, to Anitta Kynsilehto for putting together the index, to Tarja Hyrkäs for copy-editing and to Anja Reini for organizational help. We are also grateful to the editorial staff at Ashgate, in particular Valerie Rose, Sarah Horsley, Rachel Keane and Adrian Shanks, and the anonymous referees of our initial proposal.

Most importantly we would like to thank the authors for their free-flowing creativity and generous willingness to share their ideas with us, for delivering their chapters within the given time limit, for their open-minded response to our comments and suggestions and for their enthusiastic participation in the whole project.

Chapter 1

Discursive Landscapes of the European North

Frank Möller and Samu Pehkonen[1]

This volume is concerned with the European North above the polar circle and its representations in Cultural Geography and International Relations. The chapters in the book deal with cultural, geographical and political imaginations of Northern peoples and landscapes. These imaginations are not treated independently of one another. Rather, emphasis is put on the triangle of and interrelationship between culture, geography and politics. It is aspired to make self-imposed, thought-paralyzing academic boundaries permeable and to replace mutual neglect with open-mindedness and cognitive curiosity.

Yet, a synthesis of Cultural Geography and International Relations is not intended. Likewise, we are not interested in a binding definition of the North, *i.e.* a precise statement of its essential culture and nature, which we consider neither possible nor desirable. The simple working definition which we have suggested in the first sentence of the chapter, on the one hand, reflects a standard approach to the North which is as arbitrary as any other approach. On the other hand, it suggests, by its seductive simplicity, to ignore its inherent geographical limitation and to think beyond it. Pretending to know the 'real' meaning of a region inevitably means reducing complexity rather than working with it. Indeed, definitions of regions always mean territorial limitations, and territorial limitations tend to result in cognitive restrictions. Inquiries into and academic obsession with the *limits* of the North, rather than its *content*, exemplify this limitation.[2]

Regions do not exist in isolation. Geographical or political mapping of a particular region, clearly distinct from neighbouring regions, is bound to fail when

---

[1] We would like to thank Anne Buttimer, Jouni Häkli and Michael Shapiro for careful reading and critical comments of previous drafts of this chapter.

[2] Rob Shields, *Places on the Margin. Alternative Geographies of Modernity* (London: Routledge, 1991), pp. 167–72, discusses this aspect in the context of the Canadian Arctic.

1

one is interested in the social world. Regions, however defined, and adjacent areas
shape each other. People gain information from beyond their own immediate area
and incorporate this into their view on both their own locality and beyond. As
Yosef Lapid has argued with respect to 'culture', the challenge "is not to push
energetically to some consensual but arbitrary reduction, but to reflectively match
suitable definitional assets to declared theoretical missions."[3] The same may be
said with respect to the North. For this reason, the geographical definition
introduced above is treated rather generously by some of the authors in this volume
in accordance with their theoretical mission. Indeed, as Raymond Williams has
emphasized, complexity is not to be found in a particular word, but rather "in the
problems which its variations of use significantly indicate."[4] The historical and
contemporary variations of meaning assigned to the North point to real processes
which have to be studied in their own terms. To meet this goal, our approach to
'the discursive landscapes of the North' does not plainly denote the sites and levels
of discourses (be they academic, political or popular) but it takes into account the
material circumstances making the context of the European North.[5]

Meaning assigned to the North is subject to change. It varies across actors and
over time. Not surprisingly, then, the authors in this book variably depict the
North as different, as an oxymoron, as a nothing, as the periphery pure and
simple, as a sign of utmost peripherality, but also as positively-valued marginality,
as an example of an alternative way of life and so on. Yi-Fu Tuan adds that the
North may be a "homeplace". For example, the igloo is said to be a homeplace for
indigenous people, "a protected – at least partly enclosed – space" marked by "a
variegated world of shapes and colours, sounds and odours". Likewise, the
prefabricated, multi-storey buildings dominating the cityscapes of, for example,
Murmansk may be a homeplace for the dwellers. Yet, Tuan's question "Why
would anyone want to go there?" implicitly expresses scepticism about the
possibility of the North being perceived as "homeplace"[6] – while ignoring at the

---

3   Yosef Lapid, 'Culture's Ship: Returns and Departures in International Relations
     Theory,' in Yosef Lapid and Friedrich Kratochwil, eds., *The Return of Culture and
     Identity in IR Theory* (Boulder and London: Lynne Rienner, 1996), p. 7.

4   Raymond Williams, *Keywords: A Vocabulary of Culture and Society* (London: Fontana
     Press, 1988), p. 92.

5   For the elaboration of "discursive landscape" as a theoretical concept, see Jouni Häkli,
     'Cultures of Demarcation: Territory and National Identity in Finland,' in Guntram Herb
     and David Kaplan, eds., *Nested Identities: Identity, Territory, and Scale* (Lanham:
     Rowman & Littlefield, 1999), pp. 123–49.

6   Yi-Fu Tuan, 'Desert and Ice: Ambivalent Aesthetics,' in Thomas Trummer, ed., *The
     Waste Land: Desert and Ice. Barren Landscapes in Photography* (Vienna: edition selene,
     2001), pp. 69 and 83.

same time that for many people the issue is not one of going there, but one of being there. Can people who feel at home in the North be fully human? The European history of ideas gives a clear answer by having invented the "monstrous races" to "illustrate theories of the influence of climate, the assumption being that people who live in places which are too cold or too hot cannot be fully human."[7] Explorers from the continent were partly amazed and partly horrified by the stories they were told about the man-eating monsters of Sámi mythology.[8] To the visitors' amusement, Sámi people (then called 'Lapps') were displayed in continental zoological gardens in the late 19th century – with apparent success: on Sundays, more than 40,000 visitors came to watch them (and other 'exotic' peoples).[9] At that time, as an element of Northern nation-state building projects, Sámi people became otherized in their own homelands, too, dispossessed of human dignity.

Only occasionally, European intellectual history has considered permanently residing in the North admirable and worthy of imitation.[10] Respect for Northern peoples has been the exception rather than the rule. Arriving at any cost is a frequent motive and an end in itself underlying explorations and journeys to the North just as conquering and 'modernizing' the North and 'civilizing' its people. Soviet propaganda, for example, celebrated polar explorations as "milestones along the path of the Stalinist express."[11] The novelist Andreï Makine, in *Requiem for the East*, resumes the Stalinist theme, but transfers it from the level of state propaganda to the individual. One of his protagonists narrates that children whose fathers had fallen victim to one of the numerous purges, in order to avoid being stigmatized as "children of parents fallen from grace, [...] invented fathers for themselves who were polar explorers trapped by ice". This motive – polar explorers trapped by ice – is a frequent motive in literature, albeit only seldom in as sympathetic and favourable a manner as in Juri Rytchëu's novel about Roald Amundsen's deliberate passing the winter of 1918 at the shore of Chukchis and his

---

7   Peter Burke, *Eyewitnessing: The Uses of Images as Historical Evidence* (London: Reaktion Books, 2001), p. 127.

8   See Xavier Marmier, *Pohjoinen maa: 1800-luvun Lappia ja Suomea ranskalaisen silmin* (Helsinki: Suomalaisen Kirjallisuuden Seura, 1999).

9   Stefan Goldmann, 'Wilde in Europa: Aspekte und Orte ihrer Zurschaustellung,' in Thomas Theye, ed., *Wir und die Wilden. Einblicke in eine kannibalische Beziehung* (Reinbek: Rowohlt, 1985), pp. 256–7.

10   See, for example, Jean Malaurie's anthropogeographical approach to the North in Malaurie (in conversation with Jan Borm), 'Walrossuppe, Seehundblut. Die Kultur der Inuit – ein Ethnographenleben im ewigen Eis,' *Lettre International*, No. 56 (Spring 2002), pp. 80–90.

11   Aileen Kelly, 'In the Promised Land,' *The New York Review of Books*, Vol. XLVIII, No. 19 (29 November 2001), p. 48.

encounters with the indigenous people.[12]

On one hand, there is a danger of romanticizing and mythologizing Northern lands and people. In particular photographic representations of Northern lands and people as well as explorers' and travel literature are susceptible to such a danger and will be discussed in chapters three and six, respectively. On the other, there obviously are also opposite representations of the North as a site of major social, economic and ecological problems. In many parts of the circumpolar North, the post-World War II period has seen people moving away from their home villages, abandoning the traditional way of life and facing the dark side of becoming part of the world economy. State bureaucrats in planning boards addressed the North mainly in terms of social engineering and taming nature. Based on the apparent success of the Nordic welfare system and indicating the optimistic belief of the time in modernization and planned change, the 'primitive' North of dark and cold winters was pushed still farther northwards:

> Almost every year the scenery is changing in Lapland. Land is cleared, new settlements are built, houses grow bigger and wealthier. The old greyish huts have almost vanished. Houses are white, green and red in colour, fields and meadows seem well taken care of. Tractors and other vehicles are of a modern type. Settlement reaches farther North.[13]

Today, the incorporation of the North into global economic structures is to some extent equated with unemployment, poor living conditions or social networks shifting away from those based on family and work towards those influenced by drugs and alcohol. Not just human beings feel deprived of a decent way of life: the northern nature is confronted with threats emanating from collapsing industrial complexes, nuclear plants and submarines. In the Northern context, the academic debate of the centres on sustainable development becomes an irony: many of the industrial communities of the North were not even built for a permanent use in the first place, but only to help capitalism, at a particular moment of its history, turn the rich natural resources to profit. Industrial complexes in the North thus serve as a powerful illustration of what Henri Lefebvre calls "the production of space", *i.e.* that capitalism's survival in the twentieth century was only possible by

---

[12] Andreï Makine, *Requiem for the East* (London: Sceptre, 2001), p. 90; Juri Rytchëu, *Die Suche nach der letzten Zahl* (Zurich: Unionsverlag, 2001).

[13] Ragnar Numelin, 'Voittamaton Ruija: Pohjois-Norjan jälleenrakentaminen,' *Terra*, Vol. 73, No. 2 (1961), p. 87 (authors' translation). For the belief in the possibilities of modernizing the North by means of agriculture or fishing industry, see Uuno Varjo, 'Agriculture in North Lapland, Finland: Profitableness and Trends since World War II,' *Fennia*, Vol. 132 (1974), pp. 1–73.

producing space. They also illustrate the painfulness of the encounter with capitalism on the part of those living in that space because capitalism inevitably has "to destroy and rebuild that geographical landscape to accommodate accumulation at a later date."[14] Without doubt, the same may be said with respect to the Soviet, socialist approach to the Northern lands and resources, modelled after the capitalist scheme.

The remainder of the chapter is organized as follows. The section on militarized geographies of the North reviews international politics in the North during the Cold War and its analysis – and construction – in the discipline of International Relations (IR). We claim that this part is not only of historical and disciplinary interest. Rather, it exemplifies a structure of political, economic and scientific domination – a specific way to see the North – which, to some extent, lingers on even after the end of the Cold War. After all, the North remains one the most militarized regions of the world; strategists and nuclear planners continue doing 'business as usual'. The North continues to be seen as a testing-ground for theories developed in the South and social engineering. Data, collected in the North, provides the South with information that contributes to what Johan Galtung, more than thirty years ago, has called "scientific imperialism". In an era of almost ubiquitous territorial, political and ideological decolonization as well as corresponding academic discourses on postcolonialism, this term might seem to be slightly covered with dust. Yet the concept may still help "economically, politically, or militarily to maintain an imperialist structure"[15] – quite regardless of whether or not it is still called 'imperialism': elements of imperialism may survive even in an age of postcolonialism.

In the part on cultural turns and limits, we introduce the reader into the cultural turns in both IR and Cultural Geography, the one being a response to developments in international politics, the other reflecting a long-standing intra-disciplinary conflict over the concept of culture and the relationship between culture and society. Although the authors of the present volume are sympathizing with the cultural turn in both disciplines, we are not intending its idealization. Rather, a benevolent critique of the cultural turn leads us to the chapter's concluding section. Here, we sketch multiple views on the North including Northern scholars' deliberate neglect of methodological approaches composed in the centre, the re-contextualization of hegemonic academic discussions and the development of approaches from starting-points usually ignored in Anglo-Saxon social sciences.

---

[14] See Henri Lefebvre, *The Production of Space* (Oxford: Blackwell, 1991) and, for the quotation, David Harvey, *Spaces of Hope* (Berkeley and Los Angeles: University of California Press, 2000), p. 59.

[15] Johan Galtung, 'A Structural Theory of Imperialism,' *Journal of Peace Research*, Vol. 13, No. 2 (1971), p. 93.

Militarized Geographies of the North

Meaning assigned to the North varies across actors and over time. During the Cold War, however, political representations of the North were remarkably stable. In particular the superpowers did not look *at* the North but rather *past* the North at each other, using the North and its physical environment as a vehicle with which to improve their position *vis-à-vis* the rival. The North was subordinated to the national security interests of the superpowers, defined in terms of the accumulation of both military and economic power. Given the basic lines of thinking about, and the importance assigned to, 'security' at the time, addressing the North in terms of 'security' inevitably resulted in its militarization. This militarization unfolded in both a *material* sense – as a steady growth in the military potential concentrated in the area – and in a *cognitive* sense: 'security' became the prevalent way the North was dealt with, analyzed and, in part, constructed by academic and political security experts. This focus had consequences beyond the narrow margins of the security establishment. People in the North became pawns on the chessboard of superpower strategists, systematically left in the dark about, and excluded from, decision-making processes and resulting military policies, the consequences of which they nevertheless had to bear. Their security interests as individuals or groups of people mattered little.

The militarized North may also be seen as that which, during the Cold War, rendered a discursive approach to the North (a negotiation and juxtaposition of different North-views) impossible. The Cold War belief system was simple. "Differences were largely defined in ideological terms, and were superimposed upon world politics regardless of local cultural characteristics."[16] Threat perceptions and enemy images seemed to be clear, the belief in the appropriateness of one's own beliefs was imperturbable. A Manichæan distinction between 'good' and 'evil', 'friend' and 'foe' prevailed, neutral positions of those in-between, simply aspiring to be left in peace, were eyed with suspicion. Security and power, militarily, economically and increasingly technologically, were treated synonymously. The belief in nuclear power as a means to security seemed unshakable. Policies of nuclear deterrence used their own linguistic symbols and followed their own obscene logic, neither intelligible to non-'experts' nor successfully challenged from within the security establishment. An alternative look at both security and the North was neither encouraged by the political establishment nor, given the conformity pressure of the time, thought advisable and useful to one's career.

---

[16] Simon Murden, 'Cultural Conflict in International Relations: The West and Islam,' in John Baylis and Steve Smith, eds., *The Globalization of World Politics: An Introduction to International Relations* (Oxford: Oxford University Press, 1997), p. 377.

This section reviews international politics in the North during the Cold War and its analysis – and construction – in the discipline of International Relations (IR). We start the historical review with one of the most dramatic events in the North during the Cold War, described by Scott Sagan as follows:

ON THE AFTERNOON of January 21, 1968, a Strategic Air Command B-52 bomber on an airborne alert mission was orbiting thirty-three thousand feet above the Ballistic Missile Early Warning System (BMEWS) radar at Thule, Greenland. At around 3:30 P.M., the copilot turned the cabin heater dial to its maximum heat in order to combat the arctic cold. A few minutes later, one crew member reported that he could detect the smell of burning rubber. The crew searched the aircraft and discovered a small fire in the rear of the lower cabin. The flames quickly grew out of control, dense smoke rendered the flight instruments unreadable, and, seven minutes later, all electrical power was lost. The pilot immediately gave orders to evacuate the plan. Six of the crew members successfully ejected and landed safely in the snow. The seventh was killed.

The B-52 also had four 1.1 megaton thermonuclear bombs on board. The pilotless aircraft continued descending, passed directly above the Thule base, and then crashed into the ice seven miles away. The speed of the plane at impact was in excess of five hundred miles per hour, and 225,000 pounds of jet fuel immediately exploded. The conventional high explosives in all four of the nuclear bombs went off. No nuclear detonation occurred, but radioactive debris was dispersed over a wide expanse of ice.[17]

The 1968 Thule B-52 bomber accident serves well as a starting-point for sketching the approach to the North prevalent during the Cold War. Established in 1951 under total secrecy, the US military basis in Thule indicated the transition from "a period of discovery to a period of occupation",[18] both periods characterized by a profound neglect of the interests of indigenous people. Likewise, the discipline of IR has never been interested in the North *per se*. Rather, it has considered the North worthy of consideration mainly as background to the superpower rivalry. The North as such mattered little, even less the peoples of the North and their needs and desires. Here, the academic discipline of IR and international politics joined hands. Cold War politics were theoretically confirmed rather than critically reflected. In particular Realism, the leading school of thought after World War II, theorized and helped reproduce the political *status quo* without aspiring to change it. Realism rationalized, and became a legitimation theory for,

---

[17] Scott D. Sagan, *The Limits of Safety: Organizations, Accidents and Nuclear Weapons* (Princeton: Princeton University Press, 1993), p. 156.

[18] See Malaurie, 'Walrossuppe, Seehundblut,' p. 87.

Cold War politics.[19]

The story of the B-52 bomber accident, its interpretations and possible consequences is also an example of how easily the Cold War could have become a nuclear war had some ingredients and perceptions of the accident only been slightly different.[20] The crash, occurring just one day before parliamentary elections in Denmark, caused both a crisis in Danish-US relations, invisible to the public, and a domestic political crisis in Denmark, thus exemplifying the intimate relationship between international and security policies on one hand, domestic politics on the other.

As early as 1957, then Danish Prime Minister H.C. Hansen, under considerable pressure from his own party and public opinion being in a non-nuclear mood, seems to have preferred to close his eyes over the possibility of US deployment of nuclear weapons to Greenland, an integral part of Denmark since 1953. Answering to a US inquiry of 13 November 1957 "whether the Danish Government would want to be informed in case the United States stored nuclear weapons in Greenland", the Prime Minister had neither protested against possible deployment nor wished to be informed about the actual deployment of nuclear weapons.[21] This vagueness was in line with the Danish-US agreement for the 'defence of Greenland' of 1951, neither explicitly permitting nor prohibiting nuclear weapons in Greenland, thus leaving considerable space for interpretation.[22] Hansen's renunciation of insisting on being informed – perceived by the US government as a wish *not* to be informed and as a green light for deployment – revealed an élitist understanding of security policy: neither his government partners, the parliamentary foreign policy committee, parliament as such nor the public were informed. In public the Danish government claimed the non-deployment of nuclear weapons to Greenland from June 1961 onwards. Yet only after the B-52 incident the general Danish ban on nuclear weapons was officially extended to cover Greenland (and acknowledged by the US government in May 1968). Prior to that, Denmark had *de facto* two nuclear policies, the one with respect to Southern

---

[19]   See Ulrich Albrecht, *Internationale Politik. Einführung in das System internationaler Herrschaft* (Munich/Vienna: Oldenbourg, 1986), p. 33; Stanley Hoffmann, 'An American Social Science: International Relations,' in James Der Derian, ed., *International Theory: Critical Investigations* (Houndmills and London: Macmillan, 1995), p. 223.

[20]   See Scott Sagan's brilliant discussion in *The Limits of Safety*, pp. 156–203.

[21]   See Nikolaj Petersen, 'The H.C. Hansen Paper and Nuclear Weapons in Greenland,' *Scandinavian Journal of History*, Vol. 23, Nos. 1–2 (June 1998), p. 21.

[22]   Hans Mouritzen, 'Thule and Theory: Democracy vs. Elitism in Danish Foreign Policy,' in Bertel Heurlin and Hans Mouritzen, eds., *Danish Foreign Policy Yearbook 1998* (Copenhagen: Danish Institute of International Affairs, 1998), pp. 81–2.

Denmark (non-nuclear status), the other covering Greenland (tacit acquiescence in US deployment of nuclear weapons). The B-52 accident and military policies in the North in general, which resulted in the accident, are important cases for IR analysis. It raises questions pertaining to international relations theory, nuclear weapons, military and security policy in general, the domestic dimension of foreign and security policy, the degree of democracy underlying foreign and security policy decision-making and implementation,[23] legal and moral issues, majority/minority relations (Copenhagen/Home Rule),[24] weak power (Denmark)/strong power (United States) relations[25] and so on. Yet, most of these issues are absent from IR analysis. Conventionally, the discipline has dealt with Northern territories and waters in the light of military security and strategic issues reflecting national interests as defined by national decision-makers. It has largely been a state-centred analysis, equating security with availability and state control of military means and following a Realist reading of the world in terms of balance of power, strategic equilibrium, second-strike capability, nuclear deterrence, military defensibility and material capabilities.[26]

The selective blindness of IR analysis, indicative of its prevalent identification "with representing the interests of the status quo powers in the international system",[27] is not surprising. The institutionalization of the discipline of International Relations was a product of war. 'Born' in the aftermath of World War I, IR reflected the universalization of inter-societal conflicts, which had lately erupted

---

[23]   Mouritzen is discussing this aspect in *ibid.*, pp. 79–101.

[24]   For this issue, see Nikolaj Petersen, 'Denmark, Greenland, and Arctic Security,' in Kari Möttölä, ed., *The Arctic Challenge: Nordic and Canadian Approaches to Security and Cooperation in an Emerging International Region* (Boulder and London: Westview, 1988), pp. 60–68.

[25]   For Denmark as a "weak power", see Barry Buzan, *People, States & Fear: An Agenda for International Security Studies in the Post-Cold War Era. Second Edition* (Hemel Hempstead: Harvester Wheatsheaf, 1991), p. 114.

[26]   See, for example, Marian K. Leighton, *The Soviet Threat to NATO's Northern Flank* (New York: National Strategy Information Center, 1979); Clive Archer and David Scrivener, eds., *Northern Waters: Security and Resource Issues* (London: Croom Helm, 1986); Walter Goldstein, ed., *Clash in the North: Polar Summitry and NATO's Northern Flank* (Washington et al.: Pergamon-Brassey's, 1988); Falk Bomsdorf, *Sicherheit im Norden Europas. Die Sicherheitspolitik der fünf nordischen Staaten und die Nordeuropapolitik der Sowjetunion* (Baden-Baden: Nomos, 1989); Ola Tunander, *Cold Water Politics: The Maritime Strategy and Geopolitics of the Northern Front* (London, Newbury Park, New Delhi: Sage, 1989).

[27]   Martin Hollis and Steve Smith, *Explaining and Understanding International Relations* (Oxford: Clarendon Press, 1991), p. 20.

10 *Encountering the North*

in war, and policy-makers' desire to learn from the past in order to prevent mistakes and misperceptions from recurring. The discipline's history to some extent determined both the issues IR scholars were interested in and their emphasis on superpower relations. It did not, however, pre-determine the discipline's focus on military security, armament, balances of power, material factors and, increasingly, nuclear deterrence which resulted from Realism's post-World War II triumph over competitive ways of thinking about international relations.

Realism has been labelled "'the power-politics model', because of its stress on the power-political situation of a state as the central determinant of its interests."[28] Yet, more important is *how power is constituted.*"[29] According to Realists, power is an expression of material (military, economic, technological) capabilities of states. Structural Realism in particular is almost exclusively concerned with great powers, since small powers are said to lack the material capabilities to crucially affect the structure of the international system (in which Structural Realism is primarily interested).[30] It is thus consistent that the North became a relevant subject for IR analysis only in connection with its incorporation as potential theatre of war into the power-political confrontation of the Cold War. Reflecting technological and strategic developments, the North increased in significance as an area of superpower rivalry, beginning in the mid-1950s and culminating in the mid-1980s. The Soviet military build-up on the Kola peninsula and the US Maritime Strategy transformed the Arctic into one of the world's most active and important areas of military operations. The Arctic emerged as "a strategic arena of vital significance to both of the superpowers."[31]

Yet, the B-52 accident also shows that the North was becoming a militarized and nuclearized region long before, in the mid-1980s scholarly attention started to reflect this development. Like the United States, the Soviet leadership – after having achieved a genuine second-strike capability and decided, in 1966, to create a bastion for strategic submarines – directed increased attention on northern waters. While the US commitment in the North declined in the 1960s and early 1970s, the Soviet Union deployed Delta submarines in the Greenland Sea and the Arctic Ocean. Reflecting, among other things, the relative proximity to the decision-

---

[28] *Ibid.*, p. 27.

[29] Alexander Wendt, *Social Theory of International Politics* (Cambridge: Cambridge University Press, 1999), p. 97.

[30] See Kenneth N. Waltz, *Theory of International Politics* (Reading: Addison-Wesley, 1979).

[31] Oran Young, 'The Age of the Arctic,' *Foreign Policy*, No. 61 (Winter 1985–1986), p. 160. See also Gennady P. Luzin, Michael Pretes and Vladimir V. Vasiliev, 'The Kola Peninsula: Geography, History and Resources,' *Arctic*, Vol. 47, No. 1 (March 1994), pp. 1–15.

making centre in Moscow and the protection provided by the Arctic ice, the bases on the Kola peninsula were, throughout the 1980s, constantly favoured above the Kuril Islands. This, in turn, seems to have been one of the reasons for the significant increase in US interest in northern Europe in the early 1980s.[32] The Reagan administration's "emphasis on unilateral American pursuit of security" included a "call to stimulate an arms race with emphasis on high technology in new areas of competition where the United States had a lead".[33] One of the places where this arms race took place was the North. Ola Tunander describes the situation in the 1980s as follows:

> The general international tension and the Reagan administration's naval investment programme and ambitious rhetoric for Northern Europe have been followed by more provocative Soviet operations in the Nordic region, a continued re-equipping and modernization of the Soviet Northern Fleet and apparently more provocative intrusions into Swedish territorial waters.[34]

Most of these developments must be seen in the context of the global superpower antagonism. They were not directed against the Nordic states or at any one specific Nordic state. For example, the Soviet Union's activities in the North were thought of in terms of second-strike capability reflecting the superpower polarity; they were pointed primarily at the United States rather than at the Nordic states. But, of course, these developments affected international relations in Northern Europe and the way the discipline saw, analyzed and helped construct them during the Cold War.

Yet, the Cold War has come to an end in the North. Although the Kola peninsula remains one of the world's most militarized regions, perceptions and meanings assigned to (dangers emanating from) the militarized North have changed during the 1990s. For example, while the Northern Fleet has formerly been seen as a formidable military threat, the fleet – and in particular its about 100 decommissioned nuclear submarines, most of them with reactors still onboard – is perceived nowadays primarily as an equally formidable threat to health and environment. This insight paved the way for international cooperation to an extent inconceivable during the Cold War.[35] Problems, such as environmental ones, are

---

[32] Tunander, *Cold Water Politics*, pp. 28–33.

[33] Raymond L. Garthoff, *The Great Transition: American-Soviet Relations and the End of the Cold War* (Washington: The Brookings Institution, 1994), p. 508.

[34] Tunander, *Cold Water Politics*, pp. 31–2.

[35] See, for example, Steven G. Sawhill, 'Cleaning-up the Arctic's Cold War Legacy: Nuclear Waste and Arctic Military Environmental Cooperation,' *Cooperation and Conflict*, Vol. 35, No. 1 (March 2000), pp. 5–36.

increasingly perceived as *common* problems, necessitating joint strategies to deal with them. Borders in the North are seen to some extent as vehicles with which to tie communities together; cross-border cooperation flourishes. The Arctic region, up to fairly recently "a veritable tundra as far as international cooperation is concerned",[36] became formally institutionalized in 1993 with the inauguration of the Barents Euro-Arctic Region.

Of course, the restructuring of international politics in the North has not been without conflict. It has often been said that Barents cooperation, although being a 'top-down' initiative, should be rooted in the region. The involvement of representatives from the counties and *oblasti* in a separate body in the Barents Euro-Arctic Council exemplifies this ambition. Some Sámi representatives, however, displayed a rather cool attitude with respect to the Barents initiative, interpreting it as "one stage of continued colonization."[37] Barents cooperation has once even been called "a counter-strategy of the nation-states against the rising civil society."[38] Region-builders are said to have underestimated "the way in which the representational practices they have utilized in order to promote change have, in many respects, only served to reinscribe the very world they have sought to transform."[39] Pertti Joenniemi, while tapping into the European Union's Northern Dimension (ND), concedes in chapter eleven of this volume that what he calls the "neo-North" may not be appreciated wholeheartedly in the High North because different cultural readings of northernness may be incompatible. Nevertheless, says Joenniemi, the Northern Dimension, by offering an alternative representational frame, has helped northernness to gain an accepted standing within the confines of the EU as a site of power (without having to result in something societally tangible in order to count). Even in the absence of some sort of functionalist argument, the ND is important as a field of communication, a ground for genuine dialogue and an in-between space which helps to give voice to formerly suppressed actors. The ND, in Joenniemi's view, is a move away from passively accepting whatever the traditional centre defines as reality; it may be a path to self-liberation without challenging the centre in its core and thus provoking resistance. The Northern Dimension, thus, may be seen as an antidote against naturalized

---

[36] Clive Archer, 'Arctic Cooperation: A Nordic Model,' *Bulletin of Peace Proposal*, Vol. 21, No. 2 (June 1990), p. 165.

[37] Elina Helander, 'The Status of the Sámi People in the Inter-State Cooperation,' in Jyrki Käkönen, ed., *Dreaming of the Barents Region: Interpreting Cooperation in the Euro-Arctic Rim* (Tampere: Tampere Peace Research Institute, 1996), p. 299.

[38] Jyrki Käkönen, 'North Calotte as a Political Actor,' in *ibid.*, p. 75.

[39] Christopher Browning, *The Region-Building Approach Revisited: The Continued Othering of Russia in Discourses of Region-Building in the European North* (Copenhagen: Copenhagen Peace Research Institute, 2001), p. 4.

orders of knowledge.

Jochen Hille, for his part, suggests a way to break with these orders outside of, but inseparably linked with, the European Union: turning the maps of Europe around and using alternative indicators in order to determine centre and periphery (chapter eight). Exploring northern Norwegian resistance to EU integration, Hille departs from both an actor-oriented and a socio-economic interpretation: both approaches tend to ignore the underlying conceptions and conditions which generate collective interests and give social movements meaning. Rather, Hille argues, in order to explain the stability of northern Norwegian euroscepticism, the relationship between the concepts of northernness and European integration has to be taken into consideration, in particular the construction of the High North as, in Hille's words, "an alternative society to the 'central European society'", as the economically strong and politically stable "locus of 'the good life'" based on common identities and emotional attachment to the North. Indeed, even among those who, for mainly rational, functionalist reasons, are in favour of EU integration, emotional support for the EU is weak. In addition, the intra-Norwegian centre-periphery conflict is projected onto the relationship between Norway and the EU and the distance to power ('Brussels') is seen to contradict the traditional penchant for decision-making close to the people. Finally, prominent concepts of European integration like 'core Europe' and the 'model of gravity' support the image of northern Norway as a periphery which, in turn, strengthens both the resentment against integration and the northern Norwegian (self-)construction as antipole to European integration.

Cultural Turns and Limits

The end of bipolarity in the late 1980s and early 1990s was followed by critical (and sometimes not so critical) scholarly self-appraisal and the reformulation of international relations in academic writings. The way the Cold War came to an end shattered Realism's leading position in International Relations.[40] It brought to the fore constructivist and culturalist approaches which deviate from Realism by, among other things, claiming that the structures of international relations are social rather than material and that immaterial and cultural factors like ideas are

---

[40] See Richard Ned Lebow, 'The Long Peace, the End of the Cold War, and the Failure of Realism,' in Richard Ned Lebow and Thomas Risse-Kappen, eds., *International Relations Theory and the End of the Cold War* (New York: Columbia University Press, 1995), pp. 23–56.

constitutive of power and interests.[41] Like Realism, alternative approaches emphasize to some degree studying international security because much in world politics seems to be about 'security'. The intimate relationship between parts of International Relations and Security Studies is therefore not surprising. After all, in world politics the search for 'security' appears to be "universal" and 'security' is said to be "the highest end" in an international system characterized by the absence of central government and the presence of nation-states.[42] The form of modern international politics is, to a large extent, expressed and shaped in contemporary security discourses. As a by-product (or is it actually the main product?) most security discourses tend to affirm the modern nation-state *per se* and as prime subject of security, thus marginalizing alternative subjects of security and legitimating traditional security policies.[43]

Security discourses are currently displaying a variety of cultural features – and not for the first time. It thus seems to be appropriate to speak about a return of 'culture' or, better, 'cultures' in international relations and security studies. Michael Desch identifies at least two cultural approaches to 'security' prior to what he calls the "post-Cold War wave of culturalism in security studies". The World War II wave of discussions of how to deal with the Axis powers was strongly cultural or anthropological in the sense that ostensible 'national characters' of the Axis powers were linked with actual or expected conduct of war – with debatable results. During the Cold War wave of cultural approaches, US theories of warfare and strategy were said to largely ignore cultural characteristics of the Soviet enemy and to incorrectly expect the enemy to have the same understanding of rational decision-making as oneself supposedly had. Differences in strategic and organizational cultures were referred to in order to explain differences between the United States and the Soviet Union in preparing for and actually fighting war. Strategic cultural theorizing also included speculations about contrasting American and Soviet political cultures leading to different attitudes to conflict and international

---

[41] For a concise statement, see Alexander Wendt, 'Constructing International Politics,' *International Security*, Vol. 20, No. 1 (Summer 1995), pp. 71–81.

[42] For the quotations, see respectively Kalevi J. Holsti, *International Politics: A Framework for Analysis. Seventh Edition* (Englewood Cliffs: Prentice Hall, 1995), p. 85, and Waltz, *Theory of International Politics*, p. 126. 'Security' is put in inverted commas because, like 'culture', the word means different things to different people.

[43] See R.B.J. Walker, 'The Subject of Security,' in Keith Krause and Michael C. Williams, eds., *Critical Security Studies: Concepts and Cases* (London: UCL Press, 1997), pp. 61–81.

relations.[44] Later, US foreign policy analysis began to question the rationalism ostensibly underlying foreign policy decision-making in the United States by paying more attention than before to bureaucratic cultures, traditions within decision-making groups and game-playing among decision-makers and bureaucrats.[45]

Similar to Desch, Fred Halliday calls into question the novelty of many recent cultural writings in international theory. He hears "many echoes of earlier thinking in the current discussions of culture", including the *Annales school*, regarding history as medium and long-term events, among others cultural ones; the *English School* in International Relations, arguing for a view according to which a society of states is based on shared values and assumptions; or, by contrast, the prevailing 19th century view of international relations "in which culture played a central part not to promote interdependence, but as part of an explicitly *unequal and warlike* system – the right, and duty, of the white races, superior to all others, to impose their ways on the rest of the world."[46] Seeing the world through cultural eyes therefore is by no means equivalent with a benevolent and peaceful worldview.[47]

According to Desch, significantly different and potentially contradictory concepts of culture are nowadays being pursued simultaneously in IR and security studies.[48] This is not at all surprising first of all given the complexity of the term's meanings. Furthermore, 'culture' as used in International Relations is not different from other key words in the sense that the meaning of many of them is disputed. This cannot be otherwise because the meaning assigned to a word reflects political processes, including precisely the power to give words a particular, binding meaning and make others accept this meaning and take it for granted. The power of naming and the performance of hegemony are inseparable. Struggles over the meaning of words are at least struggles for scientific hegemony. Given the close relationship between the academic and the political foreign and security

---

[44] Michael C. Desch, 'Culture Clash: Assessing the Importance of Ideas in Security Studies,' *International Security*, Vol. 23, No. 1 (Summer 1998), pp. 141–8 (see p. 142 for the quotation). See also Lapid, 'Culture's Ship,' pp. 5–6.

[45] See Brian Ripley, 'Cognition, Culture, and Bureaucratic Politics,' in Laura Neack, Jeanne A.K. Hey and Patrick J. Haney, eds., *Foreign Policy Analysis: Continuity and Change in Its Second Generation* (Englewood Cliffs: Prentice Hall, 1995), pp. 85–97. Also in this regard, Sagan's analysis of *The Limits of Safety* is brilliant – and alarming since it reveals the incapacity of learning inherent in US strategic culture.

[46] Fred Halliday, 'Culture and International Relations: A New Reductionism?' in Michi Ebata and Beverly Neufeld, eds., *Confronting the Political in International Relations* (London: Macmillan, 2000), pp. 53–6 (the quotations are from p. 55).

[47] See Samuel Huntington's notorious 'The Clash of Civilizations?' *Foreign Affairs*, Vol. 72, No. 3 (Summer 1993), pp. 22–49.

[48] Desch, 'Culture Clash,' p. 152.

policy establishment, these struggles inevitably have consequences beyond the academic realm. In Ken Booth's words, to the extent that security studies has political influence, the current debate, reflecting "agreement that security is crucial, but disagreement about what security is and how it should be studied", is "part of a struggle over the next set of worldviews of Western opinion."[49]

The "oversupply of *potentially rewarding* definitions"[50] of 'culture' is neither unpolitical nor a problem. Just the opposite: definitions such as the one suggested by Desch – "collectively held ideas that do not vary in the face of environmental or structural changes" – or the somewhat colloquial "the way things are done around here" referred to by Brian Ripley; Alexander Wendt's "cultures of anarchy" in the international system, reflecting particular knowledge on the use of organized violence prevalent in a particular system, or Peter Katzenstein's sociological focus on culture, norms and identity and explanatory power assigned to them with regard to national security policies independent of the distribution of material capabilities are useful in the light of a specific research interest and declared theoretical mission.[51] Furthermore, Fred Halliday's attempt to departmentalize culture without atomizing it is helpful. Halliday has suggested four different meanings of culture:

> culture as 'artistic culture', *i.e.* literature, art, music, painting; culture as contemporary media – satellite TV, fashion, pop music, lifestyle; culture as 'civilization' – *i.e.* long-term systems of value and meaning, such as language, and religion; culture in the sociological sense, as a set of values defining community, as in identity, tradition and in legitimating or challenging systems of power.[52]

According to Halliday, the relevant questions from the point of view of IR social science are "what use existing states and contenders for state power are making of culture [as well as] how culture is an independent influence on politics, or how it is weakening the power of states."[53] Although all four meanings of culture are represented in current IR to some extent, there is a certain prepon-

---

[49] Ken Booth, 'Security and Self: Reflections of a Fallen Realist,' in *Critical Security Studies*, p. 83.

[50] Lapid, 'Culture's Ship,' p. 7.

[51] Ripley, 'Cognition,' p. 89; Desch, 'Culture Clash,' p. 152; Wendt, *Social Theory of International Politics*, pp. 246–312; Peter J. Katzenstein, 'Introduction: Alternative Perspectives on National Security,' in Peter J. Katzenstein, ed., *The Culture of National Security: Norms and Identity in World Politics* (New York: Columbia University Press, 1996), pp. 2–6.

[52] Halliday, 'Culture and International Relations,' p. 53.

[53] *Ibid.*, p. 58.

derance of the fourth, sociological meaning over the other meanings. Questions pertaining to identity, norms, values and community figure prominently among recent IR and in particular constructivist writings, while questions concerning artistic culture do not. In any case, culture may both stabilize and challenge existing forms of domination. It can contribute to hegemony and counter-hegemony and has to be analyzed with respect to both potentialities. We therefore agree with Don Mitchell's statement that "a distinction between culture wars and other manifestations of power (and resistance to it) is hardly sustainable."[54] Frank Möller outlines in chapter three an iconographical approach to Northern photography from a political point of view, which takes into consideration both photography's history as legitimacy provider for state policies and its potential for resistance, cultural counter-hegemony and indigenous self-representation. Photography, he argues, can play a role in what Ajay Heble, in a different context, has called "promoting purposeful oppositional interventions in the public sphere"[55] and altering socio-cultural formations. Photography's inter-subjective character (requiring an audience that is capable of responding) and communicative nature (constructing meaning only through the dialogue, including conflicts over meaning, between photographer and audience) makes it an apt vehicle for breaking with institutionalized orders of knowledge and envisioning an alternative reading of past, present and future.

Yet International Relations, culturally turned or not, does not seem to be particularly qualified to come to grips with spatial representations of culture in *regional* settings. Even *local* contexts may be important because "constructivism's expectation of multiple identities for actors in world politics rests on an openness to local historical context."[56] Current constructivist social science in fact places emphasis on identity construction and considers identities as being constructed on multiple levels (rather than exclusively on the level of the system). Constructivist IR should therefore be more open to influence from related disciplines, which are better equipped to deal with non-systemic issues than IR itself, and leave the systemic level of analysis more often than before. Indeed, Alexander Wendt acknowledges that "it may be that domestic or genetic factors [...] are in fact much more important determinants of states' identities and interests than are systemic

---

[54] Don Mitchell, *Cultural Geography: A Critical Introduction* (Oxford: Blackwell, 2000), p. 5. We do not agree with him in calling conflicts over culture "culture wars" (see below).

[55] Ajay Heble, *Landing on the Wrong Note: Jazz, Dissonance and Critical Practice* (New York and London: Routledge, 2000), p. 219.

[56] Ted Hopf, 'The Promise of Constructivism in International Relations Theory,' *International Security*, Vol. 23, No. 1 (Summer 1998), p. 194.

factors." He has nevertheless "systematically bracketed" non-systemic factors.[57] Here, some constructivists seem to become prisoners of their own attack on Structural Realism which, by having determined what is worth being studied for so long, still to some extent dominates the current research agenda of IR social science. By choosing political topics and empirical cases close to Realism in order to defeat Realism, the selective blindness of International Relations shifts from Realism to constructivism.[58]

"The primary task of IR social science is to help us understand world politics", writes Wendt.[59] Given the ubiquity of, and the importance given to 'culture' in parts of current IR writings, it seems to be puzzling that IR has ignored up to now questions, approaches and findings of a discipline which carries 'culture' in its title – Cultural Geography (while generously, but often without much reflection, using its vocabulary like 'landscape', 'space' or 'place'[60]). Likewise, Wendt's assessments that "[n]ature yields control only grudgingly" and "culture supervenes on nature"[61] almost invite a cultural-geographical approach to international relations. This is why, for the remainder of this chapter, we call attention to Cultural Geography. We do so through both a genealogy of the discipline and a critical appraisal of current developments even though placing 'culture' and 'Cultural Geography' in relation to each other seems to be a mission impossible. This is so since the concept of culture in Cultural Geography varies historically and geographically as different schools have attached different meanings and understandings to it. Furthermore, it is a highly negotiable matter what Cultural

---

57   Alexander Wendt, 'Anarchy Is What States Make of It: The Social Construction of Power Politics,' *International Organization*, Vol. 46, No. 2 (Spring 1992), p. 423.

58   Like *The Culture of National Security*, the present volume "is self-conscious in bringing together two fields of study usually kept apart." Unlike the Katzenstein volume, the contributors to the present book are not aspiring to "meet reigning paradigms on their preferred grounds." Thus, while our approach may be "dismissed as skirting the hard task of addressing the tough political issues in traditional security studies", our choice is a deliberate one: we are more interested in blind spots in the current research agenda than in contributing to the ongoing struggle between different schools of thought over epistemological hegemony. Likewise, we are more interested in a fruitful exchange of ideas between IR and Cultural Geography rather than in internal struggles within both disciplines. For the quotations, see Katzenstein, 'Introduction,' pp. 10–11.

59   Wendt, *Social Theory of International Politics*, p. 370.

60   For a strong critique of some political scientists' refusal to "elaborate on the material grounding for [their] incredible arsenal of spatial metaphors [...] as if the only thing that matters is getting the metaphors right", see David Harvey, 'Cosmopolitanism and the Banality of Geographical Evils,' *Public Culture*, Vol. 12, No. 2 (2000), pp. 538–41.

61   Wendt, *Social Theory of International Politics*, pp. 112 and 371.

Geography is – it certainly can and does mean different things to different generations and intellectual domains of geographers: it is, for example, a synonym for Human Geography encompassing all human action on the surface of the earth; and a subdiscipline of Human Geography with its own historically constructed field of study different from those of economic and social geographers; or is it just the recent phase of the cultural turn that gives reasons to speak about *cultural geography*?

One way to avoid entering this debate here is to take a very standard, though over-simplifying view and claim that there are two basic approaches to 'culture' in Cultural Geography which are usually seen as conflicting or even mutually exclusive. The first one reflects a 'more traditional' approach to geographical studies on culture, the 'traditionality' of which comes with the close link to Carl Sauer, the early 20th century initiator of the so-called North American landscape school. According to this school, cultural landscapes are fashioned out of natural landscapes by distinctive cultural groups. Sauer and his disciples have largely been concerned with depicting cultural landscapes in order to delimit cultural groups from one another. The visible characteristics and close links between "land and life" – that is, the material landscape in which people live – are emphasized. The focus is on tradition, and culture is equated with custom, while the key research themes are raised up from rural surroundings. Moreover, the cultural landscape created by the group is seen as a central component in producing a political community.[62]

'New' Cultural Geography, emerging as critique of the North American landscape school's approach to culture, insists on paying more attention to the political production of cultural identities, especially as they are formed with reference to other groups of people. These 'more recent' developments originated from the cultural turn in social studies in the 1980s and early 1990s and brought influences mainly from British cultural studies into Human Geography, thus making 'new' Cultural Geography a distinct British inquiry. James Duncan's article, written in 1980, was one of the first attacks on the superorganic concept of culture used in American Cultural Geography.[63] Superorganic refers to the ontological assumption that culture is a real force, an agent, existing above and independent of human intention. This, in turn, leads to approaches which reify

---

[62] See John Leighly, *Land and Life. A Selection from the Writings of Carl Ortwin Sauer* (Berkeley: University of California Press, 1993). For recent, yet tradition-based developments in landscape (nature-society) research, see, for example, Kent Mathewson, 'Cultural Landscape and Ecology II: Regions, Retrospects, Revivals,' *Progress in Human Geography*, Vol. 23, No. 2 (1999), pp. 267–81.

[63] James Duncan, 'The Superorganic in American Cultural Geography,' *Annals of the Association of American Geographers*, Vol. 70, No. 2 (1980), pp. 181–98.

culture, *i.e.* they see culture as a thing rather than as a process, as a consequence of which the sociological and psychological components of culture are being neglected. Thus, 'new' Cultural Geography theoretically challenges the Sauerian universal notions of culture, normative values and beliefs by valuing cultures more like a signifying processes, *"maps of meaning* through which the world is made intelligible."[64] Furthermore, the methodological basis of 'new' Cultural Geography, being in close engagement with poststructuralist and postcolonial theories and showing "a heightened reflexivity toward the role of language, meaning and representation in the constitution of 'reality' and knowledge of reality",[65] differs from the Sauerian tradition influenced by ethnographical field research.

Moreover, 'new' Cultural Geography has discovered the political. Peter Jackson, for example, writes in his influential book:

> That the cultural is political follows logically from a rejection of the traditional notion of a unitary view of 'culture', and from a recognition of the plurality of cultures. If cultures are addressed in the plural [...] then it is clear that meanings will be contested according to the interests of those involved. [...] The present reformulation of cultural geography should therefore be situated in its immediate political context.[66]

Cultural Geography thus has shown an increasing awareness of the intimate relationship between culture and politics, cultural politics and the politics of culture. At the same time, however, it has been rather insensitive to some basic categories of political thought such as 'power', the meaning of which is often taken for granted rather than problematized in its specific context, or too generous in stretching the meaning of particular words beyond recognition. Don Mitchell, for example, claims that "culture is politics by another name". According to Mitchell, arguments over 'culture' are "arguments over real spaces, over landscapes, over the social relations that define the places in which we and others live." Yet, by equating conflicts over culture indiscriminately with "culture wars", he disregards standard definitions of 'war' in political science, which are not just definitions but statements about the essence of war: about that which distinguishes 'war' from 'politics'.[67] Just as IR should not be content with borrowing the language used in Cultural Geography without bothering about the underlying concepts, Cultural

---

[64]   Peter Jackson, *Maps of Meaning: An Introduction to Cultural Geography* (London and New York: Routledge, 1989), p. 2.

[65]   Clive Barnett, 'The Cultural Turn: Fashion or Progress in Human Geography?' *Antipode*, Vol. 30, No. 4 (1998), p. 380.

[66]   Jackson, *Maps of Meaning*, pp. 4–5.

[67]   Mitchell, *Cultural Geography*, pp. 3–6.

Geography has to be more aware of the intricacies of key analytical categories and concepts used in IR.

Up to here, our presentation of the genealogy of Cultural Geography followed a standard way of narrating the discipline's recent developments. We made a clear separation between 'old' and 'new' approaches to the geographical study on culture – a separation, the usefulness of which in the context of this chapter and, in fact, the whole volume we now want to call into question (therefore the use of the prefix 'more' when we were discussing 'traditionality' and 'recentness' of geographical ideas on culture). We do this in particular because assuming the existence of something like 'the' new Cultural Geography appears to be highly questionable. Such dichotomizing and categorizing may easily result in an over-simplified writing of intellectual history which, in Peter Jackson's words, would deny "the significance of historical continuities, exaggerat[e] the coherence of different 'schools' of thought and encourag[e] hostile caricature".[68] The bipolar setting of 'the' 'old' and 'the' 'new' excludes other 'cultures' in the disciplinary history of Geography, for example, the cultures of spatial analysis (equating culture with human action), cultures of behavioural geography (understanding culture as a filter of perceptions) and cultures of humanistic geography (where culture forms a base for experience and meanings).[69] Furthermore, the discipline's designation as 'the' new Cultural Geography tends to hide some significant differences within and between, for example, three particular fields: landscape iconography, literary post-structuralism and cultural politics.[70] Finally, it is rather questionable if 'new' Cultural Geography is really as radical as its protagonists like to see it (and themselves): the work done in the field of landscape iconography, for example, is deeply rooted in the Sauerian concept of landscape. The 'weavers' of 'new' Cultural Geography, thus, have not entirely escaped from the ideas they intended to ban. They can be criticized of being caught up in their own pretensions.

The development of 'new' Cultural Geography – and the exaggeration of its 'newness' – can be seen as an effort to construe a distinctive British approach to

---

[68] Peter Jackson, 'Rematerializing Social and Cultural Geography,' *Social & Cultural Geography*, Vol. 1, No. 1 (September 2000), p. 9. It is somehow ironic (or self-critical?) that this statement comes from Jackson who, in *Maps of Meaning*, helped construct a clear 'old'-'new' divide in Cultural Geography. An example of a textbook which intentionally tries to avoid the division between 'old' and 'new' is William Norton, *Cultural Geography: Themes, Concepts, Analyses* (Don Mills: Oxford University Press, 2000).

[69] See Anssi Paasi, 'Deconstructing Regions: Notes on the Scales of Spatial Life,' *Environment and Planning A. Society and Place*, Vol. 23, No. 2 (1991), pp. 239–56.

[70] Peter Jackson, 'Berkeley and Beyond: Broadening the Horizon of Cultural Geography,' *Annals of the Association of American Geographers*, Vol. 83, No. 3 (1993), p. 519.

cultural geography. Of course, there is nothing to be said against it *per se*. The problems start, however, with the often blind acceptance without reflection of questions and approaches developed in and appropriate to the British context in cultural-geographical contexts different from the British. Just like the US influence on IR, the influence of British cultural studies on recent Cultural Geography may be a part of the problem of the disciplines' estrangement from the 'real world' and 'real people'. It almost inevitably results in the adaptation of marginal geographies to the dominant research canon (scientific and otherwise) and the application of research agendas developed in the centre to marginal regions. Yet, questions and approaches developed in the centre are not necessarily relevant to the margins; questions relevant to the margins are not necessarily asked in the centre. What the margins may learn from the centre is asked more often than what the centre may learn from the margins. The conformity pressure affecting scholars interested in marginal regions, people and landscapes is however enormous. Ignoring dominant agendas and approaches may result in exclusion from leading international journals, major conferences, financial schemes and grants as well as isolation within the academic community. Adaptation to the dominant discourse, however, means the perpetuation of patterns of scientific hegemony and academic centre-periphery relations rather than mutual social learning.

Multiple North-Views

The cultural turn has been met with some noteworthy and legitimate criticism. This is hardly surprising because every 'turn' aims at a radical break between the ongoing project – which becomes the target for criticism usually without both being prepared for it and believing to deserve any kind of criticism in the first place – and the forthcoming 'prophecy', the ideas of which are, however, still too obscure to present deliberate grounds on which to stand.[71] Criticizing the cultural turn, however, should not be taken as a kind of personal revenge, but rather as something that "can only do good" in the efforts to "deflate the pretensions of the cultural turn" and to explore both its strengths and weaknesses.[72] From one point of view, the cultural turn has been charged with having become too theoreticist in

---

[71] For example, although sympathetic to constructivist approaches, Jeffrey T. Checkel, 'The Constructivist Turn in International Relations Theory,' *World Politics*, Vol. 50, No. 2 (January 1998), pp. 324–48, misses, among other things, a coherent research design and engagement in theory development in current constructivist writings.

[72] Nigel Thrift, 'Introduction: Dead or Alive?' in Ian Cook *et al.*, eds., *Cultural Turns/Geographical Turns: Perspectives on Cultural Geography* (Harlow: Pearson Education, 2000), p. 2.

'turning the worlds merely into words'. Yet seen simultaneously from another perspective, the cultural turn is said to have fallen into a hard-hearted journalistic empiricism where anything 'interesting' and selling goes. Because of this, geography is surely on the public agenda again.[73] But critics wonder whether or not it has paid too high a price for that pole position. Furthermore, the cultural turn is seen as having equipped itself too easily with culturalist attitudes thus dematerializing Cultural Geography, eclipsing the realm of the social and "lacking political bite" and substance.[74] All these strands of criticism seem to point in one specific direction: the focus has shifted from questions of social justice to a concern with cultural representation, which might lead researchers to lose sight of the very real material consequences and effects of social differences and social relations of power.[75]

Does the cultural turn with its potentials and limitations bear relevance for the debate on the North? In the chapters of this volume, the reader can find a variety of different cultures and meanings attached to (the concept of) culture in the context of the North. Culture can refer to skilled human activities through which non-human nature is transformed (as the word culture initially does) but also to geographically and historically varied ways of understanding the relationship between humanity and nature on the basis of the daily life (such as the patterns of land use in Lapland). It can refer to both the material and the immaterial (Innu tents and Sámi place names), to artistic products and conventions attached to them (photography, novels and travel literature), to a way of life and a system of (presupposed) shared values (Skolt *siida* organization and *Thing* as a constitutive part of Nordic political culture). It can refer to processes where culture makes culture but also where all these meanings are hierarchically ordered.[76] Yet, Inger

---

[73] The last time the relevance of geography was considered to cover more than just the academic realm was when it produced geographical knowledge to be used for military purposes during the world wars. This does not, however, mean that the nexus of geography and military-political aspirations has ceased to exist. For American geography's and geographers' contribution to the construction of the US foreign policy in the 20th century, see Neil Smith, *American Century: Roosevelt's Geographer and the Prelude to Globalization* (Berkeley and Los Angeles: University of California Press, 2003 forthcoming).

[74] See Thrift, 'Dead or Alive?', p. 2. These remarks summarize various critical voices during the last decade, a presentation and discussion of which can be found in *Cultural Turns/Geographical Turns*.

[75] Gill Valentine, 'Whatever Happened to the Social? Reflections on the "Cultural Turn" in British Human Geography,' *Norsk Geografisk Tidsskrift*, Vol. 55, No. 3 (2001), p. 168.

[76] Mitchell, *Cultural Geography*, p. 14.

Birkeland points out that in the context of Nordic Geography "the cultural turn has been met with a remarkable silence."[77] The themes introduced by 'New' Cultural Geography do not seem to have fully entered the northern research agenda. But is this automatically a case worth worrying about?

After all, for a long time there has been keen interest in cultural themes of the North: as Paulo Susiluoto's discussion on the Skolt Sámi geographies of Väinö Tanner and Karl Nickul – two of the foremost Finnish geographers of the first half of the 20th century – shows, there is a strong tradition of Northern geography rich both in scope of themes and deepness of analysis (see chapter four). Tanner and Nickul challenged naturalized, institutionalized ways of depicting the Sámi communities, but they were largely ignored at the time by both the general public and academic geography. Susiluoto sketches the Skolt Sámi's remarkable forms of social organization and complex relationship between foreign and internal policies as well as the negative influence of border changes, conflicts between different Sámi groups and specific economic developments on the *siida* (or Lapp village). By elaborating Tanner and Nickul's methodological approaches, which included an interpretative and local-geohistorical perspective as well as extensive fieldwork and tremendous respect for their study objects, Susiluoto shows the rich heritage of early Finnish geography in dealing with Northern lands and peoples. Yet, the potential of the studies done by Tanner, Nickul and other earlier geographers such as Johannes Gabriel Granö and Ilmari Hustich remains largely unnoticed: they seem to be too far away from today's British or American mainstream discourses. Yet, one should try to turn the picture upside down and consider this difference as a positive sign. The approaches northern researchers have grown into might be more appropriate for the contexts and objects of their studies, rather than those introduced for the purposes of the centres of knowledge.[78]

That the cultural turn has been met with silence can thus also be interpreted as a strength of geographers from marginal regions (meaning remoteness in distance, language and academic culture). Researchers dealing with the North, by sticking to their 'traditionality' rather than flirting with everything 'new', have managed to avoid the trap of losing the culture-nature relationship in favour of a society-space relationship. They have succeeded in resisting the sophisticated theories written in

---

[77] Inger J. Birkeland, 'Nature and the "Cultural Turn" in Human Geography,' *Norsk Geografisk Tidsskrift*, Vol. 52, No. 4 (1998), pp. 229–30.

[78] In a different, yet related context Heble, *Landing on the Wrong Note*, p. 65, puts the questions emanating from these thoughts this way: "How and why [...] is knowledge produced and maintained? What kinds of things count (and don't count) as knowledge? Who has the power to determine what counts as knowledge? How do we understand the complex relations among knowledge, power, and history" and, it may be added, geography?

the economic centres of the globe which do not necessarily bear relevance to the everyday lives of the Northerners. Despite scholarly emphasis put on the problems of representation, certain representations will do better than others when making theories and presenting ideas to be read in academic departments all over the world. It remains for geographers on the margins to find (to borrow Ajay Heble's words) "alternative sources for knowledge production"[79] either by re-contextualizing hegemonic discussions or by composing alternative theories from starting points radically different from those internationally recognized.

Kenneth Olwig's work on landscape can be mentioned as an example of the first alternative.[80] In his writings he has criticized some of the basic assumptions made in the newly formulated British landscape studies of the 1990s. In these studies the origins of the concept of landscape are exclusively linked with certain types of representation and contents such as, for example, the continental landscape paintings of the sixteenth century with the natural scenery as their primary subject matter. Olwig redirects emphasis to alternative representations and contents of landscapes – something which he labels "substantive landscapes" – as a way of pointing out the relevance of the social and material contexts in which the politics of representation take place. Instead of understanding the origins of landscape solely in terms of visual representation, Olwig's discussion on landscape as "the organizing of things in a land" allows the interpretation of the idea of *Thing*, as analyzed by Ulrich Albrecht in chapter seven of the present book, as one of the substantive landscapes of the North. Unique features of creating public political space in the Nordic states – a culture of compromise and egalitarianism – are linked by Albrecht with the heritage of the ancient *Things* and even transferred to the recent regime transformations in central and eastern Europe, some of which were negotiated at round tables. Here, the unique features of Nordic political culture pave the way to alternative forms of political decision-making, participation and knowledge production.

The second alternative, the composition of alternative theories from new starting points, is inherent in Ari Aukusti Lehtinen's contribution (chapter two). Grasping the multilayered coexistence of a Northern geography requires keeping the door open for ideas both within and outside academic circles. Lehtinen therefore is searching for the multilayered geographies of the North in the non-academic cultural journal *N66. Culture in the Barents Region*. He finds an agenda

---

[79] *Ibid.*, p. 70.

[80] See Kenneth Olwig, 'Recovering the Substantive Nature of Landscape,' *Annals of the Association of American Geographers*, Vol. 86, No. 4 (1996), pp. 630–53 (especially p. 633); Kenneth Olwig, 'Landskabet som samfundsmæssig forførelse,' *Nordisk Samhällsgeografisk Tidskrift Debat*, at < http://www.geo.ruc.dk/NST/Debat/Debat Olwig.pdf >.

with "a multi-cultural profile" that can be read as an invitation to a lively debate
in which geographers have so far not been deeply involved.[81] The topics
presented on the pages of *N66*, while carefully listening to local voices, also mirror
a broader cultural reorientation which bears the potential for "unlearning the
(post)colonial manoeuvres" of academic research practice. Being conscious of
coexisting temporal scales, for example, the approach suggested by Lehtinen calls
into question the settled and hopelessly closing chronological periodization so
typical of scientific inquiry. Together with the other examples of the life and land
in the North drawn for the pages of *N66* such as the topography of dwelling,
Lehtinen encourages geographers and scholars from related disciplines to think
beyond the pressures of "submissive adjustment" emanating from the southern
centres of knowledge and thus to make sense of the marginal narrations which
would otherwise, *i.e.* when seen from a universalizing point of view, remain
invisible.

   In addressing these issues, the important questions to be answered concern the
terms 'culture' and 'the North'. If not the limits, what then are the contents of the
discursive landscapes of the North? Although there are already some promising
departures on this theme, emphasis has so far been mainly on the marginal regions
only as far as they are constitutive for the centres.[82] The idea here is based on the
acknowledgement that the construction of centre and periphery is dialectical: it is
through representations of the northern periphery that the southern centre
constructs itself and *vice versa*, as Juha Ridanpää's chapter in this book illustrates.
Ridanpää, discussing selected writings of the Finnish novelist Rosa Liksom,
emphasizes the clear distinction in Liksom's short stories between the North and
the South of Finland: "places and spaces of the North and the South are
constructed differently, through different discursive practices" (chapter five). Yet,
although life in the North is often depicted in terms of human misery, living
conditions in the South do not necessarily have to be better: misery everywhere.

---

[81]   For some exceptions, see Bjørn Bjerkli, 'Landskapets makt: Sted, ressursutnyttelse og
       tilhørighet i en sjøsamisk bygd,' *Norsk Antropologisk Tidsskrift*, Vol. 8, No. 2 (1997),
       pp. 132–56; Michael Jones, *Perspektiver på landskap og hvordan det kan anvendes
       i sørsamisk sammenheng* (Trondheim: Norwegian University of Science and Techno-
       logy, Department of Geography, 1999), at <http://www.uit.no/ssweb/dok/seminar/
       sorsamisk/JONES.html>; Samu Pehkonen, *"Tänne, muttei pidemmälle!"* – *Alta-
       kamppailu ja Pohjois-Norjan monikulttuurinen maisema* (Tampere: Tampere Peace
       Research Institute, 1999).

[82]   See, for example, Shields, *Places on the Margin*, pp. 162–205; Petri J. Raivo, 'The
       Limits of Tolerance: The Orthodox Milieu as an Element in the Finnish Cultural
       Landscape,' *Journal of Historical Geography*, Vol. 23, No. 3 (1997), pp. 327–39;
       Häkli, 'Cultures of Demarcation.'

Ridanpää interprets her stories as an implicit critique of some over-simplifications in postcolonialist theory. In particular her use of irony, rather than being directed only at the social inequalities between the North and the South, challenges, according to Ridanpää, the whole theoretical composition of postcolonialist theory. Addressing her writings from a socially critical, literary geographical perspective, the author reveals the extent to which the marginalities of the periphery and the centre are interrelated and contingent on each other. However, it is all too often that the periphery is studied only because it is needed in order to say something about the centre. Would we not have grounds to insist on looking at the construction of the North and still emphasize those aspects which are relevant for the North?

The problem, however, is who is able and indeed legitimate to judge over 'those things relevant for the North'? If traditional regional geography has not been particularly successful in giving a comprehensible understanding of the North, post-colonialist approaches have done no better. "The main mistake in the field of post-colonialist research has been [the] overflowing compassion and pity, which is dumped [...] over 'the North'."[83] The same attitude can clearly be observed when the climate change concerning polar regions is discussed by environmental specialists and politicians: the unstable dichotomy between indigenous and other (non-indigenous) human communities noticed by Monica Tennberg in chapter nine bears witness to the fact that the opinions and aspirations of the indigenous people are asked only seldom. Indeed, for most climate researchers the Arctic is just an interesting laboratory, but for the local and regional decision-makers climate change is an issue with real social and economic consequences. Thus, although scientific and political debates revolve around the climate change, the debates and interests are different. By 'freezing' the temporal dimensions when representing the Arctic and the effects the global climate change might bring with it, the indigenous peoples, too, become frozen in place and time. This will, as Tennberg argues, inevitably limit the possible future options people in the North – and political decision-makers there and elsewhere – can conceive of while situating the socio-environmental conditions of the Arctic and its inhabitants in a wider global framework. What indigenous people 'gain' from immaterial support and pitiful attitudes (the recipients of which they might not at all want to be) is that the real problems occurring in their everyday lives are wiped off the political agenda.

Those who criticize the cultural turn of having detached its practitioners from the realities of everyday life bring us back to the rather oppressive picture of the current situation of the people in the North painted at the beginning of the chapter. Many of the social problems are due to different and conflicting ways of framing

---

[83] Juha Ridanpää, 'The Discursive North in a Few Scientific Discoveries,' *Nordia Geographical Publications*, Vol. 29, No. 2 (2000), p. 35.

Northern realities. Even nature is not uncontested in this respect: northern nature is not an empty laboratory space, an image conventionalized by natural sciences. The strong presence of nature in the North and the central role it plays as the signifier for the life of the Northern people calls for questions relating to the cultural constructions of nature. To jargonize the case, one should pay attention to the 'spaces of nature' rather than 'natures of space'. People from the North have relatively seldom showed their resistance towards the southern monopoly of decision-making but when they have done it, it has very often been in connection with questions relating to nature. Leena Suopajärvi's chapter on the contested meanings of nature in Finnish Lapland points in this direction (chapter ten). Combining the key moments in the rebuilding of Lapland after the devastation of World War II with the individual anxieties so present in her discussions and interviews with the local population, Suopajärvi pictures a horizon of South-North debate where, on the one hand, the needs and necessities of the national economy are clearly articulated. On the other hand, however, these very same needs are hidden in the rhetoric of providing the local people the possibility of survival. Yet survival has meant survival in terms of economy (in the sense that new industries provide a certain amount of work) rather than in terms of cultural and social continuity and self-representation: much of the recreational possibilities and traditional modes of livelihood (such as reindeer herding) have been lost.

Much of the current criticism toward the cultural turn could be overcome by returning to an original strength of Geography, its tradition of fieldwork.[84] From the point of view of the North this is a welcome invitation, but only so far as it manages to diverge itself from the heroticized accounts of exploration and fieldwork practices discussed by Hanna-Mari Ikonen and Samu Pehkonen in their chapter on Northern exploration (chapter six). They review some of the central late 19th and early 20th century Nordic literature on northern exploration and reveal its close connection to the then prevailing modes of academic geographical thinking. These modes were thoroughly dichotomizing and the opposite pairs of home/abroad, safety/danger and nature/culture are argued to have formed the basis of both popular and scientific ways of knowledge production. To be able to introduce a different kind of geographical imagination the authors suggest as the starting points for revealing Northern 'realities' to go beyond the senses of sight and hearing (traditionally regarded as masculine senses) towards the senses of smell and taste (conventionally regarded as feminine senses). Today, in many parts of the North, (southern) researchers with their scientifically formulated questions have become part of the everyday life of the North without really entering 'the North'. In other words, researchers may be 'there' making insights into northern cultures,

---

[84] Neil Smith, 'Socializing Culture, Radicalizing the Social,' *Social & Cultural Geography*, Vol. 1, No. 1 (September 2000), pp. 25–8.

but they are not inside culture; they have not entered the social, cultural and economical problems of the people studied but are rather just dealing with the representations of these problems.

Is there then any positive news coming from the North? The chapters in this book point to various ways to approach the North in a non-standardized manner, challenging institutionalized ways of constructing and representing social reality and producing naturalized orders. From the international and IR point of view, the reinvented social interaction across borders is certainly opening up new opportunities to tackle the problems in terms of economy, culture, social organization and (self-) determination the people of the North are facing.[85] From the local perspective, however, these changes may not be the ones most eagerly awaited. We are, once again, confronted with the question of who defines the problems; are the new opportunities something people of the North *should* take advantage of; and if so, is it because of pity or apology that the South is allowing and promoting cooperation in the North? Or are there other interests and motives hidden behind the recent political re-invention of the North?

To borrow a term used by Andreï Makine in a different context, truths are "absolutely logical and absolutely arbitrary."[86] On the one hand, then, the discursive North is arbitrarily treated and constructed in speech, writing and pictorial representations, both past and present. On the other hand, however, the discursive North is partly logical. The word discourse, in one of its older meanings, also means to run or travel over a space or a region.[87] The "presence of North in man"[88] thus obliges the reader to undertake such a journey to the North himself or herself, to which the reader is hereby invited. With this in mind, it is easier to understand the peculiarity which can be observed while looking at the Norwegian box office movie list: the peculiarity that along the Hollywood films on global distribution, there is a Norwegian documentary film, *Heftig & Begeistret* ("Cool & Crazy"). The film is made by Men's Chorus from Berlevåg, a small fishing village at the edge of the European continent. The progressive closure of fisheries, its catastrophic impact on the local economy and the unremittingly bleak climate would, one thinks, give little reason for cheer. Since its foundation in 1917, the choir still offers an escape from the hardships of the real world, "while

---

[85] For historical and recent developments among the Sámi population, see various contributions in Kristiina Karppi and Johan Eriksson, eds., *Conflict and Cooperation in the North* (Umeå: Kulturgräns Norr, 2002).

[86] Makine, *Requiem for the East*, p. 16.

[87] *The Shorter Oxford English Dictionary on Historical Principles. Vol. 1* (Oxford: Oxford University Press, 1973), p. 563.

[88] Jack Warwick, *The Long Journey: Literary Themes of French Canada* (Toronto: University of Toronto Press, 1968), p. 47.

the camaraderie it inspires brings warmth and good humour to this most forbidding of outposts", writes a reviewer for the BBC. With their passionate, hilarious and sometimes majestic portrait of singing and survival in the North the members of the choir (the oldest one, a 96-year-old, has been with the choir since 1926) have gained huge success – over half a million spectators in Norway with a total population of 4.4 million. Since 2001 the choir has been on the road almost constantly including a visit to the USA and Australia. Everywhere these Northern strangers have been met with astonishment and surprise. But, the reviewer of *The Guardian* wonders: "Why should anyone want to see a film about a choir from a Norwegian fishing village?"[89]

---

[89]   See Berlevåg Mannsangførening, at < http://www.berlevag.kommune.no/heftig/ >.

Chapter 2

Mnemonic North: Multilayered Geographies of the Barents Region

Ari Aukusti Lehtinen

"The further to the North, all objects and beings are shedding their firm forms and contours. The strictly organized cosmos is defused. The Universe is losing its spatial and time boundaries. The space is melting away and disappears. The time also slows down and, being unable to reach the final limit, stops forever, turning into eternity." Thus formulates Nikolaj M. Terebikhin, Professor of History at Pomor University in Archangel, and derives his view from the historical experience of the Northern people. According to him, "the North in the worldview of Scandinavian and Russian nations was never a proper geographical category".[1]

The Magnetic North

Through ages the Nordic North has served as a specific source of inspiration for researchers and travellers and especially for geographers. The written and celebrated past of the North is a heroic one and, in addition, a central cause for the initial institutionalization of Nordic geography as an academic discipline during the 19th century.

The return of Adolf Erik Nordenskiöld to Stockholm in 1880, after the pioneering sail across the Northeast Passage, was the event to raise public interest in the North and, moreover, it motivated the Nordic countries towards more organized and systematic inventories of the North. The Swedish Society for Anthropology and Geography, founded in 1877, began to publish *Ymer* in 1881 and, especially during the early years, the journal grew into an important forum for arctic research. Lund, Uppsala, Gothenburg and Stockholm got their first

---

[1] Nikolaj M. Terebikhin, 'Bjarmaland. Spiritual Culture in the Barents Region,' *N66. Culture in the Barents Region*, No. 1 (1997), p. 21.

professors of geography in a short period at the turn of the century, which was, undoubtedly, a sign of the growing societal interest in locating Sweden centrally on the map of modern science regimes. During those days, geography was the science of expansion and in Sweden, as within the rest of the Nordic countries, the primary focus for areal curiosity was the North.[2]

In Norway, the Geographical Society was founded in 1889, after the return of Fridtjof Nansen from his expedition to Greenland. The founding idea of the Society was to support geographical voyages of discovery and spread information about the explorations to the broader public. The early professional geography was part of the construction of a nationalistic framework for the emerging (nation)state of Norway but it was also integrated in the expansionistic politics of the country. Polar research was part of the national identity-building and industrial-territorial transnationalization. Military leaders, diplomats and entrepreneurs were common visitors to the first meetings of the Geographical Society.[3]

The success of Nordenskiöld was also noticed in Finland and, consequently, the necessity to participate in the competition over the North became evident. In Finland, as in the western neighbours, the expeditions northwards were regarded both as a tool for solidifying the national cultural existence and a means for optional territorial expansion.[4] The provincial country, then under the Russian regime, could show her vitality by these kinds of symbolic activities. Zoologist J.A. Palmen, who for several years had analyzed the bird observations of Nordenskiöld's Vega voyage, soon made plans for a Finnish expedition to the Kola Peninsula. The Kola expedition took place in summer 1887 and it had a remarkable impact on the foundation of the Finnish Society for Geography later in the same year. Palmen was one of the founders of the Society and also its secretary for the first 29 years.[5]

---

[2] See Sverker Sörlin, *Framtidslandet. Debatten om Norrland och naturresurserna under det industriella genombrottet* (Stockholm: Carlsson Bokförlag, 1988); Anders Karlqvist and Olle Melander, 'Sverige åter i Antarktis,' *Ymer*, Vol. 110 (1990), pp. 11–21.

[3] Ann-Cathrine Åquist, 'Idéhistorisk översikt,' in Bjørn T. Asheim *et al.*, eds., *Traditioner i Nordisk kulturgeografi* (Uppsala: Nordisk Samhällsgeografisk Tidskrift, 1994), pp. 1–13; Teuvo Pakkala, 'Kaksi viimeistä naparetkeä,' in Anneli Kajanto, ed., *Namsarai* (Helsinki: Suomalaisen Kirjallisuuden Seura, 1999), pp. 46–56.

[4] See Anssi Paasi, 'The Rise and the Fall of Finnish Geopolitics,' *Political Geography Quarterly*, Vol. 9, No. 1 (1990), pp. 53–65.

[5] Two associations were initially founded in Finland, namely the Geography Association of Finland (in 1887) and the Society for Finnish Geographers (1888) but these merged in 1921 under the title of the latter. See Kalevi Rikkinen, *Suuri Kuolan retki* (Helsinki: Otava, 1980); Matti Tikkanen and Pentti Viitala, 'Pieni Kuolan retki,' *Terra*, Vol. 99, No. 1 (1987), pp. 3–12.

In Denmark, the Royal Geographical Society was established already in 1876 and one of the main interests of the Society was focused on Greenland, the vast northern island currently connected to Denmark as an independent administrative unit with home rule. Greenland research intensified in Denmark during the post World War II decades, and especially after the 1980s in the context of North Atlantic studies.[6] The North Atlantic still is a painful question to the Danes, surfacing for example in the current debates on the relations between the mainland and the Faroes. Similarly, the celebrated and widely-read novel *Smilla's Sense of Snow* by Peter Hoeg, witnesses in its own way the constantly difficult elaboration of the Danish colonial-territorial past in the North Atlantic.[7]

• • • • •

Today's geography of the European North is inevitably less heroic and often rather critical of the early master narratives of expansive modernization. However, the North continues to have the role of inspiration in geography. The last decade has witnessed several new openings northwards by geographers and colleagues within the related sciences. The new North has been approached by a broad variety of frameworks, including for example geopolitical, regional economic, eco-social and ethno-cultural perspectives.[8]

---

[6] Peter Friis and Rasmus Ole Rasmussen, *The Development of Greenland's Main Industry – the Fishing Industry* (Roskilde: Roskilde University Center, North Atlantic Regional Studies, 1989); Jørgen Ole Baerenholdt, *Bygdeliv* (Roskilde: Roskilde University Center, Department of Geography, Social Analysis and Computer Science, 1991); Jørgen Ole Baerenholdt, *Innovation and Adaptability of the Murmansk Region Fishery Kolkhozes* (Roskilde: Roskilde University Center, Department of Geography and International Development Studies, 1995).

[7] Michael Haldrup, 'Living on the Edge – Considerations on the Production of Space in the Faroe Islands,' *Nordisk Samhällsgeografisk Tidskrift*, No. 23 (October 1996), pp. 18–27; Lise Lyck, ed., *The Faroese Economy in a Strategic Perspective* (Stockholm: Nordic Institute for Regional Policy Research, 1997); Peter Hoeg, *Frøken Smillas fornemmelse for sne* (Copenhagen: Munksgaard/Rosinante, 1992).

[8] Baerenholdt, *Bygdeliv*; Ari Aukusti Lehtinen, 'The Northern Natures – A Study of the Forest Question Emerging Within the Timber-line Conflict in Finland,' *Fennia*, Vol. 169, No. 1 (1991), pp. 57–169; Yrjö Haila and Richard Levins, *Humanity and Nature* (London: Pluto Press, 1992); Kenneth Olwig, 'Eurooppalaisen kansakunnan pohjoinen luonne,' in Svenolof Karlsson, ed., *Vapauden lähde. Pohjolan merkitys Euroopalle* (Helsinki: VAPK Publishing, 1992), pp. 158–82; Baerenholdt, 'Innovation and Adaptability'; Margareta Dahlström, Heikki Eskelinen and Ulf Wiberg, eds., *The East-West Interface in the European North* (Uppsala: Nordisk Samhällsgeografisk Tidskrift, 1995); Inger Birkeland, 'The Mytho-Poetic in Northern Travel,' in David Crouch, ed.,

The early geography of expansion turned into a multitude of approaches during the 20th century, increasingly informed by a sensitive articulation of the specific time-space geographies of the North. Imperial spaces of mapping were replaced by zones of cross-border interaction, lands of otherness re-emerged as multicultural realms with shared heritages and, moreover, modern (and masculine) territorial divisions confronted the layers of mytho-poetic landscapes. Both the perspectives and focusings of research were diversified. In other words, the Northern landscapes painted by geographers became elastic and diffuse, comprehensible only from pluralistic and mobile positions. Simultaneously, the North of the geographers emerged as a critical voice from the margins; it grew into an invitation to reciprocal learning, an existing alternative to the modern 'land and life' order. Furthermore, the North emerged as a connection to pre-Christian belief traditions: the North became the contact to our early ancestors and their world views.[9]

The ongoing diversification of the academic geographies of the North has, however, had only limited success in gaining firm ground in the North. The theoretical sophistication has enabled us to construct a deeper picture of the North but, at the same time, rather paradoxically, it has kept us – as professional geographers[10] – in the role of foreigners in the daily spheres of living in the North. Especially the representativity and authenticity of research have been questioned. In other words, despite the immanent critique and radical renewal of the early geoscientific approaches, the (high) cultures of current academic geography have not yet been able to construct a satisfactory correspondence between the 'academic landscapes' painted and the 'everyday landscapes' of the Northern peoples. The elegant languages used and the detailed maps and statistics prepared by the researchers have often remained alien to the everyday means of

---

Leisure/tourism Geographies: Practices and Geographical Knowledge (London: Routledge, 1999), pp. 17–33; Jouni Häkli, 'Cultures of Demarcation: Territory and National Identity in Finland,' in Guntram Herb and David Kaplan, eds., *Nested Identities: Identity, Territory, and Scale* (Lanham: Rowman & Littlefield, 1999), pp. 123–49; Samu Pehkonen, *"Tänne, muttei pidemmälle!" – Alta-kamppailu ja Pohjois-Norjan monikulttuurinen maisema* (Tampere: Tampere Peace Research Institute, 1999).

[9]   Birkeland, 'Mytho-Poetic in Northern Travel'; Häkli, 'Cultures of Demarcation'; Pehkonen, *"Tänne, muttei pidemmälle!"*

[10] In this context, professional geography refers both to academic geography (as a discipline) and approaches within related disciplines with a specifically geographical (socio-spatial or eco-social) orientation. For the interrelations between different geographies, see Anssi Paasi, 'Kulttuuri: maantieteellisiä näkökulmia,' *Alue ja Ympäristö*, Vol. 20, No. 2 (1991), pp. 2–19.

orientation and communication among the local folks.[11]

On the other hand, the northern cultures have been able to develop their own pockets of internal continuity, and this has largely taken place behind the pressures of homogenization under (post)colonialism. The internal dynamics has sometimes even been fuelled by the adjustment demands from the South. These pockets have evolved according to their own rules and have constructed geonarratives and personal geographies of their own. This source of geoinspiration, emerging from inside the 'land and life' of the North, has gained only marginal interest among the professionals in geography.[12] Hence, the North still contains the 'treasures' that

---

[11] Here I base my argument on the specific ideohistorical criticism toward the dominating geography practices presented, for example, by Cindi Katz and Andrew Kirby, 'In the Nature of Things: the Environment and Everyday Life,' *Transactions of the Institute of British Geographers*, Vol. 16, No. 3 (1991), pp. 259–71; Paul Cloke and Jo Little, 'Introduction: Other Countrysides,' in Paul Cloke and Jo Little, eds., *Contested Countryside Culture. Otherness, Marginalization and Rurality* (London: Routledge, 1997), pp. 1–18; Birkeland, 'The Mytho-Poetic in Northern Travel'; Yrjö Haila, 'The North as/and the Other: Ecology, Domination, Solidarity,' in Frank Fischer and Maarten A. Hajer, eds., *Living with Nature* (Oxford: Oxford University Press, 1999), pp. 42–57; Kenneth Olwig, 'Landcsape as a Contested Topos of Place, Community and Self,' in Paul Adams, Steven Hoelscher and Karen Till, eds., *Textures of Place: Geographies of Imagination, Experience, and Paradox* (Minneapolis: University of Minnesota Press, 2001), pp. 95–119. The view presented here is also informed by the current critical (self)reflectivity focused on the geography/society relations under the working title 'activism and academia', see, for example, Paul Routledge, 'The Third Space as Critical Engagement,' *Antipode*, Vol. 28, No. 4 (1996), pp. 399–419; R.M. Kitchin and P.J. Hubbard, 'Research, Action and "Critical" Geographies,' *Area*, Vol. 31, No. 3 (1999), pp. 195–8. Moreover, similar worries were expressed by Anne Buttimer when she received an honorary doctorate at the University of Joensuu, Finland, in 1999. In her festivity lecture, she topically discussed, among other issues, the threat of geography becoming too sophisticated to be able to deal with the most central topics of our global and local worlds. The discipline of geography is evidently undergoing a phase of critical self-evaluation in relation to its historical legitimacy and societal dependencies – a uniform (post)colonial geoscience is gradually turning into an array of (critical) geographies. Consequently, and both for the sake of clearness and at risk of dualistic stereotypification, I systematically refer here to a general geoscientific framework which more or less shares a degree of socio-historical ignorance or in-difference, whereas 'geographies' refer to views informed by critical societal and cultural reflectivity.

[12] The repeatedly-used twin concept of 'land and life' covers here the reciprocal processes of humanity and nature in different socio-spatial scales. It also connects the concern of biodiversity to a broader dynamic of ecosocial and terrestrial differentiation, that is: geodiversity, especially characterizing the variation of lived landscapes of the (species-

are beyond the academic scope – which is, in many ways, a relief but also a place
for scholarly self-reflection. How to re-profilize the discipline of geography in
order to reach the credibility and legitimation needed to be taken seriously in the
North? How to avoid the traps of elitism? How to free us from the strains of the
(post)colonial era?[13]

This is the starting point and the basic question of this chapter. The argument
is that geographical imagination is deeply present in the North, and profoundly
represented by the Northerners themselves, and this needs to be considered a
challenge – both a source of critical self-reflection and inspiration – to academic
geographies. In this chapter, geographical inspiration is found in the pages of *N66.
Culture in the Barents Region* which is a new journal, published since 1997,
reflecting the cultural life of the European North. In the following pages, the
cultural geographies of the journal are examined against the centennial change in
academic geographies.

---

poor) North. The etymological and paradigmatic root goes back to Carl Sauer's writings
of the 1920s and 1930s, see John Leighly, *Land and Life. A Selection from the Writings
of Carl Ortwin Sauer* (Berkeley: University of California Press, 1993). See also Pauli
Tapani Karjalainen, *Geodiversity as a Lived World. On the Geography of Existence*
(Joensuu: University of Joensuu, Faculty of Social Sciences, 1986); Kenneth Olwig,
'Recovering the Substantive Nature of Landscape,' *Annals of the Association of
American Geographers*, Vol. 86, No. 4 (1996), pp. 630–53.

[13] With the 'strains of (post)colonialism' I refer to those heritages sedimented in our ways
of thinking (and doing), rooting from the era of early colonialism but persistently (and
partly implicitly) alive in our present. The (post)colonial world/view is here seen as the
initial phase of critical self-reflectivity in the West (or, in the South, when seen from
the circumpolar North) and within the interrelations between the West and the 'othered
rest'. It is the early stage of weeping out – or, unlearning – the current colonial content
in our societal practices. Derivatively, drawing a strict line between the colonial and
postcolonial eras is seen here as a premature demarcation between the past and the
(purified) present. It is for this reason I place the prefix of post in parentheses. Derek
Gregory has profoundly described the gradual process of becoming increasingly aware
of the colonial presence and "my own otherness" during his inspiring (auto)geographical
voyages, see Derek Gregory, *Geographical Imaginations* (Oxford: Blackwell, 1994).
See also an application of (post)colonialism into Arctic conditions, Juha Ridanpää,
'Postcolonialism in a Polar Region? Relativity Concerning a Postcolonialist
Interpretation of Literature from Northern Finland,' *Nordia Geographical Publications*,
Vol. 27, No. 1 (1998), pp. 67–77, where he refers to postcolonialism both as "a form
of social criticism, an emancipatory way of criticizing an unevenly developed world"
(p. 67) and as "a continuity of preoccupations throughout the historical process initiated
by European imperial aggression" (p. 75).

N66 and the Barents Dimension

Barents cooperation is the frame in which many of the new initiatives of the European North have been formulated. The new framework emerged in the early 1990s, soon after the collapse of the Soviet Union. The arctic part of Northwestern Russia can now be more easily and directly contacted from the centres of the Nordic North Calotte, and the Barents Region can be defined as the northernmost Europe stretching from the Lofoten Islands to Novaya Zemlya[14] (Map 2.1 Andrew Stevenson, 'The Barents Region').

In Western European eyes, the Barents Region, or officially the Barents Euro-Arctic Region (BEAR), is the extreme northern edge of the inhabited world, far beyond the limits of the central logistic maps of the continent. No essential reorientation to this mental setting has yet taken place due to Sweden's and Finland's memberships in the European Union in 1995. The Barents Region still functions as an area of resource exploitation within the (trans)continental economic development. It is also stuck in the military geopolitics as an interface between NATO and Russia. The colonial setting has not disappeared but, instead, become increasingly articulated and sensitive. The acute social and environmental problems are intensively reflected and criticized, and this is clearly an outcome of the comparative setting the cooperation across the borders has enabled.

The Barents framework cannot, however, solely be regarded as a tool for development critique. Rather, it has been established as an arena for formulating economically feasible alternatives to the (post)colonial regional division of labour. The regional emphasis is a sign of renewal which primarily takes place at the politico-economic sphere but, in addition, it is formulated as a cultural and communicative process with a specific regional emphasis. The way out of the current societal shortcomings is sketched and based on the shared experiences from the past: from the experiences of peripheralization and (post)colonial exploitation but also from within the deeper spheres of northern mentality and everyday life. Schooling and education, literary and publishing cooperation, performance activities of various kinds as well as research initiatives have taken shape across the old geopolitical curtains, and with the help of new electronic networks of communication. The northern space - given the physical constrains and distances of the Barents Region - is increasingly present and represented in the www world.

---

[14] See, for example, Jyrki Käkönen, *Perspectives on Environment, State and Civil Society. The Arctic in Transition* (Uppsala: University of Uppsala, Environmental Policy and Society, 1994); Ulf Wiberg, 'Regionformationer och nordeuropeiska integrationsperspektiv,' *Nordisk Samhällsgeografisk Tidskrift*, No. 21 (1995), pp. 55–67; Leo Granberg, ed., *The Snowbelt. Studies on the European North in Transition* (Helsinki: Kikimora Publications, 1998).

Map 2.1 Andrew Stevenson, 'The Barents Region,' 1999 (used with permission of Kenneth Mikko and *N66. Culture in the Barents Region)*

One essential element of this cultural turn of the North is *N66. Culture in the Barents Region*, a journal which has broadly developed the 'options' provided by the new multi-media age.

Arctic Land- and Lifescape

The editor in charge, Kenneth Mikko, outlines the basic idea of *N66*: It is "an invitation to participate in this reunion by making room for writers, artists, photographers, scholars and culture workers from the whole region – and give all the readers the possibility to take part in thoughts and feelings, visions and nightmares in the region".[15] The first issues of the journal concentrate on the historical and cultural diversification of the Northern peoples: dispersed and divided according to streams of settlement and geopolitical reorganizations. However, as part of the gradual differentiation and fragmentation, a deeper connectivity, based on shared and similar customs, attitudes and experiences, is identified. Ancient interdependencies, including similarities in archaic beliefs and interlinguistic connections, are introduced as events and phenomena present in everyday life practices. The drama of seasonal changes, the silence of the vast landscapes and the low blue light of the long winter months form the basis of shared identification, northern identity and belonging.

The common difficulties caused by the peripheral colonial location are indelibly sedimented in the Northern cultures and landscapes. The invaders and exploiters have, repeatedly, come from the South and brought with them the religions, economic models and political patterns of their own – to be adopted by the Northerners. There is distrust towards the South but – in the pages of *N66* – it is reconstructed as a basis for cooperation. The North is seen as an alternative to the modern South.

N66 concentrates mainly on the mytho-poetic dimension of the North: the 'edge phenomenon' – the lived world of the extreme Arctic – is elaborated on within the current renewal in literature, music, visual arts and multi-media performances and in connection to various ethno-religious re-formulations. Especially photography emerges as a key to the *N66*s Northern approach; the photographs grow into stories of land and life – and they also serve as a common language in the multi-lingual Barents Region. In the photographs, the Northern people stand on their own lands, they breath within their own landscapes. The 'lived edge' seems to reconnect the present to the ancient past: all generations have faced and lived within the same

---

[15] Kenneth Mikko, 'Dear Reader!' *N66. Culture in the Barents Region*, No. 1 (1997), p. 2.

edge. This archaic connection is essentially deeper than the shared experience of the historical (post)colonialism 'introduced' from the South.

The complementarity of languages has become the speciality of *N66*. Most of the articles are accompanied with a translation to one of the neighbouring languages, and often an abstract in English is attached, too. This kind of multilingualism is a sign of pluralism and richness, which characterizes both the journal and the regional world of which it is part. In the North, through centuries, the social world has been constructed by necessity as an arena of several coexisting languages, coloured by numerous socio-spatial dialects and lingual mixtures. This is inevitably the most concrete cultural outcome of the historical forms of (post)colonial domination and submission. For an individual, this has brought an option for a multilingual identification, which is, undoubtedly, a strength of the Northern peripheries. You had to learn to become sensitive to other 'word and world orders' and different ways of perceiving the world.

N66 takes multilingualism as a natural and characteristic feature of the North. It is also developed into an art. The frequent crossings of lingual borders deepen the understanding of the lands and lives of the Northerners. Moreover, the question is not only about languages but about identification: how do we define the North and its com/position as part of ourselves? The northern cultural awareness, distinctiveness and self-respect, emerging in the pages of *N66*, turn into a generative regionalistic movement: the past is not a record of intrusions from the South but, instead, a source of inspiration. A view to indigenous cultural continuity is opened, despite the humiliating processes of submission, or sometimes even, due to the shared experiences related to them. The North emerges as a multilayered land- and lifescape, simultaneously existing on several levels of collective awareness and identity.

As has become clear already, *N66. Culture in the Barents Region* is an important programmatic vision and, as such, a contact field for numerous diverging ambitions. One cannot avoid the question: how to represent the vast Barents area and its cultures by an 'umbrella' derived from the current regional political renewal in Northern Europe? How natural a regional unit is the Barents Region, when seen in a historical context or from the point of view of local cultures? Undeniably, *N66* looks 'from above', from a certain elite position down to everyday life circumstances and exciting local pasts. The journal can also be occasionally caught exoticizing the unknown Finno-Ugric East of the Barents region. The collaborative network is based on the new technology, it is stretched as electronic links between the main nodes. The fluid space of high-tech communication is not a familiar world for everyone in the North and, therefore, several of the cultural spheres become underrepresented or even excluded from the networking. The journal hence seems to face a risk of becoming abstracted from the 'real' lands and lives of the North, in a similar manner to those sophisticated geographers from the South we met earlier in this chapter.

The problem seems to appear unavoidable in this kind of multicultural and experiental collaboration across cultural borders. Programmatic initiatives turn into preliminary definitions of the content. Coordination emerges as unification. Regionalistic actions are – by some kind of historical necessity[16] – at risk of becoming either provincial adjustments to messages 'from above', or territorial divisioning of lands and lives of the people 'down there'. However, as will become clear in the proceeding pages, *N66* offers regionalism which is sensitive to the problems of representation and domination. Therefore, the message of *N66* is primarily read as an inspiration to geographers, and to scholars of the related sciences with an interest in the North. The question simply is: how to face the challenge of multicultural geographies emerging in the pages of *N66*?

Topography of Dwelling

> In every village the church rose above all other buildings. The bell tower with its blue domes was the highest building in the village and its white-stone walls with the green roof could be seen from a long distance. Near the church there were public buildings: schools, the house for village gatherings, the library.[17]

The above *N66* excerpt was originally published in Pitirim Sorokin's autobiographical memoirs under the title *A Long Journey*. In this book, Sorokin (1889–1968), a well-known Russian-American sociologist, writes about his childhood in the Komi village of Turja by recalling the memory traces of the physical and symbolic landscape of his initial home region. Turja is the embryonic place he is part of and it is the place he carried with him across the Atlantic. In *N66*/10, Dmitri Nesanelis and Victor Semjonov base their article on Sorokin's memories, and explicate further the symbolic dimension of dwelling in Komi. For them, both the traditional villages and houses presented an organized microcosmos, *i.e.* a social place of symbolic signs and arrangements which reflected the shared world view. The symbolic order was an expression of laws, customs and traditions, reproduced via generations.[18]

The interiors of Komi houses were situated around the stove and the 'red corner' of icons as circles and layers representing family life and broader social

---

[16] See Tim Ingold, *The Scolt Lapps Today* (Cambridge: Cambridge University Press, 1976); Ari Aukusti Lehtinen, 'Kalottipolitiikka ja saamelainen regionalismi,' *Terra*, Vol. 99, No. 1 (1987), pp. 13–18.

[17] Dmitri Nesanelis and Victor Semjonov, 'The Village Universe,' *N66. Culture in the Barents Region*, No. 10 (1999), p. 15.

[18] *Ibid.*, pp. 14–16.

connections. The symbolic organization of the houses, divided into different sections for the members of the family, visiting relatives and strangers, became especially significant in such rituals as weddings and burials. These rituals were the most important events within the families, as they implied remarkable changes in the family structures as well as in the arrangement and use of the house interiors. The yard around the house correlated with the sacred extension: household buildings, saunas, wells, chapels, crosses and worshipped springs all together constituted both the practical and the religious realms of daily life. The indoors and outdoors were concretely arranged as horizontal and vertical layers of family life and these layers also functioned as gates to the symbolic and sacral worlds. The larger neighbourhood units were built along rivers and tracks as compact settlements connecting the past to the present: new buildings were designed with respect to the will of ancestors and their spirits which still had a connection to their earthly home base. The village topography hence became the visual manifestation of the cosmic exchange between the present and the eternal, gradually landscaped into a spatial narration, or a collection of architectural forms as shared memories across the generations.[19]

In *N66/8*, Lauri Anttila, a Finnish architect and artist, describes in a parallel manner the lifestyle and traditions of the nomadic Innu people in Labrador, Canada. The traditional Innu tent is a product of the land and life in the North; it is a 'summary' of inherited customs and knowledge of surviving within the conditions of the extreme North. Anttila notes that the variation in the steepness of the coneshaped tent is connected to the latitude: the more northern tents are lower and less steep, giving better heating economy to the building. According to him, the nomadic Innu lifestyle is an expression of dwelling in which "the whole landscape and nature is their living room, and that their life view is holistic – the people and the nature are one".[20]

The layered interpretation of dwelling comes close to humanistic geographies of the lived and perceived worlds. Pauli Tapani Karjalainen has read Bo Carpelan's *Urwind*, a diary, or a collection of weekly letters written during one year by Daniel Urwind to his wife (who had gone far away from home), and found three different ways of reading them. Mimetic reading aims at constructing a 'physical map' of the letters, whereas hermeneutic reading sketches the contours of the person and his spatial extensions; the 'map of the mind' consists of such metaphoric oppositions as internal/external, past/present, winter/summer as well as cellar/attic. Moreover, 'textual mapping' focuses on the changing relations between the letters and the reader. The author, or subject, is replaced by a continuous re-creation by

---

[19] *Ibid.*

[20] Lauri Anttila, 'Tents and Fireplaces in Labrador,' *N66. Culture in the Barents Region*, No. 8 (1999), pp. 8–9.

the reader. Here, the reader, or the geographer, is invited to return to a nomadic, mobile position.[21]

This kind of orienteering and mapping, *i.e.* circular return to the threshold between the familiar and the unfamiliar, is part of the nomadism between such modern polarities as awakening and fall, creation and erosion, dawn and edge, known and unknown. Inger Birkeland, inspired by Julia Kristeva's psychohistorical interpretations, finds the same rhythmicity and in-betweenness in the specific context of Northern travelling. Her heterogeneous subject, composed by semiotic and symbolic dispositions, travels from home – *i.e.* from *chora* – to the outer boundaries of the subject and society. *Chora* is the beginning, the initial home, reached by spatial intuition and semiotic projections, and it comes before the entry to the symbolic world. According to her, "for a woman traveller the journey starts with a departure from home with the separation from the mother, to the discovery of the father, associated with language, the Phallus and the symbolic world".[22]

Birkeland also argues that it is possible to see the journey from home to the North as a discovery of the mythical projections. The unfamiliar is made familiar by repeating the creation myths of the home land by the rituals awakening the consciousness of the new land. According to Birkeland, the northern mythology, beyond the patriarchal monotheism of Judeo-Christian religion, is an expression of polytheism which took gradually shape during the risky habitation of the North, as part of the struggle against darkness, coldness and winter. The Northern sun stands at the centre of this eternal struggle between light and dark. The sun is associated with fertility, creation and healing. From here rises the magnetic experience of the midnight sun. This deviates from the pre-Christian sun of the peoples from the more southern latitudes which often took the shape of an authoritarian eye of the universe, holding the powers of destruction.[23]

The cosmology of the Komi villages, the traditional tents of the Innu people, the intimate layers of *Urwind* as well as the poetic nomadism encircled around *Chora* invite us to look at the layered coexistence of lands and lives of the North. Instead of fixed positions and panoptical perspective landscapes, the heterogeneous subject finds diversification in archaic repetition and continuity in variation.[24] In other words, the actual coexistence of layered and shared spheres of the North

---

[21]   Pauli Tapani Karjalainen, 'Elämä tässä talossa: lähikartoitusta kirjallisuuden kautta,' *Alue ja Ympäristö*, Vol. 24, No. 2 (1995), pp. 15–24.

[22]   Birkeland, 'The Mytho-Poetic in Northern Travel,' p. 26; see also Julia Kristeva, *Revolution in Poetic Language* (New York: Columbia University Press, 1984).

[23]   Birkeland, 'The Mytho-Poetic in Northern Travel,' pp. 28–30.

[24]   See Olwig, 'Landscape as a Contested Topos'; Jane M. Jacobs, 'Editorial: Difference and Its Other,' *Transactions of the Institute of British Geographers*, Vol. 25, No. 4 (2000), pp. 403–407.

questions the horizontal and vertical fixations (e.g. in geopolitics), and it also aims at deepening the understanding of the indigenous population as an integral part of the North. It asks us to critically reflect upon the various 'ecologies' offered from the South as guarantees of sustainability in Northern development programmes.

A Gate to the Other Side: Pendulum

"The theme of swinging in a traditional world view can be considered as one of the methods of overcoming fatal indefinition and unpredictability", argue V.E. Shaparov and D.A. Nesanelis in *N66/2*. They summarize that the act of swinging in the Northern folk cultures and shamanism can be interpreted in several different ways: inherently, it is seen as the symbol of cyclical time, or a central element in a ritual of purification. Swinging has also been connected to agricultural cults and even to symbolic banishment of evil forces. It has been one of the forms of traditional amusement and, undoubtedly, imitation of a sexual act.[25]

Shaparov and Nesanelis have examined the role of swinging in wedding and funeral rituals among the Komi people and found parallels to their traditions of the Easter swing. The construction of the Easter swing – originally stretched between two living trees – correlates with the image of gates and, consequently, the act of swinging becomes a symbolic crossing of the border between two worlds. In general, the archaic ceremonial acts of swinging – as part of the wedding games, coffin swaying in the funeral procession – were often linked to rituals of creation, birth and rebirth, or to deny death. The pendulum, rooted to the lived rhythmicity and repetition, functioned as the gate to layered space-time, a connection between self and other, here and there, motion and emotion.

The sensitivity to the layered presence of humanity/nature or past/present is connected to the respect for the forces that work beyond the reach of our senses; it also grows from scepticism toward geoscientific constructions solely based on measured entities. We can ask, for example, how 'real' the archaic beliefs – perhaps still remarkably conditioning our ways of thinking and being – are in comparison with those explanations and arguments about ourselves which have become established through exact testing and professional dispute? Evidently, no proof exists that the geoscientific scope currently accepted as the firm ground for geographical practices – the shared *chora* of geoscience – would not one day be revealed as one more cultural narration rooted to belief systems we do not fully recognize yet. Therefore, in order to save us from such disappointments, we could look forward to the layered presence of the human world, characterized by a deep

---

[25]   V.E. Sharapov and D.A. Nesanelis, 'The Theme of Swinging in Shamanism and Folk Culture,' *N66. Culture in the Barents Region*, No. 2 (1997), p. 2.

rhythmicity of earthly lands and lives. In this way, we could also get closer to understanding the immanent forces of humanity/nature. In addition, by learning to listen to the rhythms beyond the individual self and behind exclusive hereness, we might also learn to respect the variations of earthly rhythms in the different timescapes in which we are inevitably enveloped.[26]

The geography of coexistence links our corporeal rhythms – nature closest within – into broader socio-environmental timescapes. In the North, daily routines are conditioned by dramatic seasonal variations, which – in turn – are dependent on the geohistories of climatic change. Post Ice Age drama of the millenniums is challenged by the most recent human-influenced atmospheric warming which has a direct impact on the annual extension and variation of snowscapes. In the North, the advent of snow cover, the dawn of the snowlight, is awaited and debated every fall, *Aurora Borealis* decorates the clear night skies beyond the city lights, and the return of the sun above the horizon marks the onset of spring. This nature of winter cities in the North is currently under intensive contestation: the Southern model of winter cities guides towards indoor spaces and, hence, aims at freeing the humanity from the seasonal rhythmicity whereas the traditional Northern model is based on the adaptation and utilization of seasonal variation, for example in architecture, transportation and individual clothing.[27]

Amusingly, *N66*/10 introduces kick-sleds as part of the winter cities programming, debated in winter conferences in Luleå and Kiruna in the spring of 2000. *N66* reports that in Jukkasjärvi, Northern Sweden, the kick-sled has currently replaced local busses and is provided to the citizens for free. The embodiment of winter is then described as follows: "With the kick-sled, the friction is reduced to a minimum, the resistant world is softened up. At the same time, the kick-sled rider is in contact with the soil and hovers over it. The short thrust with one leg – the concentrated and firm contact with the ground, is then rewarded with a long period of relaxation – a lustful cancellation of the inevitable inertia of the natural forces."[28] This is indeed an excellent piece of the layered geography of winter.

---

[26] See Barbara Adam, *Timescapes of Modernity. The Environment and Invisible Hazards* (London: Routledge, 1998).

[27] See Jorma Mänty and Neil Pressman, eds., *Cities Designed for Winter* (Tampere: Tampere University of Technology, 1988); Jaana Nevalainen, 'Ikuisesta kesästä talvikaupunkiin – vuodenajat suomalaisessa kaupunkiympäristössä,' *Alue ja Ympäristö*, Vol. 24, No. 2 (1995), pp. 6–14.

[28] Kenneth Mikko, 'Just Kicking Around,' *N66. Culture in the Barents Region*, No. 10 (2000), p. 4.

Alternative Mappings

It can happen, suddenly, unexpectedly, and most frequently in the half-light-of-glimpses, that we catch sight of another visible order which intersects with ours and has nothing to do with it. Our customary visible order is not the only one: it coexists with the other orders. Stories of fairies, sprites, ogres were a human attempt to come to terms with this coexistence. Hunters are continually aware of it and so can read signs we do not see. Children feel it intuitively. Dogs, with their running legs, sharp noses and developed memory for sounds are natural frontier experts of these interstices.[29]

The pages of *N66* can be read as an introduction to alternative ways of mapping. Examples of historical and topographic cartography help our orienteering through the North, as do the imaginative maps, too. In addition, examples of logistic and thematic maps are regularly presented in *N66* and, even, satellite images. Furthermore, a special collection of maps of the Barents Region 'from above' – *Barents från ovan* – was published in *N66/7*. The narration starts by Abraham Ortelius' map of Bjarmaland from 1570, illustrated by a sea monster in the Arctic Sea, and continues through decenniums up to currently ongoing digitalization of a common geographical database to be monitored online on the Internet. On the other hand, ethnic mapping by Hans Ragnar Mathisen from Tromsdalen, Northern Norway, is included. Mathisen uses Sámi names in his cartographical and cosmological artworks.[30]

In *N66/4*, Mathisen tells that using maps as we know them is not common among the Sámi people but, instead, the maps became concrete in their language and common memory: "The topographic names of the Sámi are very detailed, and give a lot of specific information about a place and its qualities – height, vegetation, landscape forms etc."[31] He also underlines that Sámi shamans could construct maps of both their "spiritual and real worlds", and these were imprinted on their sacred drums. For Mathisen, drawing these maps is a way of showing the invisible history and culture of the Sámi people.

Within academic geographical practices, the cartography of the North has always been loaded with the tension of mapping-as-expansive-controlling vs. mapping-as-mutual-learning. Väinö Tanner, one of the first cultural geographers with an interest in the Sámi regions, experienced this tension while leading the field research of the Commissions of Reindeer Herding – *Renbeteskommisioner*,

---

[29] John Berger, 'It Can Happen,' *N66. Culture in the Barents Region*, No. 1 (1997), p. 28.

[30] Kenneth Mikko, 'Barents från ovan,' *N66. Culture in the Barents Region*, No. 7 (1999), pp. 10–15.

[31] Kenneth Mikko, 'Hans Ragnar Mathisen. Samernas kartograf,' *N66. Culture in the Barents Region*, No. 4 (1998), pp. 2–4.

1910–1917 – and, later, in his geological surveys in Petsamo during 1924–1931. Tanner repeatedly broadened the official geosurveys of the North to cover the indigenous Scolt Lapps and their ways of living, and this was criticized in the Annual Reports of the Geological Commission. It also caused trouble for Tanner when his scientific merits were evaluated for higher academic posts. Tanner, however, argued that the cartographers should not focus their interest solely on the physical terrain but, with a similar enthusiasm, on the indigenous place names and their symbolic and aesthetic meanings, too. Tanner received a chair in geography at the University of Helsinki in 1931, but with merits other than his 'philological-sociological' interest in the Scolt Lapps, which was then considered as non-geographical.[32]

This is still an open and aching question in geography. Practical mapping of the Barents Region 'from above' has become easy with the help of satellites and GIS technology, but the invisible landscapes of the North, only comprehensible from the inside, have been broadly left unexamined, or at best, received only limited theoretical attention. The geographical approach cannot possibly be expanded to reach all the spheres of humanity/nature of the North but, at the least, the unreflective ignorance toward the apparent cultural domination and bias caused by the expansive mapping technologies should be weeded out from the academic geographical expertise. Hence, we again face a variation of the same root question, initially formulated in the introduction: how to free ourselves from the (post)colonialistic manners inherently hidden in the narrations of the institutionalization of geography? Given the disciplinary past, is it at all possible for professional geography to learn to respect the various modes of mapping we humans have constructed in different times and places?

Wounds on the Edge

They have been screaming for ten years: in front of the Arctic Sea, at the Finnish president's reception, at the opening of the Olympic Games, outside the Eremitage in St. Petersburg. They are the Men's Choir Huutajat – 30 rather young men from Oulu, dressed in formal, dark suites with rubber ties. They fill their lungs with fresh air, then let it out with tremendous force, screaming hard – traditional folklore, worker songs, children's songs and new experimental compositions.[33]

---

[32] See Paulo Susiluoto's contribution in this book.

[33] Kenneth Mikko, 'Screaming all over the World,' *N66. Culture in the Barents Region*, No. 4 (1998), pp. 6–7.

Petri Sirviö, the conductor of the *Huutajat*, mentioned the impromptu performance at the Arctic Ocean as one the highlights of the tour. The concert without an audience raised strong emotions among the group. The scene painted in front of our eyes – a choir of formally-dressed men facing the vast polar sea masses, screaming the lores of the North – is loaded with paradoxical drama: humanity against and within nature. Strange performers standing proudly on the edge, clearly not belonging there and, therefore, separate from the harsh surroundings but, simultaneously, becoming one with the forces of the North. Extreme human voices drowning in the emptiness. "The Arctic is most likely silence. Why should we violate the prohibition and talk about it?" asked Ilya Kabakov and Pavel Pepperstein and they formulated the same paradoxical remark about the multilayered coexistence of humanity, felt so concretely and thoroughly on the edge.[34] Words are (only) derivations from the already known; in emptiness, shouting turns to whisper. The question is worth noticing again: how deeply sedimented is it in (masculine) scientific enterprise to restlessly look beyond the horizon, to posit and measure oneself against the eternal?

"It is precisely in the North that the continents are anchored to the space of the boundless white ice, where everything ceases in the face of total absence, lifeless emptiness, White Nothingness. It is precisely in the North that the Earth touches the Sky", formulate Sergei Bukaev and associates, and join the *Huutajat* choir.[35] Here, the North emerges as a gate to the other world, a 'place of power', and grows into a condition for our cultural continuity, a framework to our earthly existence and identification. It becomes a zone beyond human reach and a source of cultural renewal. It functions as a source of inspiration, the focal point of circular re-creation and continuous return – at least till the day it has lost all its secrecy in our minds. Till it has become thoroughly mapped and tamed and made a finite entity: the end of the earth.

"In the Nenets tongue, Yamal translates roughly as 'the end of the earth' and it is an apt description. The Yamal Peninsula juts out several hundred kilometres into the Kara Sea just east of Novaya Zemlya and is underlain by frozen ground, or permafrost, at depths which range up to 300 metres in places", describes Bruce Forbes in *N66/3*. Yamal is undergoing a large-scale industrial development phase due to its newly found oil and gas resources. According to Forbes, slightly over half of the 9000 indigenous people of the *Yamalskii Raion* still lead a nomadic or semi-nomadic life under the control of three large state farms (*sovkhozy*). In Yamal, the era of petroleum and gas explorations started in the late 1970s and it

---

[34]   Ilya Kabakov and Pavel Pepperstein, 'Tennis Game,' in Marketta Seppälä, ed., *Strangers in the Arctic. 'Ultima Thule' and Modernity* (Pori: Pori Art Museum, 1996), p. 152.

[35]   Sergei Bukaev *et al.*, 'Vorkuta,' in *ibid.*, p. 125.

has already – before the actual oil/gas production – completely transformed the lands and lives of the Nenets people. Drill pads, road and railway constructions, geological surveys, off-road vehicle traffic and quarrying for gravel and sand have turned into eutrophication of waterways, sand and dust blows, and they have changed the land-use patterns of the reindeer herders.[36]

According to Forbes, not all Nenets have condemned the industrial development; it has brought along modern alternatives to traditional modes of living, for example paid work, schools, health care and faster transportation. In other words, the circles of life have enlarged. However, life is on the edge again: how will the extreme North – already markedly modified by nomadic 'land and life' modes and overgrazing by reindeer *sovkhozes* – react to the expansive industrial activities? Could it still in the future be regarded as a source of cultural renewal? Forbes himself, a researcher from Rovaniemi, Finnish Lapland, concludes that "in the best of all possible worlds, any future changes will be made according to the choices of the people themselves, for whom their culture is at stake, and not imposed by the minions of remote government bureaucracies, petroleum companies, or well-wishing scientists".[37]

Both the traditional and modern communities of the North face the same condition, the challenge of circumpolar nature. The North is an oxymoron. It is characterized by unexpected changes, but within the variation already known, including spasmodic bursts from within a calm silence and radical difference in complete indifference. Everything can change rapidly, for example with the sudden rise of the Northwind and, just as quickly, return to the calm. According to Pavel Limerov, the wind (from the North) exists in the background of the mythological topography of the Komi people, determining the land of the dead to be situated in the North. Traditionally, the northern wind was considered as punishment for the violation of some prohibitions, emerging from the 'next world'. It could freeze crops and bring disease, corresponding to the transition from the realm of life into the world of death. The Northwind is the movement of the souls of the cursed and missing people; it carries the forces of abduction and it contaminates the lands and lives of the people by drawing them into the next world, permanently or temporarily.[38]

The peoples of the North have undoubtedly learned the intensive here-and-thereness of the environing nature; the aching memory of the events and accidents when some of us have been overtaken by the Northwind is shared. When in pain

---

[36] Bruce Forbes, 'Wounds on the Tundra,' *N66. Culture in the Barents Region*, No. 3 (1997), p. 16.

[37] *Ibid.*, p. 18.

[38] Pavel Limerov, 'When the Northern Wind Begins to Blow,' *N66. Culture in the Barents Region*, No. 4 (1998), p. 20.

you could only cry northwards. This sense of belonging is easily turned into a kind of fatalism, or submissive surrendering under the pressure from the outside. Life has become tenable by carefully listening to the signs of sudden change. You needed to know the breaks in weather beforehand. During storms, you had to learn to wait. Traditionally, the North was something to listen to, adapt to and shelter yourself from. However, the minds of the newcomers from the modern South were often too anxious to learn to listen and wait. The modern North was primarily defined as the storehouse of valuable resources and there was no apparent need to develop sensitivity towards the more layered North. On the other hand, the South had developed the worldview, and a self-legitimation, where the indigenous North was understood as the primitive other to be civilized. It was the white zone on the map, to be found and named according to Southern standards. The Northerners knew already how to adapt, now under the pressures from the South. They were prepared for an external invasion.

This is the same process – cross-cultural interfacing – which Kirsten Hastrup, professor of anthropology at the University of Copenhagen, has described as the emergence of *Uchronia*, a nowhere-land. The two histories of Iceland meet where the measured time of modernity overcomes the orally transmitted *Sagas* out of time; where the linear crashes with the eternally returning; where the modern time-spaces slice the layered memories of local communities. Hastrup writes about the North where the past that was familiar and always present becomes strange even to the Northerners themselves. She formulates the process of internal alienation initiated 'from outside', which gradually replaces the mnemonic past which used to keep the generations in touch with one another. From here emerges the modern 'other' – the other that now refers to the denied self, the externalization of the Icelandic *Sagas*.[39]

Hence, the North is the modern 'other' that has replaced – othered – the indigenous lived worlds in Iceland, in Bjarmaland, in the Sámi Aednam, Komi as well as in the Nenets' Yamal. There never was an indigenous North up there. The North is there only from a certain perspective, that is from the South. Therefore, the North disappears when seen from the inside, from here and, finally, on the edge it turns into the Northwind, arising from the other side. Accordingly, as Markku Heikkilä remarks in the introductory book to a photograph exhibition *Pohjoinen valokuva* in Oulu, "pure geography disappears and turns into relational one when we return into the mental world. The North becomes dependent on the

---

[39] Kirsten Hastrup, 'Uchronia and the Two Histories of Iceland,' in *Strangers in the Arctic*, pp. 36–49.

mindscapes of the individuals and communities they are part of."[40] This dimension of the North that, as Nikolaj Terebikhin has put it, is "independent of the laws of gravitation"[41] summarizes well the geo-scientific dilemma; the apparent powerlessness in front of the layered existence of human lands and lives. The North is different and this difference is invisible from the outside. It seems that we can only learn to sense it by respecting the difference and in this way, perhaps, gradually abandon the whole concept of the North by using more authentic georeferences shared by the Northern communities.

It becomes evident from the pages of *N66* that the sensitivity to difference can only be approached by turning to myths, beliefs and metaphoric vocabularies attached to the layered communities of the lived North which, in any case, guide and inform us across the generations. There seems to be no other way to restrain the dramatic internal 'othering' of individual and collective selves. This forces us to relocate ourselves, as geographers of the lands and lives of the North, to a new position within the processes we are studying. Then we might find ourselves completely embedded in the networks we have initially identified as our research fields. The North speaks through us and we speak through the North. The communities of the North or, more precisely, the lived communities beyond the (post)colonial concept of the North become one with the geographies of the lived North. Both are in a state of constant change. This transformation from modern geoscience to multilayered geographies seems, paradoxically again, inevitable – and unlikely, which is due to geography's indelible embeddedness in the early colonial modernization, where science was originally defined as the tool to de-mythicize the ancient (Medieval) irrationalities of humanity. On the other hand, along the rising postcolonial awareness, the modern scientific conceptualization of the arctic North "as the laboratory to collect empirical material or to test theories"[42] will certainly face increasing criticism from the communities under 'testing' – and from the side of humanities in general.

---

[40] Markku Heikkilä, as quoted in Arja Maunuksela, 'Kuvia pohjoisesta identiteetistä,' *Helsingin Sanomat*, 3 August 2000, p. B8 (a review on a photography exhibition 'Pohjoinen valokuva – identiteettien dokumentti' in Oulu, summer 2000, including an introductory book by Jukka Järvinen *et al.*, eds., *Pohjoinen valokuva – dokumenttivalokuvaus Pohjois-Suomessa* [Oulu: Pohjoinen valokuvakeskus and Kustannus-Puntsi, 2000]).

[41] Terebikhin, 'Bjarmaland,' p. 21.

[42] Lassi Heininen, *Euroopan pohjoinen 1990-luvulla. Moniulotteisten ja ristiriitaisten intressien alue* (Rovaniemi: University of Lapland, Arctic Centre, 1999), p. 25; see also Lassi Heininen, 'The International Situation and Cooperation in Change,' in *The Snowbelt*, p. 25.

Conclusions: From Geoscience to Geographies

As was witnessed, *N66* has produced a wide variety of alternative Northviews which have several parallels to the current renewal of academic geographies. Of course, both discourses are integrally linked to the streams of the broader society and, consequently, reflect upon the same trends and circumstances. They cannot avoid overlapping. However, *N66* has developed a multi-cultural profile and it also mirrors a broader cultural reorientation which cannot be ignored by the academic geographers of the North. On the contrary, it can be seen as a place of learning and, even, a 'vehicle' for unlearning the (post)colonial manoeuvres of our research practices. At least the following themes introduced by *N66* can be read as signals of the need of geographical updating.

First, the layered existence of the North questions the self-satisfaction of the cultural geographers laid down on the axis of 'politics, identity and landscape'. The North of *N66* is a land- and lifescape which varies within numerous scales: archaic, traditional and modern timescapes emerge side by side and, hence, not as chronological periods but as coexisting layers of which we humans are integral parts. Especially the archaic dimension is well illustrated by *N66* and, unfortunately, no corresponding elaboration can yet be found within current academic geographies. Some important views on psychohistorical geographies have been sketched but not in the context of the extreme North.[43]

Second, the North of *N66* is indeed a strange combination of marginality and centrality. The extreme edge magically turns into a source of regeneration; the land of death is concrete and existing in narrations and beliefs. The marginal places and zones of the North have attracted geographers and colleagues in the related disciplines from the beginning – as the white areas on the map to be named and developed. Only a few works can be found emphasizing the immanent being of the marginal(ized) North, e.g. in the spirit of Väinö Tanner during his 'field life' among the Scolt Lapps. Repeatedly, the geographical landscape has become established by an authoritarian and universalizing eye cast from the South and 'from above'. The socio-spatial fixations, built on the exclusive striving for correspondence between the physically real and its representation, are thoroughly criticized within today's geographies but the practical interconnection between the lived 'reals' and their intersubjective variations are broadly ignored. However, the invitation to geographical studies of collective cultural belonging in and through

---

[43]   See, for example, Gillian Rose, *Feminism and Geography. The Limits of Geographical Knowledge* (Cambridge: Polity Press, 1993); David Sibley, *Geographies of Exclusion* (London: Routledge, 1995); Steve Pile, *The Body and the City. Psychoanalysis, Space and Subjectivity* (London: Routledge, 1996). The only psychogeographical excursion towards the Arctic I am aware of is Birkeland, 'The Mytho-poetic in Northern Travel'.

rituals is already formulated[44] and the question about the 'geography of the sixth sense' introduced[45] and, hence, the road is already signposted. The North could be a perfect subject of these studies.

Third, the 'model' of mutual learning is starkly present in *N66*. In comparison, the geographical practices are still dominantly based on abstraction constructed according to scholarly interests, and this tends to distance the phenomena studied from their natural geohistorical contexts. The focusing on 'insular objects' – slices of humanity or nature – decisively limits the options for reciprocal learning. The identification and representation of 'geographical islands' as exotic others 'out there' only confirm the continuation of (post)colonial domination from the South. The study objects are defined as laboratories of land and life, or 'interesting cases' of theory testing. The question here is whether there will be imagined a study praxis that is free enough from the academic dependencies (derived for example from paradigmatic constraints, intertextual nihilism, meritocratic contestation, rules of publishing, etc.) so that it will settle down in a setting where the messages from inside the study focus can gain value of their own. This kind of immanency would naturally confuse the relations between the study subjects and objects.

Fourth, the programmatic North of *N66* turned into lived communities more or less embedded in local and shared life worlds: traditional villages and modern centres met the network communities of artists and researchers. The academic geographies have produced influential openings in community and identity research, for example in the field of historical *landskap* research and in connection to village studies, and these studies have remarkably strengthened the understanding of the historical spaces that are both socially inherited and societally produced.[46] However, much still remains unexamined, for example in the fields of socio-spatial commemoration and heritage construction as well as in connection to (post)colonial resource exploitation. Especially the contested settings between

---

[44] Paul Connerton, *How Societies Remember* (Cambridge: Cambridge University Press, 1989).

[45] Gunnar Olsson, *Lines of Power/Limits of Language* (Minneapolis: University of Minnesota Press, 1991); Gunnar Olsson, 'Heretic Cartography,' *Ecumene*, Vol. 1, No. 3 (1994), pp. 215–34.

[46] Sörlin, *Framtidslandet*; Michael Jones, 'The Elusive Reality of Landscape,' *Norsk Geografisk Tidsskrift*, Vol. 45, No. 4 (1991), pp. 229–44; Gabriel Bladh, *Finnskogens landskap och människor under fyra sekler* (Karlstad: University of Karlstad, 1995); Seppo Knuuttila *et al.*, *Kyläläiset, kansalaiset* (Joensuu: University of Joensuu, Karelian Institute, 1996); Olwig, 'Recovering the Substantive Nature of Landscape'; Petri Raivo, 'Maiseman kulttuurinen transformaatio. Ortodoksinen kirkko suomalaisessa kulttuurimaisemassa,' *Nordia Geographical Publications*, Vol. 25, No. 1 (1996).

local community preferences in the North and translocal regulations, introduced for example by the European Union or in the form of global intergovernmental agreements, would today need more careful examination.

Such a research emphasis would inform us about the background of the Northern criticism toward globalization and universalization. Local expressions from the North often seem to become labelled as provincial and outdated exaggerations in the dominating media in the South even though the problem, at least partly, rises from the experience of humiliation based on marginalization (from the central information flows) and deep, inter-generational experiences of injustice.[47] This is a direct outcome of the (scientific) insularization described above. *N66* offers an alternative to the insularization. The journal looks toward network communities but it also carefully listens to the traditional local voices and even develops new network relations through the experiences of the local communities. This kind of mutuality could be refreshing in academic geography, too. However, the academic orientation seems to follow a more 'futuristic' and, simultaneously, (post)colonial thinking by enthusiastically elaborating theories of fluid spaces built upon translocal or nonlocal lived worlds, perhaps becoming true only in the great sprawling cities of the South.[48]

Fifth, geographers of the North can no longer exclude themselves – as though 'purifying' themselves – from the symbolic and geopolitical contestation between

---

[47] See Pertti Rannikko, 'Combining Social and Ecological Sustainability in the Nordic Forest Periphery,' *Sociologia Ruralis*, Vol. 39, No. 3 (1999), pp. 394–410.

[48] Here I especially refer to the striking urbanization of disciplinary practices of geography during the last decades, which has taken place as part of the general Anglophonic globalization, see, for example, A.D. King, 'Opening Up the Social Sciences to the Humanities: A Response to Peter Taylor,' *Environment and Planning A*, Vol. 28, No. 11 (November 1996), pp. 1954–9. The geography discourses, including the rules of publishing, are formulated from and within the metropolises, which tends to annihilate the specificity of non-urban geographies of the peripheries. Non-urban peripheries become conceptualized as non-places – places that do not exist for the urbanized world at all, or at best, they have become 'othered' as zones of refreshment and recreation. See an interesting elaboration of this 'paradox of a peripheral place' by Ridanpää, 'Postcolonialism in a Polar Region?' pp. 74–5. The term 'fluid spaces' is derived from Edward W. Soja, *Thirdspaces. Journeys to Los Angeles and Other Real-and-Imagined Places* (Oxford: Blackwell, 1996), which is an inspiring and influential example of the more recent geographical avantgarde concentrating on the multilayered scenes of the postmodern city. Soja – in his own words – "re-opens the voyages of discovery" (p. 12) inside and outside the "postmetropolises" (p. 23) and develops a personal urban vocabulary of fluid cities under such signs as flexcities, cosmopolises, exopolises and simcities. These concepts turn, however, rather incomprehensible when reading them in the spheres of everyday life in the European North.

the regimes of the North. *N66* is part of the redefinition of the North from the direction of the South (via the establishment of the Barents Euro-Arctic Region), as are geographers – with their varying dependencies on the Southern-led universalization. Both the North and the geographers of the North face the same pressure of submissive adjustment under the current ideological discourse on globalization. Pressures 'from above' can easily turn into provincial adjustments or regionalistic actions, as was shown earlier in this chapter. Regionalism has been much debated by geographers, which might be due to its programmatic aura, decorating *N66*, too. The Northern regionalism grows occasionally into an *hauteur* in regard to the 'treasures' inaccessible or uncomprehensible from the South. On the other hand, provincialism is, for some reason, a rather neglected concept. In practice, however, provincialism – as a mentality characterized by a submissive and even opportunistic adjustment to the rules 'from above'[49] – is well developed in the North as well as among the geographers of the North. It grows into a painful reminder of our own dependence on the fluid networks of universalization, often becoming concrete in the form of tightening pressures to adopt a common and uniform language, English.

The neglect of provincialism in geographical literature might perhaps be due to the Anglophonic domination of the debate. It seems that the more recent southern interest in globalization and geopolitics is indelibly embedded in the still-existing (post)colonial structural networks within the English-speaking *ecumene* and, therefore, by necessity, reproducing the socio-spatial knowledge and practices of the (early) modern period. Provincialism simply does not belong to this vocabulary. On the other hand, the few references to the concept emphasize a specific interest in utopian and ideal connotations of provincialism, connected to agrarian authenticity, spontaneity, communitarianism, naturalness and antiurbanism. This seems to fit well into the vision of the universal Anglophonic word and world order supported by romantic and exotic labels given to peripheries. Initially, the 'noble savages' were of course pure constructions from the dominating centre.[50]

---

[49] Etymologically, provincialism is related to the conceding or submissive features of conquered territories, derived from the Latin *pro vincere*, see Michel Foucault, 'Questions on Geography,' in Colin Gordon, ed., *Power/Knowledge. Selected Interviews and Other Writings by Michel Foucault 1972–1977* (New York: Pantheon Books, 1980), p. 69.

[50] See Nicholas Entrikin, *The Betweenness of Place* (Baltimore: The Johns Hopkins University Press, 1991), pp. 66–78; Jouni Häkli, 'Manufacturing Provinces. Theorizing the Encounters between Governmental and Popular "Geographs" in Finland,' in Gearóid Ó Tuathail and Simon Dalby, eds., *Rethinking Geopolitics* (London: Routledge, 1998), pp. 131–51; Ari Aukusti Lehtinen and Teijo Rytteri, 'Backwoods Provincialism:

Finally, both the geographical discourse of the North and the discourse of *N66* become located rather similarly in the broader division of labour. Generally, the risks and the options are the same. However, in places, *N66* has been able to articulate the stratified character of the North far beyond the scopes of geography. I have underlined these *N66*-inspirations in this chapter and introduced them as challenges to geography to unlearn the (post)colonial heritage. As was argued, the communities of the North are different and this difference is invisible from the universalizing point of view. The universal narrations and knowledge even seem to turn into abjection when conceived from within a particular position.[51] The mythical North perhaps stands beyond the scope of de-mythicizing sciences – but, at least, the layered North invites the academic geographers and scholars from related disciplines to reflect upon the 'science myths' in which we have been immersed.

The Case of Kuusamo Forest Common,' *Nordia Geographical Publications*, Vol. 27, No. 1 (1998), pp. 27-37. Beyond the (political) geography literature, however, some analyses closely related to (internal) provincialism of the peripheries can be found: see, for example, Ingold, *The Scolt Lapps Today*, which concentrates on the ethnic assimilation problems among the Scolt Lapps, as well as Thomas Mathiesen, *Makt och motmakt* (Gothenburg: Korpen, 1982) who elaborates on the Scandinavian concept *vanmakt* (that is, essentially, a phlegmatic or opportunistic attitude towards the dominating power). See also Pehkonen, *"Tänne muttei pidemmälle!"*, pp. 31-9.

51 See Karjalainen, *Geodiversity as a Lived World*, p. 148.

Chapter 3

Shades of White:
An Essay on a Political
Iconography of the North

Frank Möller[1]

We can never understand a picture unless we grasp the ways in which it shows what cannot be seen.[2]

Imagining the Arctic: The Native Photograph in Alaska, Canada and Greenland at the British Museum and *Strangers in the Arctic – 'Ultima Thule' and Modernity* at Pori Art Museum in 1996; *The Waste Land: Desert and Ice. Barren Landscapes in Photography* at Vienna's Atelier Augarten, Centre for Contemporary Art in winter 2001/2002; *Ultima Thule and The Frozen Frontier* at Helsinki Museum of Cultures in winter 2001; *Arctic Inua – Arktisen henki* and *Pohjoinen hehku* at Tampere's Vapriikki Museum Centre in winter 2001 and spring 2002: recent conferences on and popular exhibitions depicting Arctic and Northern landscapes, places and peoples bear witness to a rediscovery of the North among museum curators and the general public. Images and pictorial representations of the North, including photographs, play a major role in most of these exhibitions. Ari Aukusti Lehtinen shows in his chapter in the present volume that photography is also a key element of approaching the North in print media. In particular in *N66/Culture in the Barents Region* photography provides a "common language" for peoples speaking different languages. In the photographs, Lehtinen argues, "the Northern

---

[1] The illustrations in this chapter were placed at my disposal by Muguette Dumont (Photothèque, Musée de l'Homme, Paris), Tanja Lintunen (University of Tampere, Department of Music Anthropology), Hannu Sinisalo (Tampere) and Jorma Puranen (Helsinki). Eija Kukkurainen (University of Tampere, Department of Music Anthropology) provided background information on Professor Erkki Ala-Könni. I greatly appreciate their interest in, and support of, the chapter.

[2] W.J.T. Mitchell, *Iconology: Image, Text, Ideology* (Chicago and London: The University of Chicago Press, 1986), p. 39.

people stand on their own lands, they breathe within their own landscapes."
Photography thus gives voices to peoples which in the course of unequal historical
processes have been successfully marginalized and are usually passed over in silence.
A political iconography of the North, like cultural studies in general, has to have an
awareness of both "the historical production of the cultural surfaces with which it has
to engage" and of that which is below the surface.[3] When it comes to pictorial
representations, it has to consider "the manner in which the visual *image*,
intentionally or unintentionally, has utilized and contributed to the way in which the
Arctic as a social, cultural and geographical space has been collectively and
subjectively *imagined*" – and an economic and political function may be added.[4]

To the viewer, pictures impart visual impressions "dissociated from and
independent of experience".[5] In photographs, the North is intellectually experienced
rather than physically. The resulting knowledge is a kind of comfortable armchair
knowledge, obtained, while browsing through catalogues, in our centrally-heated
apartments or in modern venues equipped with sophisticated technological subtleties.
Dissociated from physical experience, it is at best theoretical understanding. Yet even
then, I argue, it requires a politically and socially informed approach and, indeed,
detective work[6] in order to get behind the fascinating but often superficial, if not
misleading, impression of exotic and inconceivable, at least unfamiliar, landscapes;
temerarious, brave and determined explorers; and admirable but hardly enviable
peoples living in an environment obviously – or seemingly – hostile to human beings.
Only then is the viewer able to see behind the work of art, to see it as a "symptom of
something else"[7] – whatever this "something else" might be, the "intrinsic meaning"
of a work of art, as Erwin Panofsky argued, or its social and political function – and
to dis-cover something of which even the artist might not have been aware. Only then
it seems to be appropriate to speak of 'knowledge' in the first place. Rather than im-

---

[3]   Nick Couldry, *Inside Culture: Re-imagining the Method of Cultural Studies* (London,
      Thousand Oaks, New Delhi: Sage, 2000), p. 60.

[4]   J.C.H. King and Henrietta Lidchi, 'Introduction,' in their J.C.H. King and Henrietta
      Lidchi, eds., *Imaging the Arctic* (London: The British Museum Press, 1996), p. 11.

[5]   Susan Sontag, *On Photography* (London: Penguin, 1978), p. 156.

[6]   Stephen Daniels and Denis Cosgrove, 'Introduction: Iconography and Landscape,' in
      Denis Cosgrove and Stephen Daniels, eds., *The Iconography of Landscape: Essays on
      the Symbolic Representation, Design and Use of Past Environments* (Cambridge:
      Cambridge University Press, 1988), p. 2.

[7]   See Erwin Panofsky, *Studien zur Ikonologie der Renaissance. Zweite Auflage* (Cologne:
      Dumont, 1997), p. 34.

parting knowledge to the viewer, thus, viewing – and understanding – photographs presuppose knowledge.

A political iconography is interested in pictorial and symbolic representations and how they relate to society. With respect to historic photographs of the North, for example, a political iconography is not only concerned with the explorers' personal motives to use photography in the course of their journeys and the objects they, socially conditioned, chose to photograph. The issue is also one of what states, admiralties, navies, Geographical and Royal Societies and so on as official sponsors, supporters, financiers and orderers of explorations made of the photographic documentation of the explorations and with what aim they ordered them in the first place. Note in particular the intimate relationship between the explorers and their homelands: "Existentially, [the explorers] [did] not feel 'at home' at home, but ideologically they [did]: while almost constantly tempted by the frozen world they remain[ed] proud of their native place, nation, and culture."[8] Thus, more often than not, photographs became part of strategies to legitimate exploration, colonization, domination, existing power structures and nation-state building projects – usually at the expense of the native peoples. Yet, equally relevant is photography's potential for counter-power and emancipation. Since culture "is not an eternal, or ahistorical, force" the questions indeed are "what use existing states and contenders for state power are making of culture [as well as] how culture is an independent influence on politics, or how it is weakening the power of states."[9] How to approach these questions is the subject of what follows. Emphasis is on photographs although some remarks on other forms of pictorial representation are included. The focus is on the European North although some examples are taken from the Canadian North. The reader should keep in mind that, although parallels can be assumed, developments in the American North and the European North do not necessarily have to be identical.

The chapter is divided into four subsections: I begin with discussing historic photographs in relation to explorations and nation-state building projects. Then, I sketch the connection between culture, hegemony and photography. This section brings me fairly far away from the North to which I return in the third subsection with an interpretation of Jorma Puranen's photographic allegories of the North. Finally, in subsection four I argue for a consistent contextualization of photographs rather than re-imaging the North.

---

[8]   Yi-Fu Tuan, 'Desert and Ice: Ambivalent Aesthetics,' in Thomas Trummer, ed., *The Waste Land: Desert and Ice. Barren Landscapes in Photography* (Vienna: edition selene, 2001), p. 96.

[9]   Fred Halliday, 'Culture and International Relations: A New Reductionism?' in Michi Ebata and Beverly Neufeld, eds., *Confronting the Political in International Relations* (London: Macmillan, 2000), p. 58.

Then

Depicting Northern landscapes and peoples in photographs is not a new phenomenon. For example, Salomon August Andrée, in his endeavour to conquer the North Pole with a hydrogen balloon in 1897, carried on board with him a specially constructed stereoscopic camera with which to measure the topography of the ground for later reconstruction at the drawing board as well as the latest Eastman leaf film on rolls. Plans to carry with him a dark room and colour sensitive films were abandoned, but a portable dark room was used for cassette storage while changing the films. Films were conserved by means of brass cylinders. Instead of using the camera for cartographic work, however, the camera – no doubt skilfully used by one of the crew's members, Nils Strindberg – became an instrument with which to show the daily life and struggle for survival on the ice after the balloon was stranded after only three days of flight. Some of the photographs could even be developed after the remains of the expedition were found 33 years after their disappearance.[10]

Andrée was not the first explorer to use photography. Introduced in 1839, early photographic techniques such as daguerreotype and calotype were not especially suitable for field work, the one due to its dependency on a controlled studio environment, the other due to its slowness and inconsistency. All the same, a handful of daguerreotypes and calotypes of Arctic landscapes and people exist. In the 1850s field work was facilitated by technical further developments introduced by Frederick Scott Archer and others. Photographic (calotype) documentation of the Arctic dates probably back to the Belcher expedition of the summer of 1852. The first published work to include photographs of the Arctic resulted from William Bradford, John Dunmore and George Critcherson's excursion to Melville Bay in 1869. Given the limitations of the time such as low film speed requiring long exposures and slow emulsion speeds making enlargements impossible, it achieved remarkable technical results.[11] Fridtjof Nansen in 1888 used photographs to document his journey across the Greenland ice. The University of Oslo's Conservation of and Access to the Fridtjof Nansen Photographic Archive Project contains some 8,000 old pictures, 3,500 unique photographs, drawings and some watercolours. The picture library at Norsk Polarinstitutt in Tromsø even consists

---

[10]   Andréemuseet, Gränna, Sweden, at < http://andree.grm.se >.

[11]   Rainer Fabian and Hans-Christian Adam, *Frühe Reisen mit der Kamera* (Hamburg: Stern, 1981), pp. 63–78; Douglas Wamslay and William Barr, 'Early Photographers of the Canadian Arctic and Greenland,' *Imaging the Arctic*, pp. 39 and 42; Nicholas Whitman, 'Technology and Visions: Factors Shaping Nineteenth-Century Arctic Photography,' in *ibid.*, p. 31.

40,000 pictures from polar areas, some of them dating back as far as 1872.[12] Photographs thus have for a long time been used to fill empty spots on (not only mental) maps. Arguably, they contributed to what Yrjö Haila calls "gaining plausibility for the idea that the whole globe, including its most remote and hostile corners, is 'by its nature' subjected to human domination."[13] To this end, the globe had to be topographically scrutinized and measured; photography appeared to be an ideal and reliable instrument. Thus, it was certainly serving scientific interests in mapping places yet unrecorded, but geography also served as a pretext upon which other aims were being pursued: "Geography must be served. [...] The rhetoric was not always convincing even to those who made it." For some explorers, their "voyages into the geographically unknown [were] also voyages of self-discovery."[14] Furthermore, photographs should support the explorers' thirst for glory and contribute to their personal fame, including posthumous fame. They were used, too, to prove their claims in view of both competitors and an audience which increasingly replaced its initial uncritical adoration with scepticism. Notoriously facing difficulties to finance the expeditions, explorers used photographs as illustrations for books, magazines and slide-shows and in fund-raising campaigns.[15] Photography thus served the interests of the photographers rather than those of the objects of photography. Arguably, it resulted in a confirmation of the explorers' prejudices towards the indigenous peoples and the, probably mutual, perceptivity gap between the native peoples and the visitors/intruders. Nick Couldry reminds us that "[m]apping the 'cultures' of others is now generally understood within anthropology to be an instrument not of science but of power, heavily implicated in the practices involved in, and in some respects still continuing after the formal end of, colonialism."[16]

Photographic mapping of cultures in the service or in the name of science is no exception. It has, for example, been used in the late 19th century to "anthropologicize" the Sámi people – colonizing efforts in scientific disguise – and in the 1920s to solidify the Finnish nation-state by means of detaching 'Finns' from 'others' – thought either in the plural as a collective name for people perceived as different from oneself or, increasingly, as an undifferentiated, ontological,

---

[12] For Nansen, see <http://www.nb.no/baser/nansen/english.html>; for Norsk Polar-institutt, see <http://www.npolar.no>.

[13] Yrjö Haila, 'The North as/and the Other,' in Marketta Seppälä, ed., *Strangers in the Arctic: 'Ultima Thule' and Modernity* (Pori: Pori Art Museum, 1996), p. 115.

[14] Tuan, 'Desert and Ice,' pp. 83 and 91.

[15] Ann Christine Eek, 'The Roald Amundsen Photographs of the Netsilik People, 1903–5,' in *Imaging the Arctic*, p. 122.

[16] Couldry, *Inside Culture*, p. 92.

potentially frightening Other with a capital O.[17] It was, as Paulo Susiluoto reminds us in this volume, an age of ethnocentrism, nationalism and race thinking (not only) in Finland. Edward Said adds that in general "human societies, at least the more advanced cultures, have rarely offered the individual anything but imperialism, racism, and ethnocentrism for dealing with 'other' cultures."[18] Why should it have been different in the North?

Photographs of indigenous peoples thus may tell us relatively little about (the lives of) the indigenous peoples. None the less, they are valuable testimony of past periods. For, as Peter Burke argues, they may have something to say about the "fantasy world" and systems of norms, values and prejudices of photographers and the societies from which they came and which they represented. Thus, photographs of others often tell us more about the self than about others, however indirectly, mediated and (mis)interpretable – it is a kind of double mirror image: we see the self through the way the self sees others. All the same, "images may help posterity tune in to the collective sensibility" – or insensibility – "of a past period."[19] This is true also of photographs showing indigenous peoples in 'natural' poses or postures because, in Pierre Bourdieu's words, "the 'natural' is a cultural ideal which must be created before it can be captured." 'Natural' poses thus show first and foremost what the photographers or those who commission the photographs, socially (for instance, class and gender) conditioned, consider 'natural'.[20]

Likewise, Peter Burke insists, documentary photographers usually "concentrated on traits which they considered to be typical, reducing individual people to specimens of types to be displayed in albums like butterflies." This gaze can be observed, for example, in photographs of Sámi people taken during Prince Roland Bonaparte's expedition to Swedish and Norwegian Lapland in 1884. These photographs, representing Sámi people in front of white sheets and thus decontextualized, visually isolated and deprived of their own culture, have been described by Elizabeth Edwards as follows: the photographer's "'anthropological types', the full face and profile scientific portraits, were posed to show physical characteristics, especially the crania which was perceived as the site of biological

---

[17]   Elizabeth Edwards, 'Essay (Part 3): Anthropological (de)constructions,' in Jorma Puranen, *Imaginary Homecoming* (Oulu: Pohjoinen, 1999), p. 42; Peter Burke, *Eyewitnessing: The Uses of Images as Historical Evidence* (London: Reaktion Books, 2001), p. 123.

[18]   Edward W. Said, *Orientalism: Western Conceptions of the Orient. With a New Afterword* (London: Penguin, 1995), p. 204.

[19]   Burke, *Eyewitnessing*, p. 31.

[20]   Pierre Bourdieu, 'The Social Definition of Photography,' in Pierre Bourdieu with Luc Boltanski *et al.*, *Photography: A Middle-brow Art* (Cambridge: Polity Press, 1990), p. 81.

function, cultural development and thus of taxonomic significance. This mode of representation became one of the predominant ways in which Sámi people were portrayed" from the mid-19th century onwards.[21] (See plates 1 and 2.)

Since photographs may have different meanings for different viewers, anthropological cataloguing may nevertheless result in "exceptionally beautiful pictures",[22] expressing human dignity, which in turn may inspire artists to exceptional aesthetic achievements (more on which below). Furthermore, reading historic photographs may stimulate, initiate and catalyze the negotiation and re-negotiation of past and present among the descendants of those represented, while those represented often had "no understanding of pictures whatsoever"[23] and saw photography as "one more item [...] intended to take from indigenous communities but not give in return."[24] Yet, as Pamela Stern argues with respect to the Canadian Arctic, "the existence today of historic photographs permits Arctic peoples to repossess their histories and to reassert sovereignty over their cultures" – the more so since, as Brian Osborne shows, neither officially sponsored photographs nor artistic paintings have made the far North become a lasting symbol for all of Canada or Canadian nationhood. They were too specific and lacking continental dimension. They were overlapping with a discrete nationalism, a sometimes not so discrete racism and a distinctive Canadian sense of place only temporarily.[25] Juxtaposed to recent photographs, historic images show both change and continuity in change. In an almost dialectic twist, then, photographs taken in the late 19th century to strengthen the Canadian authorities' claim to land in the Arctic against the British, American and other competitors, a hundred years later bear testimony to the indigenous peoples' claim not only to their culture but also to the very same land of which they have been deprived in the course of the Canadian nation-state building process. Here, photography has an emancipatory and counter-hegemonic potential; it unmasks both the contingency of the nation-state's territorial possessions and the ideological character of the claim to naturalness. It supports an already existing political consciousness and helps legitimate demands.

---

[21] Edwards, 'Anthropological (de)constructions,' p. 42; Burke, *Eyewitnessing*, pp. 138–9.

[22] Jorma Puranen, 'Foreword,' in *Imaginary Homecoming*, p. 11.

[23] Eek, 'Roald Amundsen,' p. 122.

[24] Elizabeth Weatherford, 'Native Visions: The Growth of Indigenous Media,' *Aperture*, No. 119 (Early Summer 1990), p. 58.

[25] Brian S. Osborne, 'The Iconography of Nationhood in Canadian Art,' in *Iconography of Landscape*, p. 173; Pamela Stern, 'The History of Canadian Arctic Photography: Issues of Territorial and Cultural Sovereignty,' in *Imaging the Arctic*, p. 51.

Culture? Culture

Culture has (re)entered the agenda of International Studies; iconography has not.
Yet iconography, traditionally a method applied by art historians and in one way
or another used since Cesare Ripa's *Iconologia* of 1592, is increasingly being used
in human sciences. In Human Geography, for example, Trevor Barnes and Derek
Gregory define iconography as a "system of visual signs and symbols [...] that
gain[s] meaning from prevailing cultural discourse and the instantiated relations of
power."[26] Denis Cosgrove and Stephen Daniels emphasize the iconographic
method's centrality to cultural inquiry because "every culture weaves its world out
of image and symbol."[27] In Peter Jackson's words, Cosgrove and Daniels are
"arguing from a world of exterior surfaces and appearances to an inner world of
meaning and experience."[28] In History, Peter Burke argues for something beyond
iconography which nevertheless is firmly rooted in it. Necessary and possible is,
in his view, "a synthesis between elements of the iconographical approach and
elements of the alternatives to it." Burke himself, thus, is no iconoclast with
respect to method; rather, he is in search of a reconciliation between different
approaches.[29] In Literary Criticism, W.T.J. Mitchell is dealing with the
ideological foundations of different iconographic approaches, often ignored in art
historians' appropriation of the concept.[30]

Yet, political and international studies have remained somewhat reluctant to
both address their study objects in iconographic terms and to pay attention to
culture in its aesthetic features. The reason can hardly be the absence of political
icons in a broad sense, *i.e.* symbolic representations, in forms of speeches, maps,
paintings of leaders and historical events, aestheticization of party meetings and
parades and so on. As shown above, pictorial representations of the Canadian
North have been used to define and defend Canadian sovereignty – a more political
function of images is hardly conceivable. As Jochen Hille shows in his chapter, in
the current conflict in Norway over membership of the European Union fishing
boats are being prominently used as political icons, as symbols of the nation's
ostensible interest in general. The absence of iconographic approaches from

---

26 Trevor Barnes and Derek Gregory, 'Place and Landscape,' in Trevor Barnes and Derek
    Gregory, eds., *Reading Human Geography: The Poetics and Politics of Inquiry* (London
    et al.: Arnold, 1997), p. 296.

27 Daniels and Cosgrove, 'Introduction,' p. 8.

28 Peter Jackson, *Maps of Meaning: An Introduction to Cultural Geography* (London/New
    York: Routledge, 1987), p. 177.

29 Burke, *Eyewitnessing*, pp. 169–89 (see p. 183 for the quotation).

30 Mitchell, *Iconology*.

International Studies seems to be reflecting a specific understanding of culture prevalent in the discipline in the 1990s (and sketched in the introduction) from which culture in its aesthetic understanding is largely excluded. Yet, it could legitimately be a subject of political analysis: it is often used for political purposes, supports dominant world views, and helps stabilize forms of domination.[31] Furthermore, artists have often voluntarily and deliberately made themselves available to those in power. Finally, culture is said to have a counter-hegemonic potential, revealed, for example, by Antonio Gramsci in the 1920s and 1930s. For Gramsci, culture is said to have been "the sphere in which ideologies are diffused and organized, in which hegemony is constructed and can be broken and reconstructed."[32] This potentiality has recently been emphasized in Cultural Geography by Peter Jackson and in International Relations by Fred Halliday.

Peter Jackson, referring to Gramsci's concept of hegemony, stresses the "power of persuasion" as opposed to the "power of coercion", the one referring to the dominant class's capability to make subordinate classes accept, acquiesce and believe in its order of political, moral and cultural values, the other being based on physical force. Essential to Gramsci's understanding were the notions of "plurality of cultures" and permanent contest for hegemony: "[...] in capitalist societies, hegemony is never fully achieved – *it is always contested.*"[33] So are cultures since they "are *formed* in conflict."[34] The history of colonization makes this abundantly clear, combining coercion with the transfer of knowledge and culture from the centre to the peripheries. Johan Galtung has called it "cultural imperialism", meaning "the definition of that worthy of being taught" by the centre. Rather than developing counter-hegemony, the periphery, by accepting

---

[31]   For example, for the intimate relationship between Abstract Expressionism and the US "government-backed effort at cultural hegemony", see Christine Sylvester, 'Picturing the Cold War: An Art Graft/Eye Graft,' *Alternatives*, Vol. 21, No. 4 (October-December 1996), p. 407.

[32]   David Forgacs, 'National-popular: Genealogy of a Concept,' in Simon During, ed., *The Cultural Studies Reader. Second Edition* (London/New York: Routledge, 1999), p. 216. My understanding of hegemony follows that of Raymond Williams. According to Williams, hegemony "is not limited to matters of direct political control but seeks to describe a more general predominance which includes, as one of its key features, a particular way of seeing the world and human nature and relationships." Hegemony is different from ideology "in that it is seen to depend for its hold not only on its expression of the interests of a ruling class but also on its acceptance as 'normal reality' or 'commonsense' by those in practice subordinated to it." See Raymond Williams, *Keywords: A Vocabulary of Culture and Society* (London: Fontana Press, 1988), p. 145.

[33]   Jackson, *Maps of Meaning*, p. 53.

[34]   Couldry, *Inside Culture*, p. 35.

cultural transmission from the centre, even "validates for the Centre the culture developed in the centre".[35]

Counter-hegemony, Fred Halliday argues, has to acknowledge that culture is "part of the reproduction of [...] systems of [social and economic] power", rather than being separate from it. Counter-hegemony therefore has to unmask the ideological character of the 'naturalness' and 'universalism' claimed by the rulers with respect to their own cultural system because this system is in fact "of their own choice or creation". But it is precisely the appearance of naturalness which transforms culture into an ingredient of domination. And it is exactly here where counter-hegemony has to emerge, exposing the ideological character of the rulers' culture. In its stead those opposed to the existing forms of domination would have to articulate "their own hegemonic project, including in this an alternative reading of tradition". Yet, neither provincialism and parochial nationalism nor de-coupling culture from politics and economics would be promising paths to counter-hegemony. Rather than developing counter-hegemony, those opposed to the existing power fabric, by escaping into diverse forms of identity politics and self-isolation, would only validate the existing power fabric by not even challenging it any more.[36]

New technologies of reproduction, introduced in the early 20th century, seemed to facilitate the task to challenge hegemony. According to Walter Benjamin, the technical reproducibility has emancipated the work of art from its traditional manner of origin, devoted to magical or religious rituals. Benjamin hoped that the accompanying loss of aura – a kind of mirror-like mutuality between spectator and picture: the spectator looks at the picture and the picture looks back at the spectator – would be "more than compensated for by the emancipatory capacities of the new technologies of reproduction."[37] Benjamin's enemies were both more reckless and powerful in using the capacities of the new technologies. Of course, they did it in a manipulative and violent manner. Politics, in particular mass meetings, parades and movements, were aestheticized and culminated in war. The aestheticization of politics was considerably facilitated by modern techniques of taking and developing pictures. For example, aerial photography became an integral part of

---

[35] Johan Galtung, 'A Structural Theory of Imperialism,' *Journal of Peace Research*, Vol. 13, No. 2 (1971), p. 93.

[36] Halliday, 'Culture and International Relations,' pp. 59–64. It is not mere coincidence that the "narrow margin of identity politics" is one of the few places where "the Western-based art world permits indigenous visibility". See Jolene Rickard, 'Sovereignty: A Line in the Sand,' *Aperture*, No. 139 (Summer 1995), p. 51.

[37] J.M. Coetzee, 'The Marvels of Walter Benjamin,' *The New York Review of Books*, Vol. XLVIII, No. 1 (11 January 2001), p. 30.

the Italian Fascists' politics of internal colonization and disciplinization of society.[38] Of course, Benjamin was aware of the intimate connection between reproduction *in* mass and reproduction *of* masses.[39] At the time, however, fascism's aesthetical approach was more successful than Benjamin's counter-project, the politicization of the arts, because the former was in possession and ruthlessly made use of the state's means of violence.

What, then, could a political iconographic approach look like? Erwin Panofsky's classical three-stage non-political iconographic method – simplified: as pre-iconographic description, iconographic analysis in the strict sense and iconologic interpretation – proceeds from the formal description of an image to the analysis of conventional symbols to, finally, the detection of the image's "intrinsic meaning", revealing the work of art's characteristic as symptom of something else. Four things may be said at this point: first, since making and interpreting photographs is always socially conditioned, photography "should never be considered in or for itself, in terms of its technical or aesthetic qualities."[40] In a sense, a photograph does not exist: photographs have no intrinsic meaning of their own; the meaning assigned to them is always part of discursive formations.[41] A political iconography thus cannot do without a work of art's social history. Secondly, in addition to the work of art's social history, the viewer's social history is important, too. For not only the artist and the work of art but also the viewer and his/her interpretation is socially conditioned: the way a text is read is first of all "a particular reading of a particular text from within a particular institutional

---

[38] Karen Frome, 'A Forced Perspective: Aerial Photography and Fascist Propaganda,' *Aperture*, No. 132 (Summer 1993), p. 77.

[39] Walter Benjamin, 'Das Kunstwerk im Zeitalter seiner technischen Reproduzierbarkeit,' in Walter Benjamin, *Das Kunstwerk im Zeitalter seiner technischen Reproduzierbarkeit* (Frankfurt: Suhrkamp, 1963), p. 42 note 32.

[40] Bourdieu, 'The Cult of Unity and Cultivated Differences,' in *Photography*, p. 22. For example, aesthetic and ethical values involved in the decision on how and what to photograph are reflecting the collective rule exerted by the whole group a part of which the photographer is. "The norms which organize the photographic valuation of the world in terms of the opposition between that which is photographable and that which is not are indissociable from the implicit system of values maintained by a class, profession or artistic coterie, of which the photographic aesthetic must always be one aspect even if it desperately claims autonomy." *Ibid.*, p. 6. Trying to prove one's autonomy by consciously violating the value system of the group implicitly confirms this very value system.

[41] See Janne Seppänen, *Valokuvaa ei ole* (Helsinki: Musta Taide and The Finnish Museum of Photography, 2001).

position"[42] and it is hard to prove that a particular interpretation is more appropriate than others.

Thirdly, particular note should be taken of that which is absent from a picture and this should be thought of in the plural: "Interpreters of images need to be sensitive to more than one variety of absence."[43] For example, although the Sámi people in the 18th century were engaged in, among other things, reindeer herding, fishing and farming – often combining one with the other – they were depicted in contemporary illustrations mainly as reindeer herders. Later, when the Sámi had become a minority among immigrants from other parts of Scandinavia, illustrators were still focussing on them rather than on the immigrants, thereby creating the false image of the absence of change.[44] In particular images of explorations and discoveries tend to exhibit the "selective blindness of imperial history"[45] which, in turn, is indicative of a possessive habit. Fourthly, attention should be paid to that which is not immediately visible – the "communicative nature of the background"[46] – even if it seems to consist of immobilizing whiteness: wild, seemingly untouched landscapes, "land of ice and snow, [...] a dull, dreary, heart-sinking, monotonous waste, under the influence of which the very mind is paralysed, ceasing to care or think."[47] Such an assessment may reflect a culturally conditioned incapability to see what is actually there, for example different shades of white, and is thus to be included in that which is to be analyzed.

However, images are not to be seen as "perfect, transparent media through which reality may be represented to the understanding." Instead they have become

> enigmas, problems to be explained, prison-houses which lock the understanding away from the world. The commonplace of modern studies of images, in fact, is that they must be understood as a kind of language; instead of providing a transparent window on the world, images are now regarded as the sort of sign that presents a deceptive appearance of naturalness and transparence concealing an opaque, distorting, arbitrary mechanism of representation, a process of ideological mystification.[48]

---

[42]   Couldry, *Inside Culture*, p. 76.

[43]   Burke, *Eyewitnessing*, p. 175.

[44]   Neil Kent, *The Soul of the North: A Social, Architectural and Cultural History of the Nordic Countries, 1700-1940* (London: Reaktion Books, 2000), p. 155.

[45]   Jackson, *Maps of Meaning*, p. 168.

[46]   Antonin Dufek, 'Photography, Society and Time,' in The Municipal House and The Moravian Gallery, *Spolecnost pred objektivem (Society Through the Lens) 1918-1989* (Brno: Obecni dum, 2000), p. 23.

[47]   James Ross, as cited in A. Alvarez, 'Ice Capades,' *The New York Review of Books*, Vol. XLVIII, No. 13 (9 August 2001), p. 16.

[48]   Mitchell, *Iconology*, p. 8 (for both quotations).

Then, it should be appropriate to study the development of images in connection with, and as a consequence of, the development of ideologies.[49] Iconographic study in fact seeks "to probe meaning in a work of art by setting it in its historical context and, in particular, to analyze the ideas implicated in its imagery." It "consciously [seeks] to conceptualize pictures as encoded texts to be deciphered *by those cognisant of the culture as a whole* in which they were produced."[50] The issue, thus, is one of being aware of history and culture, space and time of both subjects and objects of photography and thus coming to deeper structures of an image and secret signs encoded in it. Indeed, the "meaning of the picture does not declare itself by a simple and direct reference to the object it depicts. [...] In order to know how to read it, we must know how it speaks, what is proper to say about it and on its behalf."[51] Voice(s) must be given to the (object[s] depicted in the) picture.

Now

The explorers came, but parallel with or immediately after them came the photographers. Early photographers took symbolically possession of land and people. Economic exploitation and photographic colonization joined hands and coincided with politically taking possession of the lands. The early explorers also used the camera as a protective shield to put between themselves and that which was unknown to them. They protected themselves also by covering their head in cloth. This procedure, in addition to the production of inverted images on a glass focus plate, strengthened the perception of the camera as an evil machine which, in turn, makes the idea of documentary, authentic photography seem illusive.

Further technical developments and the minimization of the apparatus removed technical obstacles and reduced the distance between photographers and subjects. Based on empathy and experience obtained by numerous journeys up North, the indefatigable folk music researcher and photographer Erkki Ala-Könni produced in the 1960s images of the people of Finnish Lapland which can be seen as a step beyond the conventional anthropological imagery. While not completely escaping from the trap of cataloguing people, Ala-Könni was deliberately aiming at displaying his objects in their familiar environment and living world, thus showing that the North can indeed be a "home place" (Yi-Fu Tuan) for the people living there. (See plate 3.)

---

[49] Jan Bialostocki, '"Studien zur Ikonologie" nach vierzig Jahren,' in *Studien zur Ikonologie der Renaissance. Zweite Auflage*, p. 14.

[50] Daniels and Cosgrove, 'Introduction,' p. 2 (emphasis added).

[51] Mitchell, *Iconology*, p. 28.

All the same, most photographers are, consciously or unconsciously, confirming to some extent their socially conditioned stereotypes and prejudices of others. Photographers thus tend to resemble modern mass tourists who are only seldom looking for that which they have in common with others; while taking photographs of people perceived as different from themselves, they are looking for that which confirms, rather than narrows, differences. Denying indigenous peoples the right to change, for example, photographic images often aim at portraying them as "living in some timeless past", cleared of anything which could call into question the presupposed pre-modern, "'authentic' characteristics of the place or people".[52] The Inuit writer and cultural activist Ann Meekitjuk Hanson puts it as follows:

> It has been our experience that some visitors expect us to be historic pieces. They expect us to always be smiling. Their romantic notion is that we still live in snowhouses and have been frozen in time. Some people have been disappointed to see the modern side of our life. But we are still Inuit – in our hearts, minds, body and soul.[53]

Denial of change often indicates a feeling of superiority but this does not necessarily have to be so. According to Neil Kent, indigenous peoples are often romanticized and idealized. They are envied "liv[ing] a free and easy existence, uncorrupted by the stresses of civilization" – or pitied because they seem to be unwilling or unable to climb the ladder towards civilization using their own initiative. Hence, they have to be educated and 'civilized'. In any case, denial of change assumes what Fred Halliday calls "transhistorical continuity". This assumption, in turn, reflects what Edward Said, referring to Gramsci, calls "cultural hegemony at work".[54]

By using many of the anthropological-colonizing images of the late 19th and the early 20th century – re-photographed, enlarged on plexiglass panels, positioned in (and thereby creating) current landscapes, and, again, re-photographed – the Finnish photographer Jorma Puranen, in *Imaginary Homecoming*, works against the assumption of transhistorical continuity and denial of change. He represents in his photographs the complex relationship between past and present. What in the 1920s was an ingredient of the Nordic nation-state building projects by means of

---

52 Nelson N.H. Graburn, 'The Present as History: Photography and the Inuit, 1959–94,' in *Imaging the Arctic*, p. 161.

53 Ann Meekitjuk Hanson, as cited in Marie Routledge, 'The North, Inuit Art Today,' in Pekka Lehmuskallio, Markku Lehmuskallio and Kati Kivimäki, eds., *Arctic Inua: Contemporary Eskimo Art from the Lehmuskallio Collection* (Rauma: Lönnström Art Museum, 2000), p. 32.

54 Said, *Orientalism*, p. 7; Halliday, 'Culture and International Relations,' p. 56; Kent, *Soul of the North*, p. 155.

cultivating the Sámi people's otherness and cataloguing them according to 'scientific' matrices now becomes, in Elizabeth Edwards's words, "a dynamic articulation of history as a continuing dialogue between past and present concerns". In some installations, historical pictures reproduced on plexiglass are held by living hands in front of which another photograph is placed as yet another layer of historical experience indicating, perhaps, both the continuation of the past in the present and change over time.[55] (See plate 4.)

These stratified photographs do not reveal themselves easily to the viewer. What is past, what is present remains opaque just as the relationship between them both. In any case, they deny being reduced to either past or present. They show the past in the present as well as the present in the past, the past's present as well as the present's past. The photographs show that, in fact, different "timescapes" are "co-existing layers of which we humans are integral parts" rather than chronological periods.[56] In another photograph, Puranen positions historical images side-by-side in the landscape and by doing so represents the possessory right of the Sámi people to their lands. At the same time, a fence cutting the landscape and the photograph in two symbolizes that the Northern territories are what can be called a "natural area unnaturally divided by national boundaries".[57] (See plate 5.)

In his photos, Puranen is capturing what Roland Barthes considers the essence of photography, namely, the superimposition of reality and the past.[58] Likewise, for Walter Benjamin the photograph "connects the past with the present by supplying the 'pulse', the rhythm and the motion of historical process".[59] Puranen aesthetically re-subjectivizes the Northern peoples and locates them in their own space and time. He maps "a geography of sites/sights"[60] that pertain to Northern

---

55   Edwards, 'Anthropological (de)constructions,' pp. 43–6 (for the quotations, see p. 43).

56   Ari Aukusti Lehtinen, 'Geography and Biopower: In Search of Authentic Land- and Lifescapes (A Nordic Interpretation).' Paper presented at the Nordic Symposium for Critical Human Geography, University of Oslo, October 5–6, 2001, p. 6.

57   The term is borrowed from Timo Kivimäki, 'Integration and Regionalization in the Northern Calotte,' in The Finnish Institute of International Affairs, *Northern Dimensions* (Helsinki: The Finnish Institute of International Affairs, 1998), p. 73.

58   Roland Barthes, *Camera Lucida: Reflections on Photography* (New York: Noonday Press, 1990), p. 76.

59   Angela McRobbie, 'The Place of Walter Benjamin in Cultural Studies,' in *The Cultural Studies Reader*, p. 82.

60   The term originally referred to Mimmo Jodice's pictures of Naples and Neapolitan culture long subjected to the colonizing gaze of the tourist industry. See Antonella Russo, with research by Diego Mormorio, 'The Invention of Southernness: Photographic Travels and the Discovery of the Other Half of Italy,' *Aperture*, No. 132

cultures, taking them seriously rather than treating them as expression of exotic otherness. Similar to the ethnologist Ernesto de Martino's approach to Southernness, focusing on peasant life and documented with photographs, Puranen's pictorial representation of Northernness includes the possibility to turn the Sámi people into "protagonists of their own world, allowing them to write their own autobiography, to become conscious of the culture they engendered."[61]

Yet, the relationship between aesthetical and political re-subjectivization is complex. As Barthes argues, the view on photography is critical only on the part of those who are already capable of criticism.[62] Photography in itself cannot bring about this capability. Moreover, what society does with a photograph and how society reads it, the photographer does not know (and can hardly influence). An image may reveal different things to different viewers. To some viewers it may reveal nothing, others may find a subject not suitable to be photographed, others, again, may find it simply interesting or beautiful or boring: the viewer in itself is a myth. For, the way an image is read, understood, interrogated and juxtaposed with other texts and other images, in a word contextualized, is socially conditioned and knowledge-dependent. According to Susan Sontag, the existence of a relevant political consciousness is the precondition for the viewer's being morally influenced by pictures. Benjamin argues that captions may become an essential part of the photograph, revealing what the photograph in and of itself does not reveal. Without lettering, legends and captions, says Benjamin, every photographic construction is bound to get stuck in the approximate. Mitchell adds that "imaging the invisible" is necessarily of verbal character.[63]

However, the relationship – or, in Mitchell's words, dialectic – between pictures and words is an intricate one. "Iconotexts"[64] – texts within the photograph, captions, textual explanations – may help read a picture but may also manipulate the viewer. Language is a very powerful means of domination without which cultural leadership is hardly conceivable. Sticking to indigenous languages while using iconotexts is one way to deal with the problem but the exclusion of many viewers/readers seems to be inevitable. Accompanying photographs with iconotexts in the English or another majority language avoids limited intelligibility. Yet it means to suppress and adapt the minority discourse to another,

(Summer 1993), p. 67.

[61] *Ibid.*, p. 60.

[62] Barthes, *Camera Lucida*, p. 36.

[63] Benjamin, 'Kleine Geschichte der Photographie,' in *Das Kunstwerk*, p. 64; Sontag, *On Photography*, p. 19; Mitchell, *Iconology*, pp. 42–3.

[64] Peter Wagner, *Reading Iconotexts: From Swift to the French Revolution* (London: Reaktion Books, 1995).

dominant one. It means relying on an ingredient of hegemony to question hegemony. Its counter-hegemonic potential seems to be limited. In Puranen's recent work English words printed on silk screens and placed into the landscape do not only create a new landscape. Using words like lost, hidden, forgotten, geographies, people, names and languages, the viewer is invited to an inquiry into lost landscapes and memories.[65] Of course, it is for the reader to accept this invitation or not. (See plate 7.)

In Lieu of Conclusions

Many pictorial representations of the North are still being done in the tradition of classical landscape photography, depicting seemingly untouched landscapes, free from human influence. They have often been criticized precisely because of their "idealized representations of landscapes" as "an unproblematic, romantic and harmless space between nature and culture."[66] However, by focusing on that which is absent from the photographs, the viewer may reveal the selective blindness of landscape photography. Furthermore, landscape photography does not necessarily have to idealize landscape. For example, Puranen's photographs avoid idealization and nostalgic romanticism. They show cars and railway tracks, electric power facilities and telecommunications installations (albeit often in the background which, again, underlines the importance of that which is not apparent at first sight). These photographs document the intrusion of modern civilization and technology into traditional Sámi lands and the resulting conflict over land use. Time in the North has not stopped. Positioning historic images of Sámi people in an economically exploited and partly devastated environment (for example, in open-pit mines or above railway tracks) may indicate the Sámi people's dispossession of their own lands. (See plate 6.) The "threat to landscape" is even said to be "a threat to memory and history".[67] The homecoming is only imaginary because, as a result of economic and technological developments and the integration of the Northern lands into global economic structures and modes of dependence, maintaining or returning to the traditional life-style of, say, reindeer husbandry,

---

[65]   Jorma Puranen, *Language is a Foreign Country. Photographs 1991–2000* (Helsinki: The Finnish Museum of Photography, 2000).

[66]   Janne Seppänen, 'Who Stole the Landscape? Some Remarks on Critical Practices in Landscape Photography,' in Leila Laukkanen, Janne Rissanen and Pirkko Siitari, eds., *Pohjoinen valokuva 91. Rajoilla – valokuva ja kulttuuri-identiteetti* (Oulu: Pohjoisen valokuvakeskus, 1991), p. 14.

[67]   Elizabeth Edwards, 'Essay (Part 4): Dichotomies and Disjunctions,' in *Imaginary Homecoming*, p. 61.

fishing or farming is impossible or at least difficult. Yet, at the same time the photographs may be read as a claim to Sámi participation in current forms of land use (including, perhaps, forms of exploitation) and in the benefits derived from economic development. They may work against the clichés that only as reindeer herders the Sámi people are 'true' and 'authentic', that old Sámi lands and new global networks are necessarily opposites and mutually exclusive. They may help locate the Sámi people in (post)modern times rather than anchoring them for ever in pre-modern times.

The ways in which pictures show what cannot be seen (Mitchell) can be understood only by those who are cognisant of the culture as a whole in which they were produced (Daniels and Cosgrove). Understanding photographs of the North presupposes understanding the North. The more one knows about the North, the more one discovers in pictorial representations – presences as well as absences, foregrounds as well as backgrounds – and the less photographs of the North are just nice or beautiful (which many conventional landscape photographs no doubt are). The issue, thus, need not be one of re-imaging the North but one of re-reading and contextualizing the images by incorporating them into both their own history as well as into the past and present of that which they represent. However, a degree of caution would not be amiss in particular when contextualization unfolds by the aid of dominant languages and in the light of a dominant culture. Integrating photographs of, for example, Northern peoples into their social and cultural history should therefore be accompanied by an awareness of the potential problems arising from learning about and approaching a minority people's culture and history through the prism of a dominant language and culture. For many viewers, it may be the only possible approach and the resulting knowledge may be more than just impressions. The approach, however, is not unproblematic; neither is the resulting knowledge: it is translated, mediated and adapted to the dominant discourse.

Chapter 4

Two Skolt Geographies of Petsamo during the 1920s and 1930s

Paulo Susiluoto

Introduction

We humans are geographical beings transforming the earth and making it into a home, and that transformed world affects who we are. Our geographical nature shapes our world and our selves. Being geographical is inescapable – we do not have to be conscious of it.[1]

Finland received the Petsamo corridor to the Arctic Ocean from Soviet-Russia in the peace of Tartu in 1920. The small indigenous Skolt Sámi nation lived in this new multiethnic 'colony'. In Finnish society they were doubly strange: first they were 'primitive' Sámi and second they were Russian Orthodox by religion. Erno Paasilinna wrote that attitudes towards the Skolt Sámi of the Petsamo region between the world wars were usually so negative that the Skolt Sámi were not accorded any real human dignity in Finnish society.[2] In order to integrate the new colony into the 'heart and body' of Finland, Finnish nationalism needed a story of Finnish Petsamo. Indigenous Skolts became the 'other' in their own homeland.

However, this is not the whole truth. Väinö Tanner (1881–1948) and Karl Nickul (1900–1980) studied Skolt society in a positive way and attached great value to their culture. They tried to understand and interpret the life of that unique, fascinating and very old small Northern nation. They also defended the Skolt Sámi way of life and territorial rights against nationalist and colonial pressures. They found *geographical beings*, whose culture and personality had developed in close contact with Northern nature and history. 'Wilderness' became a political space; landscape was composed of places thick with meanings.

---

[1] Robert David Sack, *Homo Geographicus* (Baltimore and London: The Johns Hopkins University Press, 1997), p. 2.

[2] Erno Paasilinna, *Petsamo – historiaa ja muistoja* (Helsinki: Otava, 1992), p. 328.

My aim in this chapter is to examine how Tanner and Nickul worked in describing Skolt Sámi society, life and regions during the 1920s and 1930s. My argument is that *what* an image is – in this case that of the Skolt Sámi people and region – depends on *how* it is studied.[3] I will examine the Skolt studies of Tanner and Nickul mainly from a geographical perspective. The content and the roots of these studies will be discussed, but I will focus on the methodological strategies that Tanner and Nickul used in describing the culture. I will also briefly comment on how these geographies affected Skolt Sámi life.

Peoples of the North – Sámi Land

Nowadays the Sámi are the only recognized indigenous people in Scandinavia, but they have a different status in the three Nordic countries (Norway, Sweden and Finland) and in Russia. The area the Sámi live in today extends from the Kola Peninsula to Central Scandinavia. There are about 6,400 Sámi in Finland, 15,000–20,000 in Sweden, 20,000–25,000 in Norway and about 2,000 Sámi in Russia.[4] Previously, the Sámi were called Lapps, but the currently preferred term is 'Sámi', which is the name that the Sámi people use for themselves in their own language (in Sámi *sápmi, sápmelaš*). Tanner used the Swedish term lapp, and Nickul used the Finnish term *lappalainen* (= Lapp). In this chapter, I use the term Lapp in quotations, otherwise the term Sámi is used.

At the beginning of the 20th century, the Sámi people lived mostly in a subsistence economy. There were three distinct and different forms of occupational cultures. One focused entirely on reindeer husbandry, while another was a mixed economy consisting of fishing, hunting and small-scale reindeer herding. The third culture, concentrating on dairy cattle, hunting and fishing, and partly on reindeer husbandry, took shape in usable agricultural areas, where proper farming, based on the growing of cereal crops, could not expand.[5]

In 1926 there were about 450 Skolt Sámi in the Petsamo region.[6] They lived in three Lapp villages, or *siidas*. Since 1530 these siidas had come under Russian rule and had adopted the Orthodox religion.

---

3   Susan J. Smith, 'Constructing Local Knowledge,' in John Eyles and David M. Smith, eds., *Qualitative Methods in Human Geography* (Oxford: Blackwell, 1992), p. 18.

4   Irja Seurujärvi-Kari, 'Legal Position of the Sámi Today,' in Juha Pentikäinen and Marja Hiltunen, eds., *Cultural Minorities in Finland* (Helsinki: Finnish National Commission for Unesco, 1995), p. 134; Irja Seurujärvi-Kari, 'Sámi Area and Populations in Finland,' in *ibid.*, p. 101.

5   Lassi Saressalo, 'History of the Sámi Area and People,' in *ibid.*, p. 112.

6   Väinö Tanner, 'Voidaanko Petsamon aluetta käyttää maan hyödyksi?' *Fennia*, Vol. 49, No. 3 (1927), p. 93.

Väinö Tanner's Skolt Geography

When Petsamo was ceded to Finland in the Peace of Tartu in 1920, the area was almost unknown to Finnish scientists. There was only scant and often unreliable knowledge about the inland area and its indigenous Skolt Sámi population. There was also a growing interest in the first colony of Finland, especially in its natural resources. Väinö Tanner, an engineer, geologist and diplomat had visited the Petsamo region in 1905, 1907 and 1909. The Chief Director of the Commission of Geology lured Tanner away from the Ministry for Foreign Affairs to direct the search for nickel ore in 1924, and Tanner continued this duty until 1931 when he became a professor in geography.[7] During these years he also conducted archaeological excavations, collected Skolt Sámi place names, wrote his anthropogeographical Skolt monograph, and researched the geomorphology and shorelines of the Arctic Ocean.[8]

Reindeer Nomad Studies in 1910–1912 and 1914–1917

How was it possible for a geologist to write a book like *Antropogeografiska studier*? Tanner's chairmanship of the Northern Reindeer Pasture Committees in the years 1910–1912 and 1914–1917 gave him an excellent opportunity to learn to understand the Scandinavian reindeer Sámi culture in relation to the sub-arctic and boreal environment. The frontier agreement under consideration would have limited the necessary movements of reindeer nomads between the interior of Northern Sweden and the coast of Northern Norway. The intention of the Committee was to find out how the planned frontier agreement would affect reindeer husbandry and the Sámi way of life. The report of the 1914–1917 Committee is impressive: eighteen parts and about 7,100 pages! The 1910–1912 Committee worked only in Northern Sweden, the next Committee extended its fieldwork to Norway.[9]

---

[7] Ilmo Massa, 'Problem of the Development of the North Between the Wars: Some Reflections of Väinö Tanner's Human Geography,' *Fennia*, Vol. 162, No. 2 (1984), pp. 203–204.

[8] Väinö Tanner, 'Petsamon alueen paikannimiä,' *Fennia*, Vol. 49, No. 2 (1928); Väinö Tanner, 'Om Petsamo-kustlapparnas sägner om forntida underjordiska boningar, s. k. jennam'vuölas'kuatt,' *Finskt Museum* XXXV (1929), pp. 1–24; Väinö Tanner, 'Antropogeografiska studier inom Petsamo-området. 1. Skoltlapparna,' *Fennia*, Vol. 49, No. 4 (1929).

[9] Massa, 'Problem of the Development,' p. 206; *Renbeteskommissionens af år 1909 handlingar* I (Helsingfors, 1912), p. 3; *Renbeteskommionens af 1913 handlingar* I:1 (Stockholm, 1917), p. 13.

The reports are based on extensive fieldwork: for example reindeer pastures were examined thoroughly and classified, and snow cover and its quality were studied. The Committees also interviewed reindeer herding Sámi about their annual herding system by focusing on the critical period (spring) of reindeer husbandry. Reindeer herders gave statements about the condition and usability of the pastures examined by the scientists. The Committee also used Sámi as advisors: the first Committee had two Sámi on staff, the latter one had four. According to the Swedish scholar Filip Hultblad these studies "take central place in the literature about nomadism".[10] The work of these Committees led to regulations in reindeer husbandry, and many restrictions were placed on it. Many Sámi families had to emigrate because the number of reindeer had to be cut.[11]

During his Committee years, Tanner learned to use the knowledge of local people, and to research their lifestyle from a geographical perspective. The Committees taught Tanner to see Sámi questions from a political and ecological viewpoint. He also learned to see the function and meaning of the state frontiers from a Sámi point of view.

Ecology and the Politics of Differentiation

The subject of the book written by Tanner was the social and spatial organization of the Skolt Sámi, the so-called *siida* or Lapp village. Tanner also explored Sámi history. He searched for elements common to all Sámi groups. Reindeer herding and a more or less nomadic way of life were a basis for Sámi cultures. The third element was language, which was tied to reindeer herding and a nomadic lifestyle. He also found a trend leading from transhumance – based on fishing, hunting and small-scale reindeer herding – to large-scale nomadic reindeer husbandry, although this development was not straightforward.

Tanner explained also how a nomadic way of life with long and narrow migration routes had developed in order to adapt specialized reindeer husbandry to the climate and topography. Knowledge about how to minimize risks caused by natural conditions in gaining a livelihood had accumulated from generation to generation, and this knowledge was essential in adopting a new way of life, like

---

[10]   Filip Hultblad, *Övergång från nomadism till agrar bosättning i Jokkmokks socken*, Acta Lapponica XIV (Stockholm: Nordiska Museet, 1968), p. 16.

[11]   *Betänkande. Svensk-Norsk Renbeteskommissionen av år 1997. Ruota-Norgga Boazo-guohtunkommišuvdna 1997* (Alta: Björkmanns, 2001), p. 45. See also Patrik Lantto, *Tiden börjar på nytt – en analys av samernas etnopolitiska mobilisering i Sverige 1900–1950* (Umeå: Kulturgräns Norr, 2000), pp. 88–9.

a nomadic reindeer culture.[12] Tanner said that natural conditions only put limits on freedom of choice and opportunity. They had put limits on the origins and earlier development of culture, but "cultural impulses managed to triumph over apparent constraints of nature", as Tanner said.[13]

An economy based on pure nomadic reindeer husbandry did not spread to Russian Lapland because, according to Tanner, the natural conditions in Russian Lapland were more suitable for a semi-nomadic life. But that was not the only reason for differentiation. Eastern and Western Sámi groups came under the influence of very different kinds of administrative and religious organs from the sixteenth century onward. An important part of the differentiation of the Sámi cultures was religion. Monks of the Orthodox Church were tolerant towards Skolt traditions and beliefs. The witch-hunts and aggressiveness of their Lutheran colleagues in the West humiliated the Sámi people. In the long run, the Western Sámi interested themselves in new economic activities, whereas the Orthodox Skolt Sámi followed their traditions in their economic life. Russian policies, which guaranteed the Skolts an almost autonomous position, had the same influence, and the Skolts succeeded in preserving their old habits, social organization and lifestyle based on transhumance. For these reasons, the River Paatsjoki became a deep cultural 'demarcation-line' between different Sámi groups.[14]

The geographical environment has an important role in the constitution of culture. Sámi cultures and territories did not exist and develop in a cultural and historical vacuum. In the background of the differentiation of the Sámi cultures stood history, modernization processes and active, innovative human beings. From a regional perspective he found different regional ways of life, and forces shaping cultural areas.

The Life and History of the Skolt Siidas

Tanner explored the lifestyle of five siidas, or Lapp villages. Four of these were located on the coast, one was an inland siida. Three of these were located in the Finnish Petsamo region, one in Norway and one in Soviet-Russia. His anthropogeographical studies provide an account of a spatial and temporal system, economic life, history and the prevailing conditions of these siidas. In the Petsamo region there was also one nomadic Sámi group that had immigrated from Norway and some other small Sámi groups, but in this chapter I will concentrate on the Skolts.

---

[12] Tanner, 'Antropogeografiska studier,' p. 49. At present the term *traditional ecological knowledge* is used.

[13] *Ibid.*, p. 230. All translations by the author.

[14] *Ibid.*, pp. 58–62.

Tanner showed what was common to all the Skolt groups, what was special for each and how the siidas have undergone modernization. A common feature for all siidas was their annual migration cycle with several annual dwelling places (transhumance). It was nature, and especially climate and topography, that determined lifecycles and resource use. Each siida had a different history. He studied the histories of the Petsamo (*Peahccam* in Skolt) and Paatsjoki (*Paččvei* in Skolt) siidas particularly closely. Changes in social and spatial systems were analyzed. Paatsjoki had lost about half of its territory in the setting of the boundary in 1826, but according to Tanner it was economic development, with its wage labour and monetary economy, which threw life in the siida into chaos.[15]

When analyzing the crisis of the Petsamo siida he found that conflicts between Sámi groups – Lutheran reindeer herders and Skolts – damaged siida life. Petsamo Skolts gave up rational reindeer herding and because they no longer had reindeer to carry their things, they little by little gave up migrating between inland lakes and the shoreline. Colonization by Finnish peasants also disturbed siida life. The new border between Russia and Finland in 1921 damaged their nomadic life. The pace and scale of colonial incursions were more modest in the southern part of Petsamo than in its northern and western parts. That is why people in Suenjel had more time to adjust and reformulate their life during modernization.

Tanner found many mechanisms behind the social and spatial changes of Skolt life: state borders reducing the territorial areas, conflicts and competition between different groups, the economic policies of the Petsamo monastery, colonialism and the First World War. Specialization in ocean fishing, retail trade, a monetary economy and division of labour had effects on Skolt social cohesion and spatial organization. Ilmo Massa has summed up this process described by Tanner:

> When individuals began to assess their relationship with each other in terms of money, the social fabric of the natives was soon torn to pieces. Individuals tried to become more independent of their families, families tried to loosen their kinship ties and the kinship groups to separate from their villages.[16]

The establishment of new economic possibilities, sources of livelihood and new social ties raised serious problems. The old *group sentiment* of the siidas no longer had a solid base, and the intrusion of a 'European form of culture' broke the northernmost siidas. The abandonment of a way of life based on reindeer herding and fishing was injurious, because social ties and societal institutions had developed together with the traditional economic life.[17]

---

[15] Tanner, 'Antropogeografiska studier,' pp. 158-60.

[16] Massa, 'Problem of the Development,' p. 206.

[17] Tanner, 'Antropogeografiska studier,' pp. 231-3.

Tanner found varied historical paths that led Sámi societies from a pure subsistence economy towards modernization. Sámi society was very different from agrarian and industrial societies. It was easier to understand Sámi society in the context of Sámi history instead of comparing it with other societies, argued Tanner. His attitude was relativistic and contextual. These siida accounts tell us a great deal about Skolt Sámi lifestyles, values and experiences. In the background of the analysis of the change and even destruction of the siidas stood an idea of the original siida society. In the next sections I will see how the 'original' society was build up according to Tanner.

The Socio-political Siida – Foreign and Internal Policies

> The siida protected the natural rights of individuals by maintaining order and justice, and was even centuries ago capable of conjuring up a happy society in a god-forsaken wilderness.[18]

Tanner studied the political activities and institutions of the siida. The siida had certain courses of actions and principles for 'internal policy' and 'foreign policy'. It had an administrative organ called the *norráz* or village council, which was simultaneously a social and political institution. Every family had one representative in the council. The norráz had absolute power: there was no right of appeal on its decisions. If someone was not satisfied with a decision made by the norráz, he or she had to leave the siida.

The norráz tended the integrity of siida territory in relation to neighbouring siidas. The boundaries of the siidas were usually strictly determined. In some areas there were flexible boundaries and between these there were 'fuzzy' areas, which Tanner called 'condominium' and Skolts *kueht'mierr*. These areas were under the joint control of two siidas, and neither siida had ownership of these areas. To govern these condominiums, Skolts had a practice of arranging meetings between the representatives of both siidas. The Skolts called these meeting *kueht-mierr-sobbar*. The 'condominium' was divided on the basis of the needs of the siidas. A new boundary was always temporary. When the population in one siida grew and more resources were needed, the *kueht-mierr-sobbar* met again. These practices show "that Lapps in their geopolitics truly intend to look after the needs of their neighbours as their own, in order to avoid a desire for revenge that could disturb the neighbourliness".[19]

Siidas also had separate rights based on agreements inside the territory of other siidas. Usually these rights concerned fishing, especially salmon and ocean fishing.

---

[18] *Ibid.*, p. 386.
[19] *Ibid.*, p. 348.

Although these resources might be located in one siida's territory, the governing of these concession-areas was between the two norráz. For example, siidas decided together on how to organize fishing, and sometimes they leased their fishing rights. The oldest known document on a concession is one between the siidas of Suonikylä and Nuortijärvi from 1574.[20]

All land and water belonged collectively to the siida. There was no private ownership of land and water. Families only had the right to use them. The norráz made all decisions concerning the rights to use them. Skolts divided inland waters according to their productivity in relation to the number of male family members more than one and a half years old. Ocean fishing was not regulated, except salmon fishing. Skolts allotted salmon draughts. Families (or two or three small families together) then rotated the draughts and after a full circle allotted them again. The division of land areas was not clear, but it seems that rights to the areas around the autumn dwelling place belonged to the family in question.[21]

I emphasize the importance of representing the Sámi people as political actors because Finnish historical works have until lately represented the Sámi peoples as living without any ordered society, as Kaisa Korpijaakko – who has conducted a pioneering study on the historical land use rights of the Sámi – has noted very much along the lines of Tanner.[22] Tanner started a chapter dealing with the political organization of the Skolts by criticizing earlier studies of the Sámi and the spirit of the times:

> It is one common illusion that the primitive Lapps have never been able to build an organized society but had just wandered without regard for each other here and there in the hope of food and shelter. The conclusion is soon reached that it was a great blessing when the northern neighbouring states created order and balance for their society. This conclusion is thoroughly without foundation. Lapps have never been a flock of savages leading a chaotic life, but formed well disciplined population groups having a balanced and tightly maintained social order.[23]

By stressing the political dimension of Sámi society, Tanner tried to correct wrongs and to influence attitudes and policies concerning the Skolts, whose

---

[20] The contract dealt with the *Tuallam'paihkk*-waterfall (Padun or Patuna), which was an excellent salmon draught. Nuortijärvi is the Finnish name, Tanner used the Skolt name *Njuehtt'jaur*. See *ibid.*, pp. 348–9.

[21] *Ibid.*, pp. 353–63.

[22] See Kaisa Korpijaakko, 'Saamelaiset ja suomalainen historiankirjoitus,' *Historiallinen aikakauskirja*, Vol. 88, No. 3 (1990), pp. 242–4.

[23] Tanner, 'Antropogeografiska studier,' p. 344, as cited in Massa, 'Problem of the Development', p. 205.

situation after World War One was bad. But it also seems to have been a methodological decision to put Skolts in the same scientific framework as any other nation. This part of the book is also an introduction to Skolt Sámi geopolitics. The book is an important source for those interested in the political organization of nature-based egalitarian societies.

Social Integration, the Individual and the Environment

We have seen how the social structure broke down and the ties of loyalty between individuals changed in the siidas. In this section I will examine how Skolt society was constructed, and which forces and ties kept people together in the 'original' siida.

Behind Tanner's study there was the question central to all social science: how is society possible, what is the glue that binds individuals together to form a community? Tanner talked about the forces behind society. Descriptions and analysis of lifestyles and local histories as well as an account of the politics of the siida had the same goal, to solve the main problem of his study, that is:

> to show the forces, which actively affect the life of our "Hyberboreer" in the borderland of the inhabited regions, and by having knowledge of these it becomes possible [...] to operate with these forces.[24]

Individuals lived in a world defined by many ecological and social relationships and cultural models. Tanner said that the power of the individual – or the family – was reduced by the close dependency on a subsistence economy, the support of the family, and the good will of neighbours: one day you are rich, but the next day utterly poor. That is why there were no dominating families or kinship groups.[25] People had formed a siida society in order to protect their vital interests and to avoid damaging competition. Skolts did not recognize the 'right of the more powerful' in territorial questions inside and between the siidas, but tried in many ways to protect equality between families and siidas. Natural hazards did not disturb the Skolt life, because of the multiplicity of their sources of livelihood. Dwellings were widely scattered and it was better to live in peace, because earning one's living was hard enough in a harsh environment. Natural conditions made it difficult to form larger groups, and this is one reason for peaceful living, argued Tanner.[26]

---

24 Tanner, 'Antropogeografiska studier,' p. 9.
25 *Ibid.*, p. 346.
26 *Ibid.*, p. 388.

The time Skolts spent together in their winter village also had an essential significance to the process of socialization.[27] Customs and social feelings so essential to the existence of the siida degenerated when Sámi groups no longer gathered together in a winter village. It appeared to Tanner that the Skolts realized the importance of the norráz in maintaining and elaborating the ideas of expediency and justice embedded in the wisdom and experience they had gathered from generation to generation.[28] Tanner stressed the importance of a reciprocal nature-human relationship, the role of social institutions, and shared regional interests, values and traditions in explaining social integration. He found a society based on a rational subsistence economy and equality between patriarchal families. Skolt society was not based on race or kinship; rather outsiders also could became members of the siida, which Tanner defined as based on contract:

> Recently the siida was simply a *society completely based on contract*, which constituted of certain families, mostly of ethnographic Lapps, which had the same goal: to guarantee material livelihood and balance their social and moral life.[29]

Regional Way of Life

Jouko Vahtola has written that the cultural anthropological work done by Tanner in the Petsamo region is of enduring value. Ilmo Massa has argued that Tanner's Skolt book is extremely valuable sociologically.[30] That is all true, but the book is also geographical. Tanner himself criticized earlier 'lappology' (=Sámi studies) for "neglecting the geographical factor".[31] In this section, I will examine one geographical root of Tanner's intellectual framework, namely Vidalian geography.

Paul Vidal de la Blache gave regional studies their classic form at the beginning of the twentieth century.[32] Analyses of ways of life (*genre de vie*) formed Vidal's fundamental contribution to geography. *Genre de vie* refers to a form of livelihood functionally characteristic of a human group – for example transhumance or peasant agriculturalist. Vidal showed how groups made use of the environment,

---

[27]   Väinö Tanner, *Ihmismaantieteellisiä tutkimuksia Petsamon seudulta. 1. Kolttalappa-laiset*, ed. Paulo Susiluoto (Helsinki: Suomalaisen Kirjallisuuden Seura, 2000), p. 55.

[28]   Tanner, 'Antropogeografiska studier,' p. 346.

[29]   Tanner, *Ihmismaantieteellisiä tutkimuksia*, p. 84–5. Emphasis added.

[30]   Massa, 'Problem of the Development,' p. 206; Jouko Vahtola, 'Petsamo Suomen tieteen tutkimuskohteena,' in Jouko Vahtola and Samuli Onnela, eds., *Turjanmeren maa. Petsamonhistoria 1920-1944* (Jyväskylä: Petsamo-seura, 1999), p. 507.

[31]   Tanner, 'Antropogeografiska studier,' p. 8.

[32]   Paul Claval, *Introduction to Regional Geography* (Malden: Blackwell, 1998), pp. 17–18.

and so did Tanner. He made an intensive journey into the siidas' social and political heart in order to explore how a siida functioned in maintaining and reproducing social and economic life – a way of life – in a particular geographical setting. In Vidalian geography people under study found themselves faced with new ideas, they adopted new models of behaviour and modified their way of seeing things and their manner of taking advantage of the environment. Tanner also wrote about these things, but he was realistic and sensitive enough to see the limits of cultural and ecological sustainability.

But what about region or territory? Vidalian geography referred to regions (*pay*) and brought out their 'personality'. A Vidalian researcher emphasizes the specific nature of regions, referring to the dominant natural conditions and the lifestyles of the people who lives there. Tanner made distinctions between different Sámi siida cultures dominated by different natural and historical conditions. The term *siida* referred to *the interest society* between the Skolt Sámi, who had traditional rights to use an established and autonomous area: a territory. Territory was a very important part of social existence, and social order affected how a territory was used and lived in. Tanner did not talk about the 'personality' of territories; he was more concerned with the reciprocal relationships between territory and society.

Paul Claval argues that for those who have understood Vidal de la Blache, there was never any question of regarding each region as a block closed in on itself.[33] A territory draws part of its character from its association with larger spaces. The siida had its origins in isolation, but later it was integrated into larger regional systems. Historical and natural forces shaped the regions and ways of life, but local level interests, attitudes and actions also affected the constitution and development of the siida. Tanner was interested in the formation and change of society and territory. Region or vidalian *pay*, in this study a siida, was not a category created by the scholar for theoretical or practical purposes, neither was it a framework to collect information. The siida was a real association of the region and a society constituted in human action to adopt life to northern environments.

Because of his geographical approach, Tanner was clearly aware of the historical development of regions, spatial patterns and ways of life. Distinctions and similarities between Sámi groups became clear. A picture of the forces shaping life and regions in the northern environment began to emerge. The book is a pioneering study of the history of Sámi territories. Sámi peoples and regions became part of the history of northernmost Europe, not as a race or as interesting ethnographic objects, but as social and territorial beings.

---

[33] *Ibid.*, p. 18.

Towards Interpretative Geography - Without Success

In interpretative science we play the tortuous game of seeing the world of individuals or groups as they see it. "The existence and orderings of such groups and individuals are accepted and used as the basis of examining the social world", argues John Eyles.[34] The task of interpretative research is to uncover the nature of a social world through an understanding of how people act in and give meaning to their own lives and spaces. This kind of approach is essential to all social sciences, for social researchers cannot understand what is going on in society if they are not going to succeed in finding out those meanings that the actors themselves give to social practices.[35]

J.E. Rosberg, the first professor of geography in Finland, noted in 1931 that Tanner was the first to understand the significance of the siida *concept*.[36] There was dualism in the Skolt worldview expressed in the word siida: a distinction between *sit-jennam* and *sit-olmai*. Sit-jennam means siida-land, an appropriately demarcated area or territory, a piece of landscape with land and water. It was a stable and almost unchanging base for siida society. *Sit-jennam* was the most original and most important part of their social organization. The population of the siida, *sit-olmai*, changed from time to time, and it was a secondary element in the siida.[37]

The origins of the siida's "strongly individualized population group, that maintains its integration against other siida populations" are in the intimate reciprocal relationship between a siida's people and territory. The population and area merge into 'a microcosmos' or 'state embryo' with a political shade. The 'cardinal duty' of the siida was to keep its territory (*sit-jennam*) together and secure the living conditions of its people.[38] Tanner found a society with citizens loyal to the region and nature. From that perspective we find the underlying rationale of the Skolt worldview: it explains, for example, the fact that there is no private ownership of land. This 'Skolt-dualism' was not only a conceptual expression of an abstract 'worldview', but also the most important part of the territorial thinking of the Skolts.

---

[34] John Eyles, 'Interpreting the Geographical World,' in *Qualitative Methods in Human Geography*, p. 2.

[35] See Risto Heiskala, *Toiminta, tapa ja rakenne: kohti konstruktionista synteesiä yhteiskuntateoriassa* (Helsinki: Gaudeamus, 2000), p. 213.

[36] See Paulo Susiluoto, 'Suomen ajan ihmismaantiedettä Petsamosta,' in Tanner, *Ihmismaantieteellisiä tutkimuksia*, p. 26, note 31.

[37] Tanner, 'Antropogeografiska studier,' pp. 87 and 338–43.

[38] *Ibid.*, p. 343.

What is important here is that Tanner studied how Skolts understand the word siida, and how they construct their lives according to their fundamental categories. In terms of the number of pages, the interpretation of the meaning of siida was only a tiny part of his grand study, but it held a central place. I claim that Tanner introduced a new philosophical aspect to Finnish geography, namely an interpretative perspective, but unfortunately, without any success. One reason for that was, of course, that there was only an interpretative perspective and an interpreted result, but not an explicit method to follow, or concepts and theoretical frameworks like we have today. Interpretative geography landed in Finland only in the 1990s.

Accepting and rejecting a theory – or a single book – is always a social act. At first the Skolt book received praise, but later it was rejected. A picture of a political and deeply historical Sámi country did not fit into the nationalistic attitudes so common among Finnish scientists.

Voices from a Native Land – Local Geohistories

The pioneer of fieldwork-based anthropology, Bronislaw Malinowski, noted during the 1920s the need for statistical coverage of the society being investigated; observation to discover the behaviour and conduct of individuals and the collectivity; and ethnographic description based on conversations and interviews to elicit the 'mentality' of the people.[39] In his study, Tanner used information derived from archives, literature, explorations and old maps. The use of statistics and map making was also included in the research procedure.

A base in constructing the history of the siidas and the political, social and territorial systems of the siida was oral information. Tanner spent four half a year periods in Petsamo with the Skolts (1924–1927) and had daily conversations in their own environment, in the forests and hills. "During numerous autumn evenings by a common campfire it has been a pleasure for the writer to listen to the Skolts' living descriptions of life in the siidas in bygone days", wrote Tanner. He tried to select parallel information from the descriptions Skolts gave him. That information he complemented with literature and old documents.[40] Because of Committee service, he was experienced in dealing with this kind of information. Rather than analyzing Skolt society alone, Tanner induced the Skolts to analyze their own society together with him.

Tanner's book is about Sámi history, but it also contains a *history of meaning*. "The socio-political concept of the siida would thus be at least two thousand years

---

[39]   See Eyles, 'Interpreting the Geographical World,' p. 5.
[40]   See Tanner, 'Antropogeografiska studier,' pp. 103 and 112.

old", wrote Tanner in his slightly bold final chapter.[41] The conceptual scheme of collectivism versus individualism was also a 'red thread'. Skolt societies remained more collective, whereas Western Sámi cultures were more individualistic according to Tanner.

Tanner tried to draw large synthetic conclusions by examining different spatial levels, and by watching the object of his study from sociological, political, geographical and ecological perspectives. His book progresses from facts to abstract concepts, like social feelings. What lies at the heart of this synthetic study is the idea of Skolt geography, basic forces and meanings by which social and spatial life were organized. Tanner does not only describe the Skolt conceptualization of the siida, but a new construction of the Skolt 'microcosmos' which is neither wholly Tanner's nor the Skolts': it was a product of the communication between them.

Tanner integrated two methodological perspectives: the structural macro level when discussing the 'outer' forces shaping society and territory, and the micro level when relating and analysing how Skolts thought about, acted in and experienced their life. These perspectives he merged with a historical approach. The historical dimension was lacking from traditional regional geography.[42] The new regional geography – which could be called *interpretative geohistory* – contains a historical dimension, which renders it possible to trace the social processes that are constitutive for society and territory. Tanner's approach was very modern.

Karl Nickul in Search of Collective Mindscapes

> The 'dry heart' of Australia, she said, was a jigsaw of microclimates, of different minerals in the soil and different plants and animals. A man raised in one part of the desert would know its flora and fauna backwards. He knew which plant attracted game. He knew where there were tubers underground. In other words, by naming all the 'things' in his territory, he could always count on survival.[43]

Karl Nickul (1900-1980) was a geodesian, peace activist and 'the leading Sámi friend'. He made 14 journeys to Southern Petsamo during the years 1929-1939 and

---

[41]   Tanner, *Ihmismaantieteellisiä tutkimuksia*, p. 103.

[42]   Anssi Paasi, *Territories, Boundaries and Consciousness. The Changing Geographies of the Finnish-Russian Border* (Chichester: John Wiley & Sons, 1996), p. 76.

[43]   Bruce Chatwin, *Songlines* (Berkshire: Picador, 1988), p. 300.

spent about 20 months there.[44] During the 1930s he published three articles in geographical journals dealing with Suenjel, the southernmost siida in the Petsamo region. In his place article published in 1934 there were more than 1,500 place names and dozens of place stories. In that article he discussed the role of cartography and the problem of place names in maps.[45] But there are also some ideas to be found that are still interesting from the perspective of modern geographical place studies. Another two articles deal with the Skolt Sámi way of life and rights.

Nickul's place studies are an early example of the 'cartography of the folk soul', but there were some scholars who had collected place stories before him. In Samuli Paulaharju's Skolt monograph there were a few stories related to specific 'horrifying places'. Uno Holmberg researched sacred places of the Petsamo region Skolts. Tanner explored some old underground dwelling places, probably thousands of years old – the so-called *jennam'vuölas'kuatt* – near the shoreline of Petsamo.[46] The paper is an interesting blend of archaeology, documents and traditional tales in relation to these places. Tanner also collected about 800 Skolt-language place names from northern Petsamo. Tanner wrote, that "Skolts add some symbolism to place names, which gives the names a special aesthetic charm."[47] Nickul said that it was Tanner's text that led him to think about place names in greater detail and Tanner's influence can also be seen in his Skolt texts in general.[48]

Place Tales

Fish and fishing were the most common subjects in place names in Southern Petsamo, the next big group of names was associated with reindeer – wild and tame. In this section I will examine 'cultural' places, which inform about foreign

---

[44] See Veli-Pekka Lehtola, *Nickul – rauhan mies, rauhan kansa* (Jyväskylä: Kustannus-Puntsi, 2000), p. 42.

[45] Lehtola examined also Nickul's radical views of cartography. See *ibid.*, pp. 27–40.

[46] Samuli Paulaharju, *Kolttain mailta* (Helsinki: Kirja, 1921); Uno Holmberg-Harva, 'Petsamonmaan kolttain pyhät paikat,' *Suomalais-ugrilaisen seuran toimituksia* LVIII (1928), pp. 15–24; Tanner, 'Om Petsamo-kustlapparnas sägner'.

[47] Tanner, 'Petsamon alueen paikannimiä,' p. 5.

[48] Karl Nickul, 'Petsamon eteläosan koltankieliset paikannimet kartografiselta kannalta,' *Fennia*, Vol. 60, No. 1 (1934), pp. 5–6. Later he wrote: "It is my pleasant duty to express my heartfelt thanks to Professor Väinö Tanner for his inspiring influence. His large work [...] ought to be read by every one who wants to get insight to the conditions and history of these people." Karl Nickul, *Skolt Lapp Community Suenjelsijd During the Year 1938*, ed. Ernst Manker, Acta Lapponica V (Stockholm: Nordiska Museet, 1948), p. 12.

influence, mythology and historical events.

Orthodox hermit monks have left traces in the place names in Suenjel. They visited and also stayed there in the islands and on the shores of inland lakes a 'few hundred years ago', and according to the Skolts, disturbed their life. Monks introduced a new religion. There were also many other foreign groups represented in place names, like Karelians, Inari Sámi, and Russians. One old winter village site had the name *Kattjettemsijtsaij*. The story told of a Russian official, who came from a neighbouring village. As he was annoyed at the long distance between villages, he told the Skolt men to undress and switched them with birch branches. The word *sijtsaij* means old (abandoned) winter village site, *kattjettem* means 'drop the underpants'.[49] The Skolts also had many place tales in relation to their enemies. Old Skolts have told me about *Tshuoppomsijtsaij*, another old winter village site, which was destroyed by Juho Vesainen in the late seventeenth century. Only two families survived.[50]

Aijektshoalm literally means Grandfather's strait, but *aijek* is also a euphemism for bear. In this tale a drowned bear had a money belt under its skin. Skolts believed that humans were able to transform themselves into animals.[51] Skolts had many cautionary place tales. Some of these places were avoided, while others were not. In these places certain behaviour like shouting may result in illness, or getting lost, or 'stuck in'. These places were inhabited by strong spirits, or shamans have done something unusual there, like transformed humans into stones.[52] Stories related to places with strong spirits reflect the same ultimate principle as the words *sit-jennam* and *sit-olmai*: nature (or territory) was to be held in honour of generations past and for future generations.

> Thus the genius loci of Reksjaur was considered so particularly capricious that the Skolt built his hut several hundred yards away from the shoreside, so that even the children at play would not disturb the lake, and in like manner a special vocabulary was used to conceal the names of the fish as well as fishing tackle in use.[53]

This conservative attitude towards the environment is still alive among older Skolts.

---

[49]   Nickul, 'Petsamon eteläosan koltankieliset paikannimet,' p. 43. Skolts build new winter villages after a few decades that is why there are so many sijtsaij place names.

[50]   See *ibid.*, p. 72.

[51]   *Ibid.*, pp. 19 and 33

[52]   *Ibid.*, passim.

[53]   Karl Nickul, 'Place Names in Suenjel – a Mirror of Skolt History,' *Studia Ethnographica Uppsaliensia* V (Lund, 1964), p. 223.

Making Place – or Made by Place?

In the summer of 2001 I visited former Suenjel with the 77 year old Skolt Matti Sverloff. He left his home country in 1944. When travelling by bus through the old siida he told me something about every little river and lake we saw, about fishing and who had done what there. Then we travelled by boat through his old home lakes. He told me where and how his family fished, how reindeers behave, who had lived in that place and who they were related to and so forth. Matti recounted his memories and the 'wilderness' began to live. It was easy to notice how warm his attitude towards the dwelling and other places was. In Nickul's article many place names are in a family tree expressing the bond between the dead and living generations.

According to Nickul, Skolts were so attached to their area that they did not willingly express the names of their surroundings to outsiders, and that place and self affected each other: "In this manner Skolts have been unwilling to reveal the various local place names to strangers. It seems that to reveal a name was almost the same as to uncover oneself."[54] Many pre-modern and preliterate societies are bound to their environment in the sense that they and the place seem to be virtually one.[55] In this case place obviously has about the same meaning as territory. When Tanner talked about the siida as a microcosmos, he was talking about the same thing. In this regard, we are talking about 'little' places.

Skolts mapped and mediated their relationship with the land and waters through places, translating their home region into culture. But there are practical reasons for detailed place knowledge. Landscape is used as a part of memory in an oral society that must remember everything about itself and its practices. That is why place must be a more intimately part of its culture.[56] The more technically developed a culture is, the more humans break away from the hegemony of the landscape, said Nickul. When discussing the human-nature relationship of the Skolts he used the terms *landscape based culture* and *place-tradition.*[57]

Vincent Berdoulay has said that the study of place has a strong narrative component and that "a place comes explicitly into being in the discourse of its

---

[54] *Ibid.*

[55] Sack, *Homo Geographicus*, p. 136.

[56] *Ibid.*

[57] Karl Nickul, 'Suenjel, kolttain maa,' *Terra*, Vol. 45, No. 2 (1933), p. 79; Karl Nickul, 'Eräs Petsamokysymys. Suonikylän alueesta kolttakulttuurin suojelualueena,' *Terra*, Vol. 47 (1935), p. 90.

inhabitants, and particularly in the rhetoric it promotes".[58] Nickul had no theories about place, but he studied place in the same spirit. Nickul attempted – and succeeded – to describe a Skolt worldview, living world and historical shift by using place names and stories.[59] Place became a relevant vehicle for understanding Skolt culture. Place names and place tradition in the broader sense were part of culture, but they were also essential vehicles for making that culture visible and intelligible to outsiders.

Before his place article Nickul made many journeys to Southern Petsamo. He based his studies on conversations, interviews, observations and literature. His approach was based mostly on extensive fieldwork with daily conversations. Nickul did not evaluate any theory, but his approach can be defined as *narrative-descriptive* after Yi-Fu Tuan.[60] The object of that kind of study is human experience, which is almost always ambiguous and complex. In the narrative-descriptive approach complex phenomena themselves – in this case place – occupy the front stage, but the explicit formulation of theory is not attempted.

Pure geography developed by Johannes Gabriel Granö in the 1920s described landscape as visual forms 'purified' from all human meanings and values.[61] For Nickul place names in the cultural landscape "tell us always something about people in the region, about their actions and moods." Place names also "characterize human affect, culture, in these areas."[62] When comparing Nickul's view with Granö's pure geography we can see a huge philosophical difference. "The aim is to demonstrate by this example, that the population of this region – its character, views, past, economic life, culture – is reflected in the names" argued Nickul. Pure geography aimed to be an objective science of the forms of landscape. Nickul was searching for an experienced and culturally developed landscape of meanings: "Something of the human relationship to his surrounding area is to be revealed in the place names especially. It can be seen in his mode of

---

58 Vincent Berdoulay, 'Place in French Language Geography', in John A. Agnew and James S. Duncan, eds., *Power of Place* (Boston: Unwin Hyman, 1989), pp. 134–5.

59 See Nickul, 'Petsamon eteläosan koltankieliset paikannimet,' p. 10: place names give a map a time dimension like contour lines give the illusion of topography; p. 28 the transition from a hunting economy to a reindeer economy is to be read from place names.

60 Yi-Fu Tuan, 'Language and the Making of Place: A Narrative-Descriptive Approach,' *Annals of the Association of American Geographers*, Vol. 81, No. 4 (December 1991), p. 686.

61 Pauli-Tapani Karjalainen, 'Mahdollisten maisemien semantiikkaa,' *Terra*, Vol. 107, No. 2 (1995), p. 124. See also Johannes Gabriel Granö, *Pure Geography* (Baltimore and London: John Hopkins University Press, 1997).

62 Nickul, 'Petsamon eteläosan koltankieliset paikannimet,' pp. 11–12.

making abstractions of the features of the landscape and how he separates its parts."[63] The picture we gain of the Skolt landscape is thick with meanings, memories, history and actions. "Suonikylä, his Suenjel was a real fairytale country", wrote Nickul about his Skolt friend Jaakko Sverloff.[64]

There are different descriptions of the past – and different geographies – and these can be described from varied perspectives. Nickul's place based approach reviews many interesting things about a very old egalitarian culture, where people are in close contact with nature. Nickul did not have any official rank – he was an engineer doing geodetic work – in scholarly circles and his ideas on studying place did not spread into academic geography.

Image and Method

Outside the Margins – Sámi People in Cartography

In literature and in history indigenous peoples have been marginalized. They have been on the margins in cartography as well.[65] In this section, maps will be regarded as value-laden images, which have often been used in colonial promotion, and as tool of 'civilization' and exploitation in the colonies.[66] Petsamo was more or less a Finnish colony. My aim in this section is to explore how Tanner and Nickul used maps in the context of the Sámi and the Petsamo question.

Tanner's map of different Sámi groups showed the spatial systems of many groups, but not state borders, and not non-Sámi names. The map also showed how reindeer nomadism was a kind of Sámi way of conquering new regions. In his map of the siidas of the Petsamo-region (see map 4.1) state borders are drawn with a very narrow line, but the boundaries of the siidas are depicted by a thicker line. Boundaries, regularly used seasonal family areas, and the winter village sites of the Skolt siidas had been almost unknown before. This seemingly simple map is a result and summary of intensive work. It is a cartographic masterpiece.

---

[63] *Ibid.*, p. 7.

[64] Lehtola, *Nickul*, p. 35.

[65] Evelyn J. Peters, 'Aboriginal People and Canadian Geography: A Review of the Recent Literature,' *Canadian Geographer*, Vol. 44, No. 1 (Spring 2000), p. 45. This article deals with Canadian geography, but its arguments are also valid in a Finnish context.

[66] J. Brian Harley, 'Maps, Knowledge, and Power,' in Stephen Daniels and Roger Lee, eds., *Exploring Human Geography* (London: Arnold, 1996), pp. 378–9.

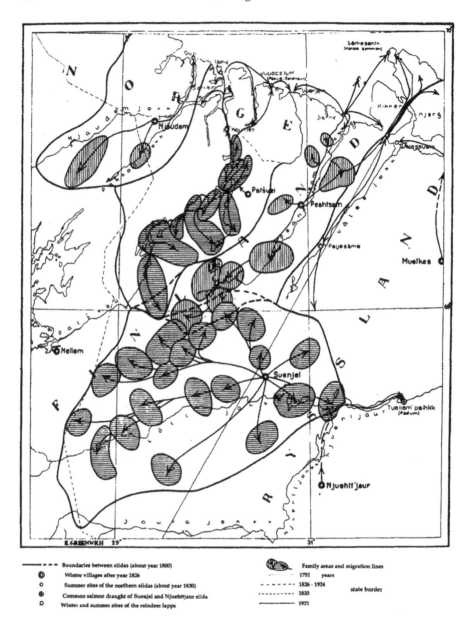

Map 4.1 Väinö Tanner, 'Resource Use and Boundaries of the Siidas in Petsamo Region,' 1929 (reprinted with permission of the Geographical Society of Finland)

Tanner used cartography in order to illustrate a Sámi way of life, history and territorial system. When discussing the problems raised by the setting of a new state border Tanner was diplomatic, but his map clearly shows how territorial 'reshaping' was harmful to the siidas. The basis for a Sámi cartography was created.

The significance of place names in relation to minority cultures has been an important question in recent times – the same as in Petsamo in the 1930s.[67] Who has the right to assign names to that 'new' region? Nickul answered: Lapps [...] have the moral right to assign names to their region."[68] The right to name places was not dependent on legal ownership of a region but had more to do with everyday life and actions. This view was radical in those nationalistic and colonial times.

For Finnish people these regions were almost uninhabited – or Finnish people wanted to see these areas as 'empty space' or 'wilderness'.[69] Actually this 'discourse of empty lands' and 'wilderness' remained strong until current times. The Finnish general and writer K.M. Wallenius saw the Petsamo region as a virgin land, God's country and man's territory, which became a "totally Finnish region in only twenty years".[70] The strategy – be it consciously or unconsciously – of proving the area to be almost uninhabited is common for a conqueror, and as a leader of the Petsamo military operation in 1919 Wallenius was a conqueror. In addition, the branding of Skolts as reindeer thieves proved useful for the purposes of representing Petsamo as a region that needed to be 'civilized' in the name of Finnishness. To create an idealized image of a unified Finnish nation in Finnish Petsamo, writers used indigenous Skolts as a negative counter image. But how did Nickul find this 'fairy-tale landscape'?

This collection of place names, done for practical reasons, appeal to me personally and became an introduction to Lappish culture. When my Skolt helpers at the campfires explained the meaning of the names and old stories in connection with them, it was obvious how intimate the relations were between the Skolts and their home tract from immemorial times. The place names fascinated me and opened a door to the Lappish world.[71]

---

[67] Pellervo Kokkonen, 'Kartan sosiaalinen todellisuus,' in Tuukka Haarni *et al.*, eds., *Tila, paikka ja maisema* (Tampere: Vastapaino, 1997), p. 68.

[68] Nickul, 'Petsamon eteläosan koltankieliset paikannimet,' p. 13.

[69] See also Lehtola, *Nickul*, p. 38.

[70] Kurt Martti Wallenius, *Petsamo – mittaamattomien mahdollisuuksien maa* (Keuruu: Otava, 1994), pp. 10 and 41. Wallenius tried to publish his book in 1948, but he was not allowed to do so because of the political situation. The book was published posthumously in 1994.

[71] Nickul, 'Place Names in Suenjel,' p. 219.

In his racist novel Wallenius presumed that Skolts have no names for lakes or "names manifest lack of imagination".[72] This sharpened Nickul's pen; he stood up for Skolt culture, and used maps and place-tradition to fight against the 'politics of empty lands'.[73] He found out that wilderness was an integral part of the worldview and history of the Skolts. In that sense there was no 'wilderness', but a deeply humanized and cultural Skolt Sámi national landscape. Cartography became a weapon against imperialism.

A map was a valuable resource in many ways, as Nickul noted: "Although a map is mainly a cultural product designed to serve practical life, mapmaking is serious scientific work. A map is not only a good guide to the area, it is also a valuable document, which, when compared with maps made earlier and later, gives a view of development, such as expansion of areas under cultivation and the road network, conquests of cultural landscapes in general, movement of populations and linguistic changes."[74] Mapmaking was more than a weapon against imperialism. Cartography had, according to Nickul, great research potential when discussing the question concerning "the great problem of the landscape-culture relationship and the relationship between culture, people and nature."[75] Forager-hunters tend to be described as "living in the midst of nature". The description is misleading, argues Tuan, for one can assert as truthfully that they live in a deeply humanized world:

> Outsiders say "nature", because the environment seems barely touched. Insiders see "homeplace" – an environment that is familiar to them, not because they have materially transformed it but because they have named it.[76]

Nickul wrote in the same spirit: "Wilderness is the Skolts' home, but for the inhabitant of the road it is a place to visit. They have a different attitude to it, they leave different traces on it."[77] By *inhabitant of the road* he meant Finnish people, for a road was an unfamiliar element to the Skolts of Suenjel. By examining the landscape through places Nickul learned to look at the 'wilderness' more as text. *Sit-jennam* became a concrete and meaningful social space full of memories and actions. Values and meanings related to the landscape are cultural contracts of what

---

72   Kurt Martti Wallenius, *Ihmismetsästäjiä ja erämiehiä* (Helsinki: Otava, 1962 [1933]), p. 18.

73   See also Lehtola, *Nickul*, pp. 53–7.

74   Nickul, 'Petsamon eteläosan koltankieliset paikannimet,' p. 11.

75   *Ibid.*, p. 8.

76   Tuan, 'Language and the Making of Place,' p. 686.

77   Nickul, 'Suenjel,' p. 82.

we can and want to see in the landscape. Nickul was ready to see and show us what Skolts could and wanted to see in *their* landscape. Nickul proved place to be an important subject of research, and his article introduced a new way of writing ethnography. This representational strategy seems to be very fruitful in researching so called nature dependent cultures.

Instead of representing empty lands or Sámi populations as plots on the map, Tanner and Nickul showed how all the useful areas had been in regular use. Cartography served to show indigenous territorial arrangements and 'national landscape'. They clearly took up an anti-colonial view. Cartography is an important part of research, and Tanner and Nickul showed how to use it.

Homo Geographicus in a Political Space

The intellectual atmosphere concerning the Sámi minority was characterized by ethnocentric historical interpretations originating in the ideas of Kalevala romanticism, nationalism and a Greater Finland ideology.[78] The well-known Finnish scholar Väinö Voionmaa wrote in 1918 about the Lapps, who were 'unchanging like nature' and about the battle between the races.[79] Nationalist ideas were also reflected in the discussion about Petsamo, as the following quotation from Voionmaa's *Suomi Jäämerellä* shows:

> The Arctic Ocean is no longer a Lappish sea, nor does the problem of the Arctic Ocean concern only the fish soup and lichen lands of a few Lapps. There is no sense or right in giving to few hundred reactionary and backward people great economic advantages which hundreds of thousands of people and whole states can make use of.[80]

Tanner and Nickul saw the Skolt as active agents of their own life. The Skolts did not oppose the demands of development, but they were not trained enough to compete against other groups. As Tanner put it: "Nomadic Skolt society was not completely unchanged, as has been assumed. Multiple relevant factors of the period of transition indicate that Skolt societies have tried to adjust themselves to economic development when getting acquainted with it."[81] Through their geographical research on Skolt society Nickul and Tanner challenged prevailing perceptions of indigenous societies as static and frozen in time.

Part of Finnish nationalism in the 1920s was a deep hatred of Russians, regardless of whether they were communists or not. Scholars have talked about a

---

78 Massa, 'Problem of the Development,' pp. 204–205.

79 Pekka Isaksson, *Kumma kuvajainen* (Jyväskylä: Kustannus-Puntsi, 2001), p. 220.

80 Väinö Voionmaa as cited in Massa, 'Problem of the Development,' p. 204.

81 Tanner, *Ihmismaantieteellisiä tutkimuksia*, p. 72.

cult where Russians were stereotyped as enemies, barbarians, vermin and so on.[82] Skolts were branded twice over: as 'a product of Russian culture' and as primitive Sámi in an agrarian and developing country.[83] Erno Paasilinna has said that it is almost impossible to understand how deeply Finnish people despised Skolts.[84] Tanner was informed that the Skolts are all a degenerate, drunken and excessively untidy rabble, who do their best to avoid honourable work. When he got acquainted with the Skolts more closely, he could notice that these slanders were based on a total misunderstanding. Tanner became convinced with his friends that, in spite of their poverty, the Skolts were a talented and morally developed part of the population and not some kind of pariah class.[85] They tried to correct the mistaken image of the Skolt Sámi, and research offered a means for their mission.

Kaisa Korpijaakko has argued that the Sámi people are often represented as a nation without history and rights.[86] The Skolts, however, knew their traditional rights. Tanner mostly appreciated the clear thinking and argumentation of the Skolts. I think political consciousness and the art of argumentation of the Skolts had roots in a transparent and democratic political system with open village council meetings. The siida was an ideal subject of study and easy to approach. Many Skolt men had been in the Russian army in World War I, and had seen the dark side of civilization. That is why they saw their own societies 'microcosmos' more clearly as a counter image to the great powers. Maybe original Skolt society became an ideal for themselves?

> The socio-political culture of semi-nomadic Lapps, "everyone's government for everyone", straight, vigilant and a strong form of government [...] was not only developed and passed down from father to son, but even thrived for centuries. It was in olden times the highest possible culture in the region in question.[87]

There is no reason to underestimate the role Skolts have in these studies. As 'professional politicians' Skolts taught Tanner and Nickul to understand the functions of an egalitarian society. As a matter of fact, the proposal to protect Suenjel was originally made by Skolts in 1930.[88]

---

[82] Outi Karemaa, *Vihollisia, vainoojia, syöpäläisiä* (Helsinki: Suomen Historiallinen Seura, 1998), pp. 190–200.

[83] See also Lehtola, *Nickul*, pp. 50–51.

[84] Paasilinna, *Petsamo*, p. 328.

[85] Tanner, 'Antropogeografiska studier,' p. 7.

[86] Korpijaakko, 'Saamelaiset ja suomalainen historiankirjoitus,' pp. 241–6.

[87] Tanner, *Ihmismaantieteellisiä tutkimuksia*, p. 97.

[88] Lehtola, *Nickul*, p. 58.

Tanner and Nickul conducted extensive fieldwork; both got into close contact with ordinary people. They seem to have had only male informants, and this is one of the most important gaps in their research. They talked and interviewed Skolts on specific, often political, social and territorial questions. They appreciated the local way of knowing, they had open enough minds to understand a worldview so unfamiliar to us. They also learned to see macro scale history from a local micro level perspective. That is something we can call an archaeology of local knowledge. They found dynamic and willing geographical beings. They went even further by planning new minority politics; they wanted to influence the future of the Skolts and promote justice in minority questions.

We would like to live in the same way as our ancestors lived, maintaining respect for their ways and traditions. In their time we had a lot of wild reindeer and fish and our little nation was happy. But now we are afraid that our rights are no longer respected.[89]

They both had the ability to call the taken for granted – and very negative – image of the Skolts into question. They examined how Skolts had organized their society, how they experienced and understood their place on Earth, and what they felt in the storm of modernization. They learned to understand this foreign culture and found positive ways of describing it. Between earth and culture there lived a geographical being, *homo geographicus*.

The Right to be What You Are – in Your Own Country

Tanner and Nickul were anxious about the future of a unique culture and about the denationalizing of the Skolts. Tanner discussed whether it was possible to merge the traditional form of governing of the Skolts with the deep social tendencies of that time. He also asked if it was possible to maintain some educational and conciliatory aspects of Skolt society in order to avoid the social problems so common in the communities where the siida ties had broken. Tanner was aware of the possibilities of research affecting politics because of his Committee work.

Tanner actually seemed to have planned some kind of minority policy based on his research. He was not explicit in arguing for a certain kind of minority policy. Tanner saw the interdependence between the way of life and the institutional practices of the society. The analysis of the constitution and functions of the siida also had practical goals: "the knowledge of the forces that nurture the life of Skolts makes it possible to work with them without falling into naive a priori opinions and

---

[89] Tanner, 'Antropogeografiska studier,' p. 189, as cited in Massa, 'Problem of the Development,' p. 206.

sentimentality which may harm more than benefit the welfare of the Skolts." He also made a radical proposal: about 90 per cent of the area of Petsamo should be left for use by the reindeer economy – in practice in the hands of Sámi people.[90] I like to ask questions concerning Tanner's influence on 'Skolt affairs', because it seems like that question needs to be cleared up. Veli-Pekka Lehtola has written a fine book about Nickul, but he neglected Tanner's ideas and questions concerning the protection of Skolt culture and rights in Finnish society. It was Tanner who laid down the general outlines for these questions.

It was Nickul who really started to act in order to find practical solutions for the questions asked by Tanner. Nickul focused his research work on the Suenjel siida and soon started to act in order to integrate Suenjel into Finnish society. He proposed that Suenjel should be made into a protected Sámi cultural territory, where the Skolts would have an opportunity to live according to their own traditions and develop their culture in Finnish society 'in the Sámi line' so that social changes would not bring about moral and material deterioration.[91] Finnish authorities planned an agrarian settlement model with farming and cattle raising in order to 'develop' Skolt life – and to reduce their territorial rights. Ideas of ecological and social sustainability steered the way to improving and protecting Skolt culture. During the 1930s a proposal for protecting Skolt territory was under consideration, but World War Two stopped the process.[92]

The ways in which we make use of geography as a means of investigating the world and of recommending change – or maintaining something – is highly political. It reflects a particular kind of understanding. Geography – as a discipline – is never value free or innocent.[93] Tanner and Nickul made a clear decision to advocate Skolt culture. The geographical approach suited to illustrate the intimately intertwined relationship between territory and society, between human and nature. The representational strategies used transformed the Skolts and their regions conceptually. The 'Wilderness' or Lapland was originally a mosaic of siidas – which were "like little folk islands in the Ocean of wilderness" – constituting together *Same-jennam*, the Sámi country.[94]

---

90 See Tanner, *Ihmismaantieteellisiä tutkimuksia*, pp. 75 and 97; Tanner, 'Antropogeografiska studier,' pp. 9–10. About the reindeer economy area see *ibid.*, pp. 452–3; Tanner, 'Voidaanko Petsamon aluetta käyttää maan hyödyksi?' p. 104.

91 Nickul, 'Eräs Petsamokysymys,' pp. 103–104; see also Nickul, *Skolt Lapp Community Suenjelsijd*, p. 11, and Nickul, 'Suenjel'.

92 More detailed information is to be found from Lehtola, *Nickul*, pp. 57–78.

93 Stephen Daniels and Roger Lee, 'Editor's Introduction,' in *Exploring Human Geography*, p. 12.

94 Tanner, 'Antropogeografiska studier,' pp. 9–10.

There are always competing definitions and meanings. Which description will achieve its 'place in the sun'? Tanner and Nickul took a stand against nationalist and agrarian policies by proving the Skolt way of life to be rational and sustainable in those northern circumstances. These pioneering works challenged earlier Sámi studies and defined a new place for the Sámi people and their regions. There were no empty spaces, no mute landscapes, but intensive communication between humans and nature. Their studies gave voice to a tiny nation, and made the landscape speak. They traced the Skolt story of 'us and our country'. The voice tells of Skolt history, rights, hopes, indigenous philosophy and territorial thinking. These topics are again at the heart of Human Geography.

Chapter 5

Rosa Liksom's Literary North: Traditional Confrontations or New Discursive Practices?

Juha Ridanpää

Introduction

In current day discussion and especially in the media definition the North has been approached from two different general views. Foremost, the North has been related to problems around the Northern Dimension, questions simply and solely based on European and global politics. When the Cold War came to an end the whole question of limits and boundaries in the northern context changed. New kinds of questions and new kinds of answers were needed and the Northern Dimension was this required answer. The Northern Dimension was a new, hazier version of those strictly delimited northern borderlines.[1] Secondly, the North has been approached, as if it was a sort of battlefield for economic contradictions, mainly because of its still not completely conquered natural resources. The alternative view from which the concept of the North has, now and then, been approached is associated with the continually changing characteristic features of the northern culture. Although the position of the cultural North has partly risen in recent discussions in the media, its importance can still be considered to be marginal rather than general.

There have been several ventures trying to set a regional definition for the concept of the North. Many of these are suggesting that the North was a combination of different physical aspects, for example geological or biological determinations, a line behind which some vegetation does not grow anymore.[2] The

---

[1] Pertti Joenniemi, 'Changing Politics along Finland's Borders. From Norden to the Northern Dimension,' in Pirkkoliisa Ahponen and Pirjo Jukarainen, eds., *Tearing Down the Curtain, Opening the Gates. Northern Boundaries in Change* (Jyväskylä: SoPhi, 2000), pp. 114–32.

[2] See Juha Ridanpää, 'The Discursive North in a Few Scientific Discoveries,' *Nordia Geographical Publications*, Vol. 29, No. 2 (2000), pp. 31–2.

other quite simple manner of defining is to draw a line on the polar circle. One can also say that the North is a point of the compass and therefore is always relating to a hand holding a compass, but culturally the North is much more. When perceived from this perspective, it is not so important to ask what the region referred to as 'the North' contains culturally, but more important is to ask how and why it is culturally defined as being such. Regional lines between the South and the North are seen here as sliding, bluring and therefore becoming irrelevant, while the main attention is paid to how these kind of confrontations are culturally constructed and how complicated these definitions can actually be.

In this chapter the main intention is to look at the cultural, particularly literary practices of constructing something we generally call 'the North'. The attention is on how these kinds of geographical totalities are experienced, interpreted and produced by different literary practices, groups of signs, 'discourses', especially through postcolonialist ideology. The focus in this chapter is centred on the Finnish author Rosa Liksom and her three collections of short stories concentrating on the difference between Finnish northern and southern cultures, *Yhden yön pysäkki* (1985), *Unohdettu vartti* (1986) and *Tyhjän tien paratiisit* (1989). The attention is on how Rosa Liksom's North is spatially constructed against her South and how certain images following the universal ideology of postcolonialism are attached with this construction. Firstly, the focus is on the differences in human misery between northern and southern spaces, and their representations through different discursive and symbolic features. Secondly, the focus is directed on Liksom's ironic way of describing colonialism between the Finnish northern and southern everyday lives. Her collections of short stories are interpreted as not only reflecting different forms of discursive distinctions between the South and the North, but also as reconstructing something new through their social critic. The ambivalence of her work arouses many new questions when trying to get answers of what the culturally discursive North actually contains.

Postcolonialism in the Discursive World

'Postcolonialism' has been one of the most frequently repeated phrases in current cultural and social sciences. Mostly this phrase has been referred to as some kind of theoretical background for research, but on the other hand, postcolonialism can also be regarded as a form of social criticism, the emancipatory way of criticizing the unevenly developed world, rather than a theory of it. To be precise, the majority of the research done in the field of postcolonialism is directed at resisting and criticizing the discourses totalizing society.[3] The actual theoretical context is usually borrowed from different deconstruction theories, like those of Jacques Derrida, Mikhail Bakhtin, Antonio Gramsci or Michel Foucault.[4] What is common in their theories is that they all comprehend the world as some form of a rhetorically structured system.

Understanding the world as a rhetoric system, a multiple collection of different discourses is academically based on structural linguistics and especially on the trend's predecessor Ferdinard de Saussure's (1875–1913) linguistic definition for the sign. The most significant innovation in Saussure's theories was when the sign was divided into two separate parts, into the conceptual part of the sign, 'signifier', and into the meaningful part which this signifier is signifying, that is 'signified'.[5] The difference between a nature's object and a sign marking it came to the fore and following this structural linguistics focused on how different social and cultural ways of expression are in fact socially and culturally conventional.

Michel Foucault has approached this question of difference between a nature's object and a sign marking it with the phrase 'discourse', a phrase with one of the most continuously and contextually varying contents, but still his definition of this phrase can be somehow argued as being a constant one. He approaches discourse with such an abstraction as 'words and things', "the ironic title of a work that modifies its own form, displaces its own data, and reveals, at the end of the day, a quite different task". Although discourses are groups of signs, composed of signs, it is very important to understand when trying to follow Foucault's basic arguments, that discourses work as "practices that systematically form the objects of which they speak". Discourse can be easily entitled as a process of "the ordering of objects".[6] In geographical and especially in the social context this means, that social theory and everything associated with it can be named as 'a discourse', for all communication, signs, symbols, practices, through which anything at all can be communicated, is so easily crammed under this same label.[7] Following this, discourse becomes more a practical aid for research, while contem-

---

[3] Jonathan Crush, 'Post-colonialism, De-colonialism, and Geography,' in Anne Godlewska and Neil Smith, eds., *Geography and Empire* (Oxford: Blackwell, 1994), pp. 334–5.

[4] Sinikka Tuohimaa, 'Pohjois-Suomen unohdetut naiskirjailijat,' in Katja Majasaari and Marja Rytkönen, eds., *Silmukoita verkossa. Sukupuoli, kirjallisuus ja identiteetti* (Oulu: University of Oulu, 1997), p. 45.

[5] Ferdinand de Saussure, 'The Nature of the Linguistic Sign [1916],' in Lucy Burke, Tony Crowley and Alan Girvin, eds., *The Routledge Language and Cultural Theory Reader* (London: Routledge, 2000), pp. 26–7.

[6] Michel Foucault, *The Archaeology of Knowledge* (London: Tavistock, 1972), p. 49.

[7] Derek Gregory, *Geographical Imaginations* (Oxford: Blackwell, 1994), p. 11.

poraneously being a methodological aid, too.

Edward Said's piece of work, *Orientalism* (1978), has been generally considered as one of the most important examples of how Michel Foucault's basic ideas about the World being structured discursively can be applied in geographical research, though Said's work is probably a more historic than geographic one. Said follows Giovanni Battista Vico's basic idea of how mankind is always writing its own history, giving contents and naming everything, including places and regions. Geographic concepts, such as a world divided into 'East' and 'West', 'Orient' and 'Occident', exist only because of their human cultural innovations. Said particularly concentrates on how different manners and cultures are artificially kept alive on the ideological stage and how at the same time features with no correspondence with reality are attached to the concept of Orient.[8] Although Said bases his research into orientalism totally on Foucault's way of perceiving the world as a culturally, socially and politically structured totality, there are also some differences when compared to Foucault's theory of discourse. While Foucault regards separate texts and their producers more in relation to their structural relevance, Said stresses the importance of a single work and the author in relations to the social discourses in question.[9]

Said's ideas about mankind writing its own reality does not mean that geographic categories, such as 'Orient', would not have a correspondence to reality, although this 'correspondence to reality' is actually a very complicated expression, when reality is understood as 'written by mankind'. In spite of this, Said believes in the existence of some form of a 'real reality' and in this way, for example, distinctively draws a separating line between 'pure' and 'political' knowledge.[10] According to Said, people's way of comprehending reality is not based on nature, but precisely on culture and often on untruthful images created by cultural domination. For example, cultural descriptions produced or for some reason non-produced of the 19th century's Orient, were often under harsh political supervision. Such European novelists as Flaubert, Nerval and Scott were often advised as to what kind of descriptions they were allowed to write, or if they were allowed to write at all. Describing East was often wholly denied, so the comprehension of Orient was actually politically determined.[11] For new cultural geography Said's work has been a magnificent example about the discursive constructing of geographic totalities but in particular Said's work has been a pioneer in the field of postcolonialist research.

---

[8]   Edward Said, *Orientalism* (London: Routledge, 1978), pp. 4–5.

[9]   See *ibid.*, pp. 23–4.

[10] *Ibid.*, pp. 9–10.

[11] *Ibid.*, p. 5 and pp. 43–4.

When following the main lines of postcolonialistic ideas, applied to the case of Finland, the relationship between the North and the South is not only a relationship between two spatial contraries with their different cultural features but also a relationship between discursively constructed totalities, one being marginal and other being a centre. One of the main and most simple ideas in postcolonialist theories has been that there cannot be a margin without a centre and no centre without a margin. To be marginal there has to be something central to define your existence. This relationship between centre and margin is about using power, about domination in relation to bureaucratic, economic, political and cultural activities, which form all kinds of inequalities and subordinate relationships.[12]

From the semantic perspective the phrase 'postcolonialism' is simply referring to the process in which the world and society are separated into periodical categories of colonialism and the period following it. Thus, postcolonialism is defined through the different national cultures of previously colonized countries, whose cultural contents were re-formed only after imperialist states – such as Great Britain, France, Portugal and Spain – left them behind.[13] In this way post-colonialism can also be comprehended as the re-writing of colonial history through different theoretical perspectives.[14] From this point of view, the postcolonialist theories and the discussion concerning the North may sound an odd combination. When northern Finland and its past is compared to countries, whose historical background has continually been associated with the tradition of imperialism, the whole question of northern colonialism may possibly not feel so 'imperialized' or 'colonialized', after all. On the other hand, the postcolonialist way of perceiving history can be applied to any centre-margin relationship. Comparing the marginal literature of northern Finland to the history of urbanized southern Finland, in spite of some lack of global colonial relevance, is a relevant subject for postcolonialist research as well.[15] From this point of view, post-colonialist ideology works as a socially theoretical context in northern research, against which texts under

---

12 Matti Savolainen, 'Keskusta, marginalia, kirjallisuus,' in Matti Savolainen, ed., *Marginalia ja kirjallisuus: Ääniä suomalaisen kirjallisuuden reunoilta* (Helsinki: Suomalaisen Kirjallisuuden Seura, 1995), p. 12.

13 Bill Ashcroft, Gareth Griffiths and Helen Tiffin, *The Empire Writes Back. Theory and Practice in Post-Colonial Literatures* (London: Routledge, 1989), p. 1.

14 See, for example, Robert Young, *White Mythologies: Writing History and the West* (London: Routledge, 1990).

15 See Savolainen, 'Keskusta, marginalia, kirjallisuus'; Veli-Pekka Lehtola, *Rajamaan identiteetti. Lappilaisuuden rakentuminen 1920- ja 1930-luvun kirjallisuudessa* (Helsinki: Suomalaisen Kirjallisuuden Seura, 1997), p. 26; Tuohimaa, 'Pohjois-Suomen unohdetut naiskirjailijat,' pp. 46–7.

interpretation can be placed.[16]

Postcolonialist perspective and structural analysis with its criticism cannot still be considered as some form of a final answer to these questions concerning reality and its representations. Its way of articulating these questions is still not without any serious contradictions or paradoxes. According to Jacques Derrida, all deconstructive methods are "trapped in a sort of circle", "for the signification 'sign' has always comprehended and determined, in its sense, as sign-of, signifier referring to a signified, signifier different from its signified".[17] It has also often occurred that postcolonialist research cannot emancipate those totalizing discourses, which were under criticism from the first, but the efforts usually remain captured by themselves.[18] As Said himself has mentioned of his own work, the critic in *Orientalism* (1978) is directed at the difficulties of representing the other, but in spite of a critical perspective, his work actually repeats the same models of totalizing structures and theories, which the work was eventually supposed to be criticizing.[19]

Literary Space

In discursive methodology it is heavily stressed that culture is written through literature and literature is written through culture. New cultural geography is more and more concentrated on the variety of manners about how social and cultural reality is actually structured. This rise of interest towards discursive methodology should mean, at least in principle, that interest towards language as an object of study as well as interest towards everyone's own way of using language should rise as well.[20] Still it can be argued that literary geography possesses a quite marginal status in the field of human geographic research.

Literary geography can most easily be defined as research into literature through interpretation of different geographic concepts, such as landscape, space,

---

[16]   Lehtola, *Rajamaan identiteetti*, p. 28.

[17]   Jacques Derrida, 'Structure, Sign and Play in the Discourse of the Human Sciences,' in Robert Con Davis, ed., *Contemporary Literary Criticism* (New York: Longman, 1986), pp. 482–3.

[18]   Crush, 'Post-colonialism, De-colonialism, and Geography,' p. 335.

[19]   Young, *White Mythologies*, p. 11.

[20]   Marc Brosseau, 'The City in Textual Form: Manhattan Transfer's New York,' *Ecumene*, Vol. 2, No. 1 (1995), p. 89.

place or region.[21] The relationship between literature and geography has been changing when different theoretical contexts have been used in geographic research. The history of this relationship goes back as far as the year 1907, when William Sharp in his research, simply titled "Literary geography", presented a map of regions appearing in different authors' novels. Between the 1930s and 1970s mainly British geographers were eager to prove, how artistic pieces of work were actually very valuable sources of information when trying to describe a region.[22] During the 1960s fictive literature was more or less used as a synonym for positive data.[23] The main idea was simply that with help of these positive details that literature was offering, the geographer could easily find a personality for each region or landscape.[24] Geographical research would probably never have been interested in literary material if the factual information and documentary values of literature had not been noticed. Literature was considered as if it was factual fiction about real reality and to be precise, 'facts in fiction' was the main criterion for literature ever becoming a geographical research material. For example, E.W. Gilbert did not consider Anthony Trollope as a proper author for research, because the places Trollope depicted were actually not real.[25] On the contrary, J.H. Paterson categorized Walter Scott as a regionalist on the basis of his truthfulness, although the regional tradition was not even born in 1900s.[26]

---

[21] Allen G. Noble and Ramesh Dhussa, 'Image and Substance: A Review of Literary Geography,' *Journal of Cultural Geography*, Vol. 10, No. 2 (1990), p. 49.

[22] *Ibid.*, pp. 50–51.

[23] Marc Brosseau, 'Geography's Literature,' *Progress in Human Geography*, Vol. 18, No. 3 (1994), pp. 336–7.

[24] This way of using literature as material for study is mainly based on regionalism, the idea that the author can produce a form of a synthesis of place and its inhabitants through her/his writing. This way the reader can experience the personality of a region, even though they have never even visited the place. See E.W. Gilbert, 'The Idea of Region,' *Geography*, Vol. 45, No. 3 (1960), p. 168; J.H. Paterson, 'The Novelist and His Region: Scotland Through the Eyes of Sir Walter Scott,' *Scottish Geographical Magazine*, Vol. 81, No. 3 (1965), pp. 146–52.

[25] See Gilbert, 'The Idea of the Region,' p. 163.

[26] According to Paterson, first, "although his work was not so localized, [Scott] has been credited by many admirers with the self-same ability as the later regional novelists – the ability to add, by his descriptions, what has been called 'a new dimension', in the imagination of the observer, to the landscape itself", secondly, "Scott was a writer who paid great attention to the details of his scenery, even when it served only as the simplest background to his story", and thirdly, "Scott deserves a place among the regional novelists because, for thousands of people both during his lifetime and after his death, it was his writing which provided their introduction to the Scottish landscape".

Along with new humanistic, cultural and structural interests in the 1970s and in the 1980s the attitude towards literature changed as well. Literature was not only data anymore. Literature worked as a methodological and ideological tool for research.[27] Humanistic geography was born at the beginning of the 1970s against technologically orientated scientific models dominating human geography.[28] Instead of scientific accuracy, humanistic geography concentrated on 'sense of place' – human values, meanings and subjective feelings connected to place – values that were supposed to reopen through art.[29] Literature was considered as a medium depicting concrete experiences, including everyday life, and also experiences about concrete things. In the same way normal talk, chatting with neighbours, has a function of reconstructing everyday experiences. In literature this reconstruction of life is only produced more formally, that is, written.[30]

The social and socially critical perspective in literary geography was born in the 1970s and 1980s as a critic towards the humanistic perspective and its subjectivism and excluding of social context. In both perspectives, humanistic and socially critical, the object of study was the meaning of place. From the critical point of view, however, it was not important what people were experiencing, but why they were doing so. The questions were directed at the structural sides of the meanings and roles of literary experiences and on the other hand, on how these experiences have their own effect on literature.[31] According to the socially critical perspective, geography should not only describe social space, but also try to theoretically criticize it. Geography should steer and lead the social development towards a more just society while literature should be a tool through which this criticism and

---

See Paterson, 'The Novelist and His Region,' pp. 146-7.

[27] Douglas C.D. Pocock, 'Geography and Literature,' *Progress in Human Geography*, Vol. 12, No. 1 (1988), p. 96.

[28] Yi-Fu Tuan, 'Humanistic Geography,' *Annals of the Association of American Geographers*, Vol. 66, No. 2 (1976), pp. 266–76; David Ley and Marwyn S. Samuels, 'Introduction: Context of Modern Humanism in Geography,' in David Ley and Marwyn S. Samuels, eds., *Humanistic Geography: Prospects and Problems* (London: Groom Helm, 1978), pp. 1–2.

[29] Lester Rowntree, 'Cultural/humanistic Geography,' *Progress in Human Geography*, Vol. 11, No. 4 (1987), p. 560; Edmunds V. Bunkśe, 'Saint-Exupéry's Geography Lesson: Art and Science in the Creation and Cultivation of Landscape Values,' *Annals of the Association of American Geographers*, Vol. 80, No. 1 (1990), p. 96.

[30] Yi-Fu Tuan, 'Literature and Geography: Implications for Geographical Research,' in *Humanistic Geography*, pp. 195–6.

[31] John A. Silk, 'Beyond Geography and Literature,' *Environment and Planning D: Society and Space*, Vol. 2, No. 2 (1984), p. 151.

prospects for a better future were presented.[32] Socially interpreted, the author is always reflecting a certain social ideology and literature is always a reflection of its society.[33]

If regional geography is using literature as a factual source for defining and describing geographic categories, the new cultural geography regards literature as one of those social and cultural tools through which the whole of reality is comprehended. Constructing a geographic object through literature or by using literary tradition as an advantage can be a very concrete process,[34] but mostly the constructing of literary regions, landscapes, places or spaces is more or less somehow hidden behind its discursiveness. The previously mentioned Edward Said's work *Orientalism* is a splendid example of how literature can act as a cultural, social and especially political tool through which places are discursively constructed. Although literature is quite often concerned with different kinds of human feelings, it is important to notice that these feelings are also always attached to some social context. Whatever they are reflecting, they are always reflections of their own social and cultural space. Discursively speaking, however, literature is not only a reflection of some cultural background, but it is more like a channel through which this background is written. Therefore one should not forget those huge amount of different literary conventions through which the textual world of the piece of art is constructed.[35]

Reality cannot ever be seen 'as such'. Although reality is textually or discursively structured, no text can ever reach such description, where the possibility of different aspects to comprehend, interpret or represent reality would have been already passed. All perceptions and descriptions are made from some point of view, where different social and cultural traditions constantly have their effect. Reality is always seen through some lens.[36] Literature describing and constructing places is socially and culturally coloured even before its formation. Nothing can be created in a vacuum. The world is not a vacuum. As it has been mentioned, "places are the stories we tell. Writing and creating go hand in hand.

---

[32] Brosseau, 'Geography's Literature,' pp. 342–3.

[33] Pocock, 'Geography and Literature,' p. 95.

[34] For example, Kairanmaa is an ideal Finnish case of 'a region', which is concretely structured in base of Pentti Haanpää's literary fiction and tradition. See Juha Ridanpää, 'Pentti Haanpää's Kairanmaa in Cultural Transformation,' in Sirpa Leppänen and Joel Kuortti, eds., *Inescapable Horizon: Culture and Context* (Jyväskylä: University of Jyväskylä, 2000), pp. 133–54.

[35] Stephen Daniels, 'Arguments for Humanistic Geography,' in R.J. Johnston, ed., *The Future of Geography* (Cambridge: Cambridge University Press, 1985), p. 149.

[36] Joanne P. Sharp, 'Locating Imaginary Homelands: Literature, Geography and Salman Rushdie,' *GeoJournal*, Vol. 38, No. 1 (1996), p. 121.

So do reading and creating."[37] If anything was not already written, comprehension of reality would be impossible. Therefore one should take writing
seriously too.

Rosa Liksom: The Northern Author?

When reading and trying to comprehend Rosa Liksom, it is very easy to read the
text with the author, or in particular the image of the author, in mind. Staying
inside the written and still forgetting the author's cultural image is actually
impossible. The question is, should the analysis of Rosa Liksom's way of
perceiving the North go only through her production, analyzing her short stories
in relation to the different social and cultural discourses there possibly are or
should analysis relate everything to the author's person? The point is that in
discursive analysis both are continuously done. An author cannot escape reality's
discursiveness, simply because this discourse is what she or he is actively
producing.

When starting to analyze Finnish literature, the first thing I notice is that
northern literature has a very minor part in its history. This does not mean that
there were not any capabilities of producing artistic works in northern Finland, but
what it means is that culture and literature and everything associated were defined
from the South. The northern manner of perceiving what art really is just did not
fit into southern literary categories. All publishers were in the South. Kuopio and
Kajaani were the most northern points on the map where nationally known writers
had come from by the end of the 19th century. In the beginning of the 20th century
such names as Pentti Haanpää (Piippola) and Ilmari Kianto (Suomussalmi)
expanded the group of nationally known writers further to the North, but still there
was no room for northern authors.[38] With a few exceptions, Lapland, northern
Finland and northern authors have been systematically left out when listing what
Finnish literature is all about and reaching the list of nationally known authors
from the North has been very exceptional, even today. Rosa Liksom has been one
of those exceptions.

When following characteristic features of colonialism, a strong horizon of
exotism and strangeness has been laid over the image of Lapland. Trying to exceed
or break this horizon has been basically an impossibility for northern authors. Timo
K. Mukka it can be argued was the first writer to earn a reputation as a nationally

---

[37]   Pauli Tapani Karjalainen, 'The Significance of Place: An Introduction,' in Pauli Tapani
       Karjalainen and Pauline von Bonsdorff, eds., *Place and Embodiment* (Helsinki: Lahti
       Research and Training Centre, 1995), p. 12.
[38]   Lehtola, *Rajamaan identiteetti*, pp. 29 and 234.

known author despite being from the North.[39] Liksom follows this tradition, a tradition where authors can be thought of as trying to continually criticize and emancipate the unevenly developed world through their literary work, in this case, the inequality between northern and southern Finland. Rosa Liksom or Anni Ylävaara as her name used to be, was born on the 7 January 1958 at Ylitornio, northern Finland, but when viewed from a more northern perspective, Ylitornio and the whole Tornio valley resembles more a wealthy centre than a poor periphery. It is actually quite comic, how marginality and misery is most perfectly or at least most famously described in those works of art coming from the most wealthy part of northern Finland.[40]

In this chapter, the analysis of Liksom's way of perceiving the North is based on three collections of short stories: *Yhden yön pysäkki* (1985), *Unohdettu vartti* (1986), and *Tyhjän tien paratiisit* (1989). Along with these there has been published an English assortment of these three collections, concentrating on southern life, *One Night Stands* (1993), which being already translated should help my work a little. Liksom's literary production can be, and has been, understood from several perspectives. Liksom writes mostly short novels of only a few pages and her style of writing the North can be considered as being more or less "post-modern".[41] The aggressive style of describing such themes as sex, alcohol, drugs and violence automatically divides readers' and critics' opinions of Liksom, but at least she has earned some form of international reputation, while being one of the few translated Finnish authors. What makes Liksom a very tempting author to approach is her manner of dividing her short stories into two clear geographically separated categories: those situated in the North and those in the South. From their counterbalance it should be quite easy to construct a description of her way of seeing the North as a contrast to the South. What makes this figure even more tempting is Liksom's continuous way of using irony in her descriptions. By means of ironic expressions in her art she is either consciously or unconsciously emancipating different forms of social, cultural, political and economic inequities that may exist in the relationship between northern and southern Finland. Secondly, it is also possible that Liksom's manner of using irony can bring all new dimensions and nuances to the theoretical discussion about postcolonial questions. On the other hand, not all of her short stories are written from an ironic point of view, some are actually quite serious and as always, the separating line is more than a blur.

---

[39] Olavi Jama, 'Haaparannan lukiosta sipirjaan. Torniolaakson kirjallisuus kahden kansalliskulttuurin marginaalissa,' in *Marginalia ja kirjallisuus*, pp. 120–121.

[40] *Ibid.*, p. 97.

[41] *Ibid.*, p. 127.

In Liksom's literature the clearest manner of separating the North from the South is expressed by differences of linguistic features between these geographical totalities, small dialectal features, which unfortunately are basically impossible to translate into English. Dialectal features are not the only way in which she separates the North from the South, but they play a major role. When starting to read any of Liksom's short stories, one can immediately place them in either a northern or southern context by means of small dialectal features in the case of the North, or slangy and English expressions in the case of the South. The only solution for the translator is to cut the corners, as seen in this example, where circumstances of northern nature are bothering life in the South:

> "Is this reality?" he asked, in English, when the wind ripped the cigarette out of his mouth. I glanced at the sleeping village by side of the road, lost in its friendly and sterile dreams, unaware of what went on elsewhere in the world.[42]

It is not only Liksom's original sentence, "Is this the reality?" that has changed somewhat, but in this translation many other features of the language have been lost as well. Using English expressions while speaking English, has a totally different meaning to using English expressions while speaking Finnish. The point here is that in case of Rosa Liksom the translators have actually insurmountable difficulties and responsibilities when trying to keep the text alive but still somehow in the same form, avoiding loosing the essence but still not giving a false image. I would actually agree that in Rosa Liksom's case this may be impossible, so the only way for translations is to cut the corners and try to explain it later.

Spatial Distinctions with Misery?

In Rosa Liksom's literature the North is constructed formally as a clear contrary to the South and so, when trying to define Liksom's North it is important to go through them both, in relation to each other, as contrary totalities with their different features. From Liksom's short stories there can be found several distinctive lines which are separating the North from the South. Cultural and social features and differences are often quite obvious. But going through these different features does not only mean that stories were situated in different cultural and social contexts or landscapes, but more that Liksom's short stories are written through different spatial discourses. Places and spaces of the North and the South are constructed differently, through different discursive practices. Liksom often relates space with human misery, but this misery is linguistically differently

---

[42] Rosa Liksom, *One Night Stands* (New York: Serpent's Tail, 1993), p. 6.

structured in the North when compared to the South. Descriptions of northern life are usually situated in a rural, almost primitive environment where cruel forces of nature are often unavoidable source of misery for people living there:

> [...] It didn't rain during the whole summer, it was lurking over the village as a pining beast ready to attack, but no attack eventually came. Flowers were drying on the hall's window sill as well as in front of the veranda, and the hay in the yard was rustling.
> In the first week of August we started haymaking and then it started raining. Black clouds came from behind the lake in the early morning and until eight o'clock in the evening they wandered around the village. It rained for three weeks without stop. Drops punched heavily on the house's window and gathered splendid pools of water in front of the yard and sauna. The sand in the yard turned into figures of leopards and flowers were being broken and rotting along the grass. During a few days at the end of summer the sun was shining hot and lively, but days with fair weather never came one after another. Hay became yellow, turned dark brown and eventually it was black, smelling and sad trash around wet and decaying hay poles or else it was along the fields as black lumps. The family was sitting all days inside the house and waiting gloomily for the rain to end. And when it ended, it had no meaning anymore. Everything was lost. Hay was gathered reluctantly and salt was thrown over it.
> During the war cows were fed with the newspapers, this is a little bit better, said grandfather when the last horse load was taken inside the barn with a shingle roof and broken walls.[43]

As it is so often portrayed, nature in the North is cruel and merciless in Rosa Liksom's short stories. Even if some days are more merciful, all northern life with its contents and meanings still revolves around the questions concerning nature. The southern landscape for living is totally different:

> My dad bought me this apartment in Eira, it wasn't cheap, but back in Haukilahti we had these constant fights about hygiene. There was dust everywhere, and god, you should have seen the bathtub. I had to disinfect it every single time I used it. I couldn't use the toilet before spreading a sterilized cover on it. [...] I started cleaning the place and had no time for anything else. Luckily, Dad brought me food from Stockmann's every day, so I didn't have to tear myself away and go out.[44]

In the South most of the stories take place inside, and nature, as well as everything related to it, has a relatively small role when the spatial environment for the story is constructed. In Liksom's writings the South means the same as Helsinki while the countryside always acts as a representation for the North.

---

[43] Rosa Liksom, *Unohdettu vartti* (Espoo: Weilin+Göös, 1986), pp. 24–5. All translations for *Yhden yön pysäkki*, *Unohdettu vartti*, and *Tyhjän tien paratiisit* by the author.

[44] Liksom, *One Night Stands*, p. 71.

Liksom uses plenty of actual place names (Eira, Haukilahti, Stockmann) from Helsinki which helps the reader to situate events in the South. But this is not only a contextual background for stories. This usage of actual place names is how Liksom's South is discursively constructured in a certain cultural language and in cultural expressions where these geographical associations with certain place names are exploited. Place names are a necessary part of a literary event, but it is their particular use, which makes the whole story a 'southern one'. In the northern discourses place names are not used in the same way, they are irrelevant and spatial content is more built around descriptions of nature outside:

Hard frost came at the beginning of November. One month before usual. There was no snow at all, but the ground was all frozen up.

Reindeer, calmed down by hunger, gathered along the sides of roads and close to villages. Many of them ended their misery during dusky afternoons by rushing under lorries carrying logs. Bloody and bony carcasses were lying messed up in road ditches and at the side of the fields.

Then came snowfalls. It snowed almost two weeks without stop. Sometimes the whole landscape was one white line. Drifts grew over the houses and in front of snow fences. Carcasses of reindeer were left under high snow banks to wait next spring.

At the beginning of January the temperature went down to minus 40 degrees celsius. People in the village were recalling times of the war. A dog was howling outside with its fur and snout frozen.[45]

The story follows almost the same route as in the previous example from Liksom's North. An unexpected cycle of nature turns people's and animals' lives into a misery. Better times are expected and when they seem to come, everything turns even worse. The only solution for this misery is to recall the war: then, it was even worse. Time and its passing play a central role. Everything is signified in relation to time. Time and space get their significance always out of each other and in those discourses through which Liksom constructs her North, time is always an important element. People in the story are physically inside, but spatial elements of the story as well as the main part of its plot are painted from outside. These spatial elements, which make these events possible, are in this manner, discursive contrary to those of southern descriptions. When the physical descriptions of nature between the South and the North are read, the differences are found easily and are quite obvious, but less important. What is more important is how their nature's contents is expressed and emphasized in those spatial descriptions. Many of Liksom's short stories situated in the North are nothing but forces of nature. Naturally all sort of things happen there, but these things are almost always discursively wrapped around spatiality and its mercilessness:

---

[45] Rosa Liksom, *Yhden yön pysäkki* (Espoo: Weilin+Göös, 1985), p. 114.

Night was lucid and bright, as spring nights of Lapland always are after harsh winters. He had been walking forward from late evening to night frost and fallen asleep in an old summer cow stable, standing grey and tired on the edge of cultivated peatland. [...] He slept a while under damp hay and woke up along with the bright six o'clock morning sun. He felt inside himself the suffering of the whole of humankind and sins of his grandfathers. All of it was poured on him without any reason. No one felt pity for him, not even that low-minded postmaster.

When the red sun was wrenched behind the hills he moved on. Soon it would be the time of final decision. The surface of the road was soft, partly covered with gravel. He walked by a birch grove appearing worried again without a glance aside. Wind swayed light birch branches slowly in the sharp breeze. So many years lived, one thin life. It hurt, it scratched deep. Why all these moments filled with misery. A sunbeam flashed on his face, and a diver cried twice. Endless road and awakening landscape. That day he yet intended to last.[46]

The main idea of this story, the red line is not a northern landscape, but human misery. Human misery has an important role in Rosa Liksom's literature, but what is more important here is how this misery is written through northern nature and how northern nature is written through descriptions of misery. Spatial reality becomes constructed through the descriptions of social reality and vice versa. The wrenching red sun, a diver crying and the rough gravel roads are all part of some kind of a store of expressions through which this misery can be symbolically expressed. When this is understood discursively, northern nature is not automatically a symbol of misery. It becomes a symbol of misery only after it is somehow – culturally, politically or socially-expressed as such. Rosa Liksom's short story is a part of those stories in which it is not symbolically but constructively expressed, that 'northern nature is a symbol of misery'. To maintain this symbolic meaning, it has to be expressed and expressed, again and again. According to some interpretations, Liksom's North has no difference from her South, because in both worlds the human is an outsider and living alone,[47] but still, misery in southern everyday life is constructed differently:

After the funeral she locked herself in the bedroom. She did not eat, drink, or sleep for twelve days and nights. The sun rose and set but she paid no attention to it. She felt like a stranger to herself, and everything she saw and thought seemed fragmented. She did not reminisce about the past, but she could still smell her husband in the sheets of their double bed. There had been moments of happiness. There had been stout infants babbling on their living room rug on Sunday mornings, the man still sleepy next to her, she herself beautiful and fragrant in the front seat of the car on her way to work. All

---

[46] Liksom, *Unohdettu vartti*, p. 116.

[47] Tuohimaa, 'Pohjois-Suomen unohdetut naiskirjailijat,' p. 55.

that had been a long time ago, or perhaps only yesterday. Either it happened, or else it was just a fantasy.

Naked, she climbed onto her bed. She looked at herself in the mirror on the bedroom ceiling. She looked hard but could not see anything. Not even her own face. She picked a heavy vase off the floor and tossed it at the ceiling. The mirror shattered and rained down on her. The splinters tore bleeding wounds into her body. Some of the wounds were deep and gaping and blood streamed out of them. The blood soaked into the sheets, her body throbbed with heat, and the blood smelled like old, hard life. She jumped down to the floor. Mirror splinters pierced the soles of her feet but she did not feel any pain. Excited, she danced from wall to wall, humming an unknown melody. The white-flowered wallpaper became stained with red. Her breasts and thighs were steaming with blood. When she finished the dance and collapsed on the floor, the last drops of blood gathered in a puddle in her own lap.[48]

This short story is full of detailed descriptions of human suffering, bleeding wounds, steaming and smelling blood, but the real misery in this story, is that she will not respond by any means to what is currently happening. She had already ended her meaningful life before this story even started. If northern misery was symbolically expressed through various objects of nature, southern misery is expressed, not only through the indifference of staining blood, but also through ignoring all exterior but her own bed. When the world outside is not meaningful, then the whole life is without a meaning. "The sun rose and set but she paid no attention to it." What makes a difference between the South and the North is how basically the same kind of human misery is discursively constructed through different symbols, that is, contents for the South and the North are defined differently, made look different. The South cannot be represented as written through harsh winds, cold nights, old sheds or some bird crying twice, because they are what the North is. But they would not be so, if Rosa Liksom or someone else did not say so.

Postcolonial Discourses with Irony?

One easy way to approach Liksom's literature is following the same ideological lines of postcolonialism as Edward Said. Simplified, there could be two different ways to perceive the relationship between the main points of Said's theory and Liksom's art. The first would be that Liksom is one of those authors who, without their own attention or by being systematically forced, are making descriptions of the North, that Liksom was contemporaneously reproducing general images of what the North is and what it should be about. The other version would be that she is

---

[48] Liksom, *One Night Stands*, pp. 82–3.

systematically breaking down this colonialist and subordinating relationship between the South and the North by producing new alternative representations of what the North could be about. Still, Liksom's position and especially her literature's position within these two versions presented can not be interpreted that simply.

When the author's style of writing is more or less ironic, it is never that simple to make any sure conclusions about possible discursive meanings of the author's output. In fact, irony can always prevent the reader from getting inside the possible intentions or purposes of the text. Of course in Liksom's case, irony could act as a gate through which the author is trying to break down this colonialist wall between the South and the North, but even this is not at all simple. Liksom's critical irony is often directed at the questions considering the social inequality between the North and the South, but on the other hand, her irony is also directed at the whole theoretical composition that has been constructed along with recent postcolonialist discussions. Although Liksom's literature is not written either straight after or straight against the postcolonialist theory, this does not mean that perceiving her writings through this model would be about taking steps to detour. It is rather about finding some new possible aspects of the confrontation between cultural definitions of the North and the South. To be precise, when trying to perceive any literature of northern Finland together with the postcolonialist perspective, there can be found an innumerable amount of new puzzling questions complicating that same difficulty,[49] which does not necessarily mean taking steps to detour.

Liksom's simple division between the North and the South is highlighted through separating northern short stories from southern short stories into their own distinctive categories. These two categories are published in the same collections, so the reader compares them automatically in relation to each other and only after this comparison, makes an interpretation of what her North or South is. Stories are narratively isolated from each other with their own contents, but cannot be read without relating them to each other. Liksom's North has its own rhythms of time and life, but what makes them northern is a silent comparison with the southern rhythm of life:

> The woman went to the cow shed after six and I went to the sauna. The oldster stayed sitting in the room with a cat. It was damn hot in the sauna just as it is supposed to be. I threw water like hell on the rocks and then put on clean pants in the damn cold sauna chamber. I swear it was minus 30 degrees over there. But so what, it only hardens.

---

[49]   See Juha Ridanpää, 'Postcolonialism in a Polar Region? Relativity Concerning a Postcolonialist Interpretation of Literature from Northern Finland,' *Nordia Geographical Publications*, Vol. 27, No. 1 (1998), pp. 67-77.

Makes tougher. The window of the sauna chamber was frozen over so you couldn't see that lump of a moon, which lighted the whole village. It was probably that Christmas sauna. The oldster went after me and I stayed with the cat in the room. I emptied a coffee-pot and was supposed to leave to the house. The weather report was coming from the radio and there it was said that it will become colder. Swedish radio has a more accurate report, but you just can't understand a thing of it. Valkko was nodding under a Christmas tree. I don't understand how it always goes sitting there. The woman came out from the cow shed with her side ahead and dragged a water bucket behind.[50]

In the previous example it is described how northern life and everything belonging to it derives its contents from its own, monotonous and dull everyday existence. The question of whether this boring and monotonous life is described here as a consequence of southern domination is a puzzling one. Life in the northern countryside, where Liksom herself was born, at least according to the back cover of *One Night Stands*, is monotonous, hard and dull. Certainly, this story could not happen without something opposite, which makes this description seem like 'life in the marginal'. Liksom describes the North as if it was a form of an icon for human suffering, but what makes this composition even more puzzling, is Liksom's ambiguous manner of exploiting ironic impressions in her short stories: "during the war cows were fed with the newspapers, this is a little bit better." If the reader comprehends all these examples more or less realistically, the interpretation is that because northern life is hard, people living there have to find a new way to get through their everyday misery. This 'new way' is to make a joke of it. From this point of view, northern irony would be as a local matter of fact and Liksom's art a depiction of it. On the other hand, Liksom uses irony as an artistic effect making serious subjects more comfortable for the readers to get through with. But this does not make these subjects less serious. In fact, Liksom's ironic depictions stress all the misery, cruelty and inequality there is in northern everyday life. The wholly other question is how effectively Liksom's manner of writing ironically is capable of removing misery from there. At least her irony is often directed at those frequently repeated untruthful images of the North, constructed by southern dominance.

The South stands automatically as some form of a contrary to Liksom's northern descriptions, but is the South necessarily a dominating part along with these cultural differences? In fact, how different are these descriptions of the South and the North? Is life in the southern city less monotonous? Maybe the previous northern example was social criticism towards the South. Or maybe it was not. In every case, there seems to appear some form of a continuing silent juxtaposition between the northern margin and the southern centre going on in Liksom's art. In

---

[50] Liksom, *Unohdettu vartti*, pp. 47–8.

most of her southern short stories the North is not acknowledged by any means. The South is always indifferent to anything concerning the North, while in some of Liksom's northern short stories references to southern life appear much more frequently. References describing the North may be ignoring the South as well, but the difference is that while southern ignorance of the North is always silent, almost oblivious, northern ignorance of the South is at least mentioned:

I have never wanted to move to the South. I would always like to live here, but these circumstances are so hard. If there would be even one man of my age within ten kilometres or closer, but there isn't. I am alone among these older people. They sure are good people and talk to me and try, but they still have a different comprehension of the world than I do.

I would not want to live in a city. I love nature and like to follow all its movements during different seasons. This nature kills me somehow when there ain't other people. But at the same time it kills and gives power. It is strange. I have become used to being alone. Just this, that I'm so scared, helps me with being alone. But still I need a friend. It is dark this life in these outlying villages, everywhere just people dying. And there you try to maintain your good mood.[51]

This example is mainly about the feelings of a northern man towards his own close environments. There are only a few somewhat bitter references to the South, but without these few references from the marginal to the centre, from the North to the South, this whole story could not have happened. These references and what they are referring to construct the North as being marginal. Southern stories do not have this kind of references, as in the next example the South is more like a gate away from Finland, a gate one step closer to the real world:

I had this tremendous itch to get out of town. At the ferry terminal, I made a couple of phone calls to the guys and told them I was leaving for a while. Micke gave me a long lecture about muggings and herpes. At first, when I told him I was going abroad, he thought I was kidding, but then I pointed the phone receiver at the terminal loudspeaker, and he believed me all right.

I got on the ferry at half past five, well ahead of its departure time, checked my backpack and went up on deck to look back at the city. I had never felt so free since the time when I celebrated finishing trade school with a trip to Roskilde. It felt like the whole world was open to me, and there was nothing to be worried or depressed about.

[...] I hadn't been sitting there for more than a couple hours when the cops arrived. They started telling me to move on. In Danish. They thought I was a Dane, and I was unable to tell them anything, I completely paralyzed. Then one of them checked my breast pocket and found my passport. They realized I wasn't one of the locals, so now they told me in English: Move On!

---

[51] Liksom, *Yhden yön pysäkki*, pp. 124–5.

I didn't budge. They looked at each other and told me, in cool and precise English, that I could choose between jail or the next train back to Finland. I would have preferred jail, but I couldn't tell them that, I just sat there squinting at them. I realized that I'd lost it. They practically carried me to platform number five. According to the digital clock, the next train would leave at seven minutes to one. One of the cops stayed with me on the bench, the other guy got my backpack from the locker. At the time, it seemed like a pretty scene to me. When the train pulled in, they carried me inside and said have a good trip. In Danish.[52]

This story is yet again one of those descriptions of misery, this time southern misery, constituted when hopes of gaining something better than a boring life in Finland through changing the environment crashes down. Hopes of forthcoming happiness clash with rules of social power and the only possibility is to go back home. Southern misery has nothing to do with the North. Northern bitterness towards the South and southern ignorance towards the North can be interpreted as a postcolonialistically set distinguishing line between these spatial totalities. Liksom's dichotomy between two socially and culturally constituted spatial categories, 'the North' and 'the South' becomes more complicated if her short stories are observed from that same ironic perspective they are written with. Liksom writes seriously, too, but making a distinction between her serious and ironic writing is sometimes quite hard. For example, her stories of settled manners of northern living can be comprehended as realistic descriptions of the North, but when perceived ironically, this realism of her short stories turns into something else:

I get on well when alone. Too many people just make me feel strange, as if a vacuum is under my breast and that doesn't feel great. The best for me is to be with my wife or alone in the backwoods. The oldster is complaining so much of his fate and he is still bitter that he had to go to war when he was young, that even I have tired of it. Man just won't stand it all the time.[53]

Realism of northern loneliness or just an ironic description of it or realistic impressions of northern irony? No matter how it is interpreted, the question will finally end up with a conclusion that northern misery is something to be taken seriously and something to be dealt with. And these same conclusions we have also when we read through Rosa Liksom what 'the South' is all about. But what changes along with this and forthcoming examples of reading about northern life through the ironic perspective is how these theories of postcolonialism should actually be approached:

---

[52] Liksom, *One Night Stands*, pp. 21 and 25.

[53] Liksom, *Unohdettu vartti*, p. 8.

Ever since lumber work came to a sudden end I have done this every single Wednesday afternoon. I gather empty beer bottles from the sink and walk with a plastic bag in my hand across the yard to the icy road and from there to the milk scaffold. The day looks exceptionally great, frost hardly to mention, the sun is shining diagonally from the Southwest on the snowy fields. Two old women and an oldster are standing on the milk scaffold. The oldster has a bag of empty bottles too. We stand quietly side by side as we are used to doing. No one has any news. What really could happen during a week. During a week, a month, a year. [...] The shopkeeper counts the money and I feel satisfied that life feels somehow more secure when there are sixteen uncorked beer bottles in a plastic bag. The shopkeeper looks at me: "There are fifteen pennies missing, how come you have counted it this wrong, has your head already started to act irregularly?" I go down the road, to search out those missing pennies at home. The shopkeeper does not nod but maintains that same expression on his face. I understand, it's a question of life or death. I find the missing coins in a coffee can, I take the rifle from over the bed and go back to the store bus. Old women are fingering rye bread. I give the missing coins to the shopkeeper, who glares maliciously. I rise the rifle, I see the distressed but selfconfident glance of shopkeeper. I shoot before any of the customers have time to notice. Women load their baskets full, hardly glancing at the bloody body of the shopkeeper, and go down the road. I take a beer bag, lift the rifle on my shoulder and walk home. It is time for afternoon coffee, but because the day has been exceptional, I decide to open a beer bottle on the front of the table and enjoy the whiz of the cork and the strong full taste of malt.[54]

In the previous description the story begins with ordinary settled manners, which have their origin in the history of the Finnish forest industry. When the forest industry came to an end in the North, human life turned into monotonous routines. In the beginning of the short story, nothing really happens and no one is really interested if something happens. Then comes a dramatic turn, a man takes a rifle and shoots a shopkeeper who maliciously prevents the man from taking a few rare joys out of his life. Discursively speaking the whole story could be a social criticism towards southern economics, politics and decision-making, with or without its dramatic turn. But in the last sentence the story turns interesting: the man decides to open a beer bottle on the front of the table and enjoy the whiz of the cork and the strong full taste of malt. Just like a straight quotation from some beer commercial. The beer, which man buys in the short story is called by himself a "Lapp" referring to the Finnish beer *Lapin kulta*, translated as the *Gold Beer Lapland*, which strengthens the already ironic impression. This ending turns the whole socially set discursive confrontation between the North and the South upside down. Misery and drama turns actually into a comedy, but who is laughing? The northern readers, the southern readers or both?

---

[54] Rosa Liksom, *Tyhjän tien paratiisit* (Porvoo: WSOY, 1989), pp. 99-100.

Conclusion: Traditional Confrontations or New Discursive Practices?

Postcolonialist theories, as well as Liksom's art, are about the continuing confrontation of different forms of contraries, about stressing how there could not ever be a margin without a centre and no centre without a margin. They are about emphasizing how there has to be something marginal to define the existence of something central and how important it is to criticize and try to systematically break down the different forms of activities constructing all kind of inequalities and subordinate relationships. Therefore, Rosa Liksom is evidently a postcolonialist artist. Liksom's North can be defined as being something that her South is not, a rustic contrary to urban life. This generalization defines cultural features of the South and the North in relation to each other, as opposites to each other. Although both of them have something culturally of their 'own', their definitions come also from the lack of what the other part has. In the case of the North this means the lack of possible fortunes which are taken away from them by the South. The South does not have features from the North, but just as postcolonialist theories go, it is no problem for the centre.

Does this necessarily make Liksom's art a form of postcolonialist theory, which is consciously or unconsciously put into practice? Sinikka Tuohimaa has interpreted Rosa Liksom's writings and especially the human types she describes as being all the time marginal no matter in what environment they are situated.[55] Tuohimaa says that Liksom's marginality is the same marginality everywhere and that this would be a main point where Liksom is cracking the traditional northern identity, a point where Liksom turns from a traditional northern author to a critical author, an author breaking some boundaries. Circumstances and the main causes for northern marginality as well as for southern marginality are naturally different, but both are marginal. Their human lives are always miserable when compared to something else, compared to what it could be. The marginal North, as Liksom so often describes it to be, has maybe its ruling centre in southern Finland, but the main point seems to be that this does not make southern life any easier. What is so simple in postcolonialist theory is not so simple in real life. It seems as if Rosa Liksom is an author who is systematically fighting against simplified descriptions of the North and contemporaneously reproducing general images of what the North is, and what it should be about. Still, it seems as if she is not trying to systematically break down this colonialist and subordinating relationship between the South and the North, but more trying to break down this whole composition of it, as if trying to stress how naïve these forms of generalizations actually are.

'The North' is a spatial concept with different kinds of, more or less ambivalent definitions and associations. 'The North' can be defined through different cultural

---

[55] Tuohimaa, 'Pohjois-Suomen unohdetut naiskirjailijat,' p. 53.

features and materials, through literature, but even cultural and literary approaches towards the North are actually full of this ambivalence. It seems that Rosa Liksom's literary world was written as if it was a pure form of some post-colonialist criticism towards the unevenly developed colonial reality. But only after one notices the meaning of her irony in this simplified dichotomy, does it become clear to what Liksom is actually pointing. The point is made against simplifying. It does not mean that cultural features, human feelings, for example human misery, would not be somehow spatially constructed. In Liksom's art northern misery is always different when compared to southern misery, no matter if her irony was comprehended or not. But what it does mean, is that these cultural features, feelings, misery, which discursively define what the North is, are remaining in the same form of the colonial composition, unless someone defines something different, something new. Rosa Liksom does not actually say anything new about the North. This new content to the North comes, when she puts aside the common comprehension of what the North should be, pushes the old, general views into such a peculiar position, that the reader must either consciously or unconsciously start to question if there is something wrong, something strange in her writings. She certainly repeats the traditional discourses of what the North is, but when the reader thinks, at least for a moment, whether she is serious or not, the traditional two-part definition of the North and of its cultural contents turns upside down.

Chapter 6

Explorers in the Arctic: Doing Feminine Nature in a Masculine Way

Hanna-Mari Ikonen and Samu Pehkonen

The journeys with a purpose of exploring the regions covered by the eternal polar ice have [...] always been considered prestigious and pure. This is partly due to the white fields of snow and the amazing light phenomena in the sky, but it is also because of the pure, unsoiled idealism motivating these expeditions. Except for the journeys made by hunters and fishermen (though polar research owes much to these journeys as well), we can be fully convinced of the fact that even a hopeless dreamer has never travelled to the polar glaciers in pursuit of money and wealth.

Science alone may claim the credit for the numerous unrivalled ventures to overcome the greatest obstacle ever standing in the way of human research, the endeavours to conquer the eternal ice which surrounds the secrets of the North Pole like a wide, impenetrable wall.[1]

This quotation taken from the Norwegian polar explorer Roald Amundsen depicts the period at the turn of the 20th century, when attempts were made to solve the Arctic mysteries and there was still hope of finding new territories. In addition, the quote illustrates an important phase in the reciprocal development of Norwegian nationalism and the discipline of Geography. At the end of the 19th century, the polar regions were merely rare white spots on the map – and, according to the ideals of the time – waiting to be discovered and claimed by the 'civilized' world. Just as the Pacific expeditions of James Cook are said to have laid the ground for the development of the British Geography,[2] the efforts of the Finn-Swede Adolf

---

[1] Roald Amundsen, *Luoteisväylä. Kertomus Gjöan matkasta 1903–1907* (Porvoo: WSOY, 1908), pp. 1–2. All the quotations in this chapter are translated by the authors.

[2] D.R. Stoddard, *On Geography and Its History* (Oxford: Basil Blackwell, 1986), pp. 28–40. For a critical reading of Stoddard's "unorthodox, eurocentric and adventurous history of modern geography", see Derek Gregory, *Geographical Imaginations* (Cambridge and Oxford: Blackwell Publishers, 1994), pp. 15–33.

Erik Nordenskiöld navigating through the Northeast Passage on the *Vega* (1878-1880), the Norwegian Fridtjof Nansen skiing across Greenland (1888-1889) and later trying to reach the North Pole on board the *Fram* (1893-1896) and the Norwegian Roald Amundsen sailing through the Northwest Passage on the *Fram* (1903-1907) formed the basis for geographical research and thinking in the Nordic countries.

The Nordic Geography found a natural field of study in polar regions. However, the traditions of polar research vary among the different Nordic countries. Sweden was the most active country between the 1860s and the 1920s, whereas Norway first became engaged in polar research in the 1890s.[3] The Finnish research concentrated on the Kola peninsula and Siberia, although some arguments were put forward also in favour of launching 'truly polar' research programs.[4] Arctic regions close to the Nordic countries were almost completely unmapped. There was an uncertainty about the climatic conditions in the Arctic, and even such a fundamentally geographical question as whether or not there was unknown land between Alaska and the North Pole was not answered before 1926. In addition, the reasons behind the interest in the polar regions were strongly culture-bound: Norway, Sweden and Finland were gradually creating an image of the cold, white and virgin landscapes of the North as a basis for their nationalistic projects. Competition as to who of the Northern men would manage to bear the even colder and harsher climate in the white margins of the world was highly intense. The national importance of polar exploration in the Nordic countries could be seen in the ways the nations appreciated their polar explorers: the men returning from the Arctic were celebrated as national heroes.

A closer look at Amundsen's opening words reveals certain attitudes of the time towards the exploration and research of the Northern regions. Amundsen stresses the special, valuable and pure character of polar research. For him, the Arctic is a visual pleasure and exploration represents the highest form of human activity in that landscape. The Arctic is not a place suitable for daydreamers, fortune hunters or huntsmen without the right gaze, unable to notice the visual miracles and to measure nature with scientific accuracy. According to Amundsen, scientific research alone is motivated by pure idealism not foiled with too concrete and

---

[3]   See Gösta H. Liljeqvist, *High Latitudes. A History of Swedish Polar Travels and Research* (Stockholm: Swedish Polar Secretariat & Streiffert, 1993); Sverker Sörlin *et al.*, *Det nordliga rummet: Polarforskingen och de nordiska länderna* (Umeå: The International Research Network on the History of Polar Science, 1995); Øystein Kock Johansen, ed., *Norske maritime oppdagere og ekspedisjoner gjennom tusen år* (Oslo: Index, 1999).

[4]   See J.E. Rosberg, 'Napaseutututkimus Lapinmeren rannoille, lähinnä Frans Josefin maassa ja Novaja Zemljassa,' *Terra*, No. 37 (1925), pp. 57-74.

mundane objectives. Here, scientific ideals contain strikingly gendered symbolism. The North is, time after time, associated with virginity and considered something untouched but also dangerous.

In this chapter, we will discuss the processes and objectives of knowledge production in the 19th and early 20th century polar explorations. Further, we will discuss the impact of the tradition of exploration on the geographical modes of knowing in those days and today. We call this process *geographical imagination*, and we claim that this imagination has defined the legitimate way of doing geography. Here, we find Gillian Rose's ideas on the aesthetic and scientific expressions of masculinity a fascinating starting point.[5] According to Rose, aesthetic masculinity refers to being charmed by the spectacular characteristics of the nature and the visual pleasures offered by it. The Northern nature is an object of mysterious desires and passions, a provider of (mental) subsistence, a visual pleasure but also an exciting element of possible danger. Considering these notions, the aesthetic masculinity is not very far away from scientific masculinity, *i.e.* subordinating nature by measuring and modelling it. Nature is seen as a feminine object, aesthetically pleasing, which is to be subordinated. The feminized Northern nature is subordinated both by praising (aesthetic masculinity) and by measuring (scientific masculinity) it. Our guiding analytical principles in this chapter are the emphasis on the visual and its relation to the idea of aesthetic masculinity, and the links between scientific masculinity and the acts of measuring and observation.

The next section highlights the dualistic thinking in science when it refers to Geography as a field science. After this, a more detailed picture will be given of the things we believe are central for understanding the popular geographical thinking which made (and partly still makes) explorations such a popular endeavour. The next four subsections – entitled 'going there', 'being there', 'finding it' and 'coming home' – will take the readers to the Arctic seas and lands in the search of the *Terra Feminarum*, the feminized North located somewhere beyond the known world. During this *imagined journey* we will analyze some samples taken from Nordic exploration literature, describing the journeys actually carried out. As a conclusion, we will aim to elaborate our own geographical imagination so as to be able to encounter research issues in an alternative way.

On Dualisms: Intertwining Scientific and Aesthetic Masculinity

The human perception of the world tends to be dualistic. Scientific work is by no means an exception. One of the basic dualisms – and the key to understanding

---

[5]   Gillian Rose, *Feminism and Geography. The Limits of Geographical Knowledge* (London: Polity Press, 1993).

many other dichotomies – is the division between feminine and masculine. The opposite pairs most relevant for understanding the nature of the Arctic explorations are especially nature/culture, home/away, everyday/exceptional, safety/danger and passivity/activity. These are gendered and hierarchically valued attributes.[6] It is more appreciated to travel away from home, to be active, to put oneself in danger and to show that by acting in a particular manner one is a part of culture, someone apart from and better than nature. These are clearly masculine features. The heroic men heading towards the extreme North were living out precisely the masculine side of the above mentioned opposites. They were actively and at the risk of death travelling away from the stable and the known, towards the unknown and dangerous. What made them successful was their single-mindedness or the fact that they were ruthless, arrogant, insensitive and self-serving.[7]

The 19th century polar research stands as a brilliant example of the masculine character of Science. The growing interest in polar regions was tightly linked with the superiority of natural scientific methods and the re-invention of the philosophical ideas of Francis Bacon. An explorer, an ultimate scientist, was the universal man making his way into unknown territories and finding the means to tame the wilderness. To make an explorer fight the forces of nature was a far more reliable and effective way to proof the superiority of man against nature than to have a scientist working in the library.[8]

Bjørg Evjen gives a brilliant example of how notions of home and abroad are still connected with the ideal masculine and feminine fields in science. In North America, where archaeology also includes anthropology, women have been concentrating on the anthropological side, working with already collected material, while male archaeologists usually head off to carry out excavations in polar regions. "Men are polar heroes, while women take their research home with them", Evjen writes.[9] Also in Geography it is the field and the fieldwork which are seen to make the geographical research real. The several existing accounts of the origins of Geography as a scientific endeavour – when, where and why there

---

[6]   See Evelyn Fox Keller, *Reflections on Gender and Science* (New Haven: Yale University Press, 1985); Rose, *Feminism and Geography*; Nancy Duncan, ed., *BodySpace. Destabilizing Geographies of Gender and Sexuality* (London and New York: Routledge, 1996); Constance Classen, *The Colour of Angels. Cosmology, Gender and the Aesthetic Imagination* (London and New York: Routledge, 1998).

[7]   Pierre Berton, *The Arctic Grail. The Quest for the North West Passage and the North Pole, 1818–1909* (New York: Viking, 1988), p. 512.

[8]   Tore Frängsmyr, 'Nordenskiöld och polarforskningen – den idéhistoriska bakgrunden,' *Ymer*, Vol. 100 (1980), pp. 7–38.

[9]   Bjørg Evjen, 'Women in Polar Research – Exotic Elements, Intruders or Equals?' *Ottar*, No. 226 (1999), p. 35.

developed a discipline we today call Geography – indicate that an acceptable understanding of the basic characteristics of providing geographical knowledge requires understanding the fact that geography originates from observing, mapping, cataloguing, classifying and ordering the surface and the inhabitants of the earth. The leading figure of early 20th century North American Cultural Geography, Carl Sauer, saw geography to be "first of all knowledge gained by *observation*, that one *orders* by reflection and re-inspection the things one has been *looking* at, and that from what one has experienced by *intimate sight* come comparison and synthesis."[10] The way of meeting these criteria demanded active participation in the research carried out in the field. In relation to producing geographical knowledge, a clear link was drawn between a vision and the field: becoming and being a geographer demanded engaging oneself working in the field, and in the field it was the sight, which was the prerequisite for the highest art of geography.

The visual emphasis on the field gave and still gives Cultural Geography a clearly visual character. Sauer was influenced by the current discussions in Anthropology, and his landscape school of Cultural Geography continued working in the spirit of 'visual anthropology' privileging sight long after many other disciplines became circumspect, notes Derek Gregory.[11] It was not before the recent work on gender and fieldwork that the complexity and the contested character of fieldwork have been recognized. This has been done by problematizing, for example, the triangle of seeing, knowing and credibility: 'field' is seen as something "always in the process of being constructed" rather than something out there waiting to be discovered by the gaze of the researcher, and 'fieldwork' becomes understood as a variety of embodied spatial practices, such as movement, performance and encounters.[12]

Despite these recent developments, Cultural Geography has still not fully succeeded in liberating itself from the spheres of male domination, and the efforts

---

[10] Carl Sauer, 'The Education of a Geographer,' *Annals of the Association of American Geographers*, Vol. 46 (1956), p. 295 (emphases added by the authors).

[11] Gregory, *Geographical Imaginations*, p. 16.

[12] Felix Driver, 'Editorial: Field-work in Geography,' *Transactions of the Institute of British Geographers*, NS Vol. 25, No. 3 (2000), pp. 267–8. See also Special Issue on 'Women in the Field: Critical Feminist Methodologies and Theoretical Perspectives,' *Professional Geographer*, Vol. 46, No. 1 (February 1994), pp. 54–102; Karen M. Morin and Lawrence D. Berg, 'Emplacing Current Trends in Feminist Historical Geography,' *Gender, Place & Culture*, Vol. 6, No. 4 (December 1999), pp. 311–30; Karen Nairn, 'Doing Feminist Fieldwork about Geography Fieldwork,' in Pamela Moss, ed., *Feminist Geography in Practice. Research and Methods* (Oxford: Blackwell, 2002), pp. 146–59.

to draw away from traditional fieldwork-based research are considered as threats.[13] Valuing masculine features over feminine ones has lead to the exclusion of women from the 'real' practices of science.[14] The exclusion has largely been caused by the imagined connection of women with nature, emotions, love and the body. Masculine characteristics, which also stand for scientific ideals, are different: they are filled with power. "The fact that the ['rational'] 'male' senses of sight and hearing were classified as 'distance' senses and the ['corporeal'] 'female' senses of smell, taste and touch were characterized as 'proximity' senses, was interpreted to mean that men were suited for 'distance activities' such as travelling and governing, while women were made to stay at home", notes Constance Classen.[15] Northern explorations were dominated by masculine senses. Sight, "the master sense of the modern era",[16] was the primary sense of observation and positioning, taken as the truth especially when linked with technical equipment to calculate and scientifically prove 'the farthest North'. Hearing and sight were absolute senses in the way that everyone in the crew could share the same scenery and sounds in the same measurable manner while tastes and smells were personal (*i.e.* feminine) senses not discussed in a scientific context.

Learning and Writing the Geographical

Despite the fact that Nansen, Amundsen and other Nordic polar explorers of the late 19th century were not geographers (since the discipline was just starting to become professionalized in the Nordic countries), we would claim that a strong connection, the so called geographical imagination, existed between exploration and geography. This co-existence has been reproduced in the educational practices of Geography where it has, until recently, largely been handled unreflectively. Here, we deal with two elements of reproducing the myth of geographical knowledge production – namely, fieldwork and literary practices – in order to shed light on the question why explorations to the Northern regions were so popular and highly valued in the 19th and early 20th century Nordic countries.

In his essay on geography, exploration and discovery, David Stoddard provides an account of the elements of action and adventure in the disciplinary history of

---

[13] See, for example, Robert A. Rundstrom and Martin S. Kenzer, 'The Decline of Fieldwork in Human Geography,' *Professional Geographer*, Vol. 41, No. 3 (1989), pp. 294–303.

[14] Keller, *Reflections on Gender*, p. 7.

[15] Classen, *The Colour of Angels*, p. 66.

[16] Martin Jay, 'Scopic Regimes of Modernity,' in Scott Lash and Jonathan Friedman, eds., *Modernity and Identity* (Oxford: Blackwell, 1992), pp. 178–95.

Geography. He quotes Joseph Conrad (1875–1924), a Polish-born American writer, who was unhappy to see the contrast between the world of action and the world of the academia. For Conrad, education had come to be controlled by:

> persons of no romantic sense for the real, ignorant of the great possibilities of active life; with no desire for struggle, no notion of the wide spaces of the world – mere bored professors, in fact, who were not only middle-aged but looked to me as if they had never been young. And their geography was very much like themselves, a bloodless thing with a dry skin covering a repulsive armature of uninteresting bones.[17]

Becoming a real geographer – both in spirit and in practice – demanded learning about nature out in the wild. This knowledge was gained during field excursions where the eyes of the observer were able to verify the theories learned in the lecture hall. Excursions were carried out both in the vicinity of the university and into remote locations. For British geography students, excursions often headed towards Eastern Europe – in particular Russia, the Carpathians, Montenegro and Poland are mentioned in this context. The participants of the excursion put up with primitive conditions and expected to conduct a full survey of the physical setting, the vegetation, the ethnography, the villages and their life and sociological structure.[18]

During these fieldtrips, "the vodka would flow"[19] – or, in the words of Gillian Rose who expresses this masculine character less heroically, "fieldwork also involves the necessary amount of drinking in order to prove how manly the fieldworker is".[20] It still holds true, as even Stoddard admits, that such trips tend to promote the education and enjoyment of the participants rather than to inspire original contributions to knowledge. "Given the absence of precise objectives, the brief periods involved in the studies, and the cultural and language barriers encountered in the area and communities studied" these excursions were doomed to fail from any 'scientific' point of view, and the only thing learned was the tradition of going to the field.[21]

---

[17] Joseph Conrad, *Last Essays* (1926) as cited in Stoddard, *On Geography*, p. 143.

[18] *Ibid.*, p. 145.

[19] *Ibid.*

[20] Rose, *Feminism and Geography*, p. 70.

[21] Stoddard, *On Geography*, pp. 145–6. At the end of the 1990s when the authors of this chapter were studying Human Geography in the University of Joensuu, the students were taken on a field excursion to Poland, a typical country for Western observation. All the distinguishing marks of the fieldwork tradition were met, and a geographical report was produced – a report representing the scientific mode of writing which means, knowing the reality of the trip, that the 'field observations' were finally abandoned.

The contents of the fieldtrips highlight extremely masculine action in the extremely feminized nature. The nature in fieldwork has been an object of geographical observation and subordination. Although the Geography of today is less dominated by surveying and sketching than it was a hundred years ago, it still holds true that every new generation of geographers will be reminded of the great discoveries and advances made by men who went into the field to look and to think. It is still true that fieldtrips made by undergraduate students, for example, act as initiation rituals of the discipline where the students become geographers – stronger men, who challenge the nature and face the wild for the sake of knowledge.[22]

Yet, it would be an insufficient endeavour to look for the 'geographical' solely inside the discipline of Geography. For those outside the academia, there were other channels to get in touch with nature. Simultaneously with expeditions to unknown places, reading and writing adventure stories began to be popular. The story of Robinson Crusoe, written by Daniel Defoe in 1719, and, later several books of Jules Verne are seen to reflect authors' ambitions to mentally conquer the world. By drawing together all the information the modern science had gathered, Verne, for example, aimed at rewriting the history of mankind. The stories told about European men not only trying to control the new territories and environments discovered but also to devote oneself, one's body, mind and life to these exploratory aims.[23]

Many young boys were eager to read about these heroes who were constantly put under pressure by the nature but who, time after time, managed to overcome all the obstacles. In addition to purely entertaining the reader, adventure tales and travel accounts provided school teaching, especially in geography, with powerful tools. Anecdotes on Nansen's journeys were collected into Nordahl Rolfsen's book, which was to become an obligatory reader for generations of Norwegian schoolchildren. In Sweden, Selma Lagerlöf was asked to write an exciting text to be used in teaching swedish geography. The author came up with the story of Nils Holgersson, a boy making his way Northwards (sic) on the back of a gander. Nils travels the length and breadth of Sweden with the wild geese. Looking at the Swedish way of life from the air and on the ground, the book gives the clearest possible view of the land and the people. Another book – entitled *Polar Explorers: Accounts of Brave Men and Their Adventurous Journeys to the Most Deserted and Inaccessible Regions of the World* – concentrating on polar travel was introduced

---

[22] Rose, *Feminism and Geography*, pp. 69–70.

[23] Mikko Lehtonen, *Pikku jättiläisiä. Maskuliinisuuden kulttuurinen rakentuminen* (Tampere: Vastapaino, 1995), pp. 50–51.

to Finnish schoolchildren in 1929.[24] These attempts of educating future geographers seemed to be effective. Just as David Harvey, one of the today's leading critical geographers, who, as a young lad, dreamt of sailing the seas of the British Empire on board a naval destroyer,[25] many young boys in the North were turning their admiration towards polar heroes. Finally, some of these boys made their dreams come true. Roald Amundsen decided to become an explorer after having read about the tragic fate of the British Franklin expedition (1845-1847) in search of the Northwest Passage, and Fridtjof Nansen's lifelong passion for the far North was kindled in his student days when, at a tutor's suggestion, he left for a passage aboard a sealing vessel to the Arctic Ocean in 1882.

Although many of the adventure stories were written by men who had never even visited the places they were writing about, the readers seemed to consider travel books as faithful accounts of reality. Direct observations and collections of exotic specimen were required as evidence of the closest possible touch with nature.[26] Not long after explorers had safely found their way back home a book was written and published to provide the public – and the sceptical colleagues – with all the amazing stages of the journey. In polar travel, the story of A.E. Nordenskiöld's sailing through the Northeast Passage (1878-1880) was published in two volumes in 1881, and the Norwegians Nansen and Amundsen published several books (translated into several languages) on their journeys already within the year of their return.

Books written by arctic travellers are an interesting blend of scientific and literary fashion. They are astonishing adventure tales filled with descriptions of both ineptitude, boredom, frostbite, isolation and hunger and, on the other hand, of victory, discovery, the sublime and human capability. Moreover, to show that 'these things really happened', the works contain detailed (scientific) descriptions – both literal and pictorial – of landscapes already modified or still untouched by the man. In addition to Nordenskiöld's original two-volume travel report of over 800 pages, the scientific results of his expedition through the Northeast Passage

[24] Nordahl Rolfsen, *Læsebog for folkeskolen*, 5 vols. (Kristiania: Jacob Dybwad, 1892–1895). Between the years 1892 and 1950 more than eight million copies of this book were delivered to Norwegian schools; Selma Lagerlöf, *Nils Holgerssons underbara resa genom Sverige*, 2 vols. (Stockholm: Albert Bonnier, 1906-1907); Hannes Salovaara, *Naparetkeilijöitä. Kuvauksia uljaista miehistä ja heidän seikkailurikkaista matkoistaan maapallon autioimmille ja vaikeapääsyisimmille seuduille* (Helsinki: Otava, 1929).

[25] Jeff Byles, 'Maps and Chaps. The New Geography Reaches Critical Mass,' *The Village Voice*, Vol. XLVI, No. 31 (August 2001), electronic version at <http://www.villagevoice.com/issues/0131/edbyles.php>.

[26] Gregory, *Geographical Imaginations*, pp. 22-3.

- representing the 'ultimate scientific' value - were published in 1882 in five volumes including almost 3000 pages which formed the scientific value of the expedition.[27]

The call for objectivity also influences how and what is written under the name of Geography. Firstly, not every account is accepted as geographical. It is "the quest for systematic and objective knowledge that distinguishes the traveller, inadvertent or otherwise, from the scientific explorer and discoverer", argues David Stoddard.[28] Thus, feminine research themes become restricted or excluded because not all themes are equal. The obsession of Geography with distance has left the spaces of proximity (such as home) outside the appreciated and dominating themes. When exploring the history of Geography, one notices that not everyone was deemed capable of creating masculine geography of objectivity. One century ago, the Royal Geographical Society in London, for example, could not admit that even white women could produce proper geographical reports of their travels (which were, of course, not as numerous as were the journeys made by men), and this was one of the reasons why women could not, in the first place, be admitted to the Society (the other reason being simply that they were women).[29] In cases that women geographers did manage to get their reports published, it was because of the way they were written, not because they would have been manifesting feminine voices.[30] Feelings were omitted from writing geography, since the ideal to be followed was a precise account of the seen and heard. "Geography has a particular written voice", argues Gillian Rose. It is "unextravagant, unembellish-ed, unpretentious, unexceptional".[31] In sum, literature on exploration is a site for negotiating (hu)man-nature relations. It is precisely this position - feminizing

---

[27] See Leena Miekkavaara, 'A.E. Nordenskiöld Jäämerellä,' in Markku Löytönen, ed., *Matka-arkku. Suomalaisia tutkimusmatkailijoita* (Helsinki: Suomalaisen Kirjallisuuden Seura, 1990), pp. 104–37.

[28] Stoddard, *On Geography*, p. 152.

[29] Mona Domosh, 'Towards a Feminist Historiography of Geography,' *Transactions of the Institute of British Geographers*, NS Vol. 16, No. 1 (1991), pp. 95–104.

[30] This is evident in the writings of Ellen Semple, the foremost female geographer of the early 20th century and an exponent in environmental determinism. "Man is a product of the earth's surface. This means not merely that he is a child of the earth, dust of her dust; but that the earth has mothered him, fed him, set him tasks, directed his thoughts, confronted him with difficulties that have strengthened his body and sharpened his wits, given him his problems of navigation or irrigation, and at the same time whispered hints for their solution." Ellen C. Semple, *Influences of Geographic Environment on the Basis of Ratzel's System of Anthropogeographie* (New York: Henry Holt, 1911), p. 1.

[31] Rose, *Feminism and Geography*, p. 8.

nature in a masculine way – we want to address next. This will be done during an imagined journey into the arctic lands and waters.

An Imagined Journey: The Only True Voyage of Discovery ...

Going There

It was midsummer day. A dull, gloomy day; and with it came the inevitable leave-taking. The door closed behind me. For the last time I left my home and went alone down the garden to the beach, where the *Fram*'s little petroleum launch pitilessly awaited me. Behind me lay all I held dear in life. And what before me? How many years would pass ere I should see it all again? What would I not have given at that moment to be able to turn back; but up at the window Liv was sitting clapping her hands. Happy child, little do you know what life is – how strangely mingled and how full of change. Like an arrow the little boat sped over Lysaker Bay, bearing me on the first stage of a journey on which life itself, if not more, was staked.

[...] Then the *Fram* weighs anchor [...] The quays are black with crowds of people waving their hats and handkerchiefs. But silently and quietly the *Fram* heads towards the fjord, steers slowly past Bygdö and Dyna out on her unknown path, while little nimble craft, steamers, and pleasure-boats swarm around her. Peaceful and snug lay the villas along the shore behind their veils of foliage, just as they ever seemed of old. Ah, "fair is the woodland slope, and never did it look fairer!" Long, long, will it be before we shall plough these well-known waters again.

And now a last farewell to home. Yonder it lies on the point – the fjord sparkling in front, pine and fir woods around, a little smiling meadow-land and long wood-clad ridges behind. Through the glass one could descry a summer-clad figure by the bench under the fir-tree....

It was the darkest hour of the whole journey.[32]

The amount of the people bidding Nansen and his crew farewell shows the tremendous popularity of polar explorers. Still, it is not the feeling of being popular but that of loneliness, which is emphasized by Nansen. It is an inconsistent feeling of being alone with the nature, similarly in control of it and at its mercy. This is further proved by the fact that Nansen does not find it necessary to address the feelings of his crew. Already at the moment of departure he is narrating the myth of a polar patriot into being.

Leaving the home behind – rather literally – is not easy, but, in any case, it is still a decision somehow worth taking. There is a (hu)manly inevitable force – and this might be the reason for Nansen not to discuss what his team-mates thought about the departure – which makes the man leave his wife and children despite the

---

[32] Fridtjof Nansen, *Farthest North* (New York: Modern Library, 1999 [1897]), pp. 38–9.

possibility of maybe not seeing them ever again. Here, we argue, it is the same unknown force that freezes women at home. They stay there to provide shelter and to take care of the daily routines while men are fighting for their (country's and science's) glory. By detaching from their home, the polar explorers left behind the domestic femininity as well. Nansen's obsession with his wife was replaced by his obsession with the Arctic. On the other hand, Roald Amundsen and Salomon Andrée, to mention two more examples, were never married. In fact, Andrée saw the spheres of the romantic and the scientific rather incompatible. Still, he had a secret affair with a married woman, Gurli Linden, who was left to live several years in uncertainty about whether her true beloved was alive or dead. One or two – even ten – years was a short time to be abroad battling against the arctic forces of nature, but a long time for those battling with the daily routines at home.[33]

Nansen's description of the *Fram*'s departure is full of detailed analysis of the scenery. The garden, the villas, the pine and fir woods and the smiling meadows are parts of the landscape he is detaching himself from. Also men may have difficulties in detaching from home and the familiar landscape – "Behind me lay all I held dear in life" – but a true man is brave enough to break down emotional bounds and throw himself into an adventure with the unpredictable nature. It is, thus, from this very first moment of Nansen's journey that the contrast between the 'civilized nature' and the 'uncivilized, empty nature' is constructed. Detailed description continues through the book, but the nature the crew encounters in the Arctic is different, with no reference to the one they have known before.

If the voyage itself lasted for several years, preparing the expedition took no less time. Provisions, ammunition and instruments for taking scientific observations were all carefully selected and checked several times, since no unnecessary things could be taken aboard. The history of arctic exploration is full of examples of men with no competence or experience whatsoever of the climate and conditions of the arctic. Therefore, it was crucial to hire the right type of crew members. In addition to physical strength, scientific background was preferred, since Science was indeed one of the motives behind the desire to conquer the North. This is why Nansen was desperately trying to make scientists join his staff. In 1893, he contacted the Scottish geologist Archibald Geikie, asking for a

> geologist, who is desirous to join us, and who is especially adapted for the purpose[.] You will easily understand that besides a clever and experienced geologist he must be

---

[33] See Mick Conefrey and Tim Jordan, *Icemen. A History of the Arctic and its Explorers* (London: Boxtree, 1999), p. 70. Here, Gurli makes an exception: during the thirty years Andrée was lost in the Arctic she did not just wait passively but took an active role in preserving (and constructing) the hero image of Andrée's expedition. See Sven Lundström, 'Andrée-expeditionen i vår tids ljus,' *Ymer*, Vol. 117 (1998), p. 40.

an exceedingly strong and enduring fellow who is able to stand any fatiguing and exhausting exercise in the shape of dragging sledges, boats, etc. I think the age between 25 and 35 is the best. He must also have a fine character of course, if possible he ought to be good humoured jolly fellow.[34]

There were not many candidates meeting these strict criteria. It was usually white European men from well-to-do families who set off for expeditions to conquer unknown regions and to make new discoveries.[35] Here, the biological sex becomes central in understanding the role of women in polar exploration. On one hand, women were not allowed to participate in these games partly because this would have risked the serious objectives of scientific work by turning the expedition "into a sex-romp" and partly because taking along participants with poorer physical and mental abilities would have limited the freedom for other (male) participants to carry out their daily practices or, even worse, made it more likely to be attacked by wild beasts because "girls have periods [which will] attract polar bears."[36] On the other hand, some explorers, such as Robert Peary, saw female companionship "as a matter of both mental and physical health and the retention of the top notch of manhood" without which it would ask "too much of masculine human nature to expect it to remain in an Arctic climate enduring constant hardship". Peary's wife, Jo, was a member of his expeditions, but she stayed safely on board the ship while her husband conducted 'ethnological research' including photographing naked Inuit women (Peary's research resulted in one of the Inuit women carrying Peary's child).[37]

---

[34] Letter to Archibald Geikie, 17 May 1893 as cited in Svetlana A. Khorkina, *Russia and Norway in the Arctic 1890–1917. A Comparative Study of Russian and Norwegian Traditions of Polar Exploration and Research* (Tromsø: University of Tromsø, 1999), p. 151.

[35] Lisa Bloom, *Gender on Ice* (Minneapolis: University of Minnesota Press, 1993).

[36] Evjen, 'Women in Polar Research,' p. 39. Some exceptions remain however. Hanna Dieset, a botanist, was among the participants of a Norwegian expedition for Svalbard in 1907. Juliet Jean-Saussin, a geologist, and Erminiya Zhdanko, taking the place as the expedition's doctor, were the two female participants of the Russian polar expeditions between the years 1900 and 1917. All these women seemed to have managed their duties well and deserved being praised – except Jean-Saussin, of whom we can not tell, since the expedition did not return from the Arctic. See *ibid.*, p. 31; Khorkina, *Russia and Norway in the Arctic*, pp. 158–60.

[37] Conefrey and Jordan, *Icemen*, pp. 36 and 44.

Being There

[N]ext morning [...] winter had come. There was white snow on the deck, and on every little projection of the rigging where it had found shelter from the wind; white snow on the land, and white snow floating through the air. Oh, how the snow refreshes one's soul, and drives away all the gloom and sadness from this sullen land of fogs! Look at it scattered so delicately, as if by a living hand, over the stones and the grass-flats on shore! [...] Suddenly, a flash of light hit our eyes. And, as if created by some magical powers, a wonderful view opened before our very eyes. [...] There was that strange Arctic hush and misty light over everything – that greyish-white light caused by the reflection from the ice being cast high into the air against masses of vapour, the dark land offering a wonderful contrast.[38]

The sea was covered by newly frozen ice, too thin to bear a man's weight but strong enough to prevent the ship from sailing forward. As far away as one could see there was tightly packed drift ice which, together with the newly formed ice, made it hopeless to move on.[39]

All around us the ice was shifting, tilting, and bobbing. The sea was chaotic, shattered into a mishmash of unstable pans. When we encountered a zone of open water and broken ice hundreds of miles long to the north the next day, we had no choice but to retreat. No sooner had we worked our way back to our starting point near Cape Arkticheskiy, however, than we were pinned down by a blizzard for eight days.[40]

The first quotation shows the friendly side of the Arctic nature. The Northern nature is a wonder, a creation of a sublime power, the aesthetic pleasure human beings want to experience when watching something beautiful.[41] It is something that paralyzes your mind in a way that it is beyond anything that the human mind

---

[38] Amundsen, *Luoteisväylä*, p. 25; Nansen, *Farthest North*, pp. 79 and 81.

[39] A.E. Nordenskiöld, *Vegas färd kring Asien och Europa* (Stockholm: Biblioteksförlaget, 1960 [1880]), p. 100.

[40] Will Steger, 'Dispatches From the Arctic Ocean,' *National Geographic*, Vol. 189, No. 1 (January 1996), p. 81.

[41] The intellectual roots of this description can be traced back at least to 18th century Europe and to the writers such as Immanuel Kant and Edmund Burke. It is especially the sphere of the sublime that plays an important role here. The sublime can be understood as the outermost or the extreme, the highest and ambivalent form of experience, but, at the same time, it is connected with the attractive, the transcendent, the physiological-sensual, the dreadful, the convenient and the intense. See, for example, Eli Høydalsnes, 'Bilder i nordnorsk ramme: Kunsthistoriske betraktninger om essens og konstruksjon,' in Trond Thuen, ed., *Landskap, region og identitet: Debatter om det nordnorske* (Bergen: Norges forskningsråd, 1999), p. 51.

could fully understand. The nurturing side of the feminized nature (Mother Nature) is present in a form of "white snow floating through the air." However, as Gillian Rose reminds us, there is also an element of danger included, since it is not within the human capability to explain the phenomenon. This time, however, the last sentence of the quotation offers a scientific explanation: "caused by the reflection from the ice being cast high into the air against masses of vapour". There is a scientific explanation, that it is no more than a reflection. In the exploratory descriptions of the Arctic nature, the line between Rose's ideas on aesthetic and scientific masculinity becomes blurred.

Ice, for that matter, is an inevitable natural element in the polar regions. There are different and sometimes contradictory meanings attached to ice, since there is no such thing as a singular ice but, rather, many different kinds of (cultural conceptualizations of) ice. For Nordenskiöld, when sailing via the Northeast Passage, the observation of the ice conditions was the foremost prerequisite for a successful journey. Although the ice and snow were beautiful natural elements, there was a also a touch of anxiety in Nordenskiöld's encounter with the newly frozen ice. For Nansen, whose whole plan to reach the North Pole rested on freezing the ship solid, and for those travelling by foot, the lack of proper ice was the omnipresent nightmare. So, when the Arctic showed its unfriendly side – such as in the last quotation – the place could turn into a prison, "a dull, dreary, heart-sinking, monotonous waste, under the influence of which the very mind is paralyzed".[42]

Nevertheless, during the expedition, the leading thought was "forward stretching the muscle, the heart beating, and the blood running through the vessels, and straightening the back".[43] Turning back home before reaching the goal was meaningless and signified nothing but failure. It was not allowed to think about retreating, not even during the most difficult moments. To explore was to be active and to be active was to be masculine. To become passive would have meant becoming an ordinary traveller without the legitimate mission of scientific endeavour.

There is no time of passivity to be found in the explorers' texts. There was always something to do during the long Arctic winter. One could repair and take care of the equipment; observations were carried out, be it night or day, storm or sunshine; and at least one could always amuse oneself by hunting. Sometimes it was, of course, a necessity, but sometimes it was pure entertainment. The white, open landscape offered little shelter for the game and was, thus, an ideal ground for hunting. When a walrus, seal, polar bear or a reindeer came into sight there

---

42   James Ross, as cited in A. Alvarez, 'Ice Capades,' *The New York Review of Books*, Vol. XLVIII, No. 13 (9 August 2001), p. 16.

43   Amundsen, *Luoteisväylä*, p. 488.

was no lack of interest in chasing it, when the joy of hunting took hold of the men. Nansen, for example, depicts how "some of us went ashore in the evening for some shooting." The results of this shooting trip are described at some length. Nansen, who was considered a very moderate man and who was against any kind of brutality, suddenly felt "the wild beast in [him]", when "[t]he passion of the chase vibrated through every fibre of [his] body."[44] In addition to these types of episodes, zoological expeditions form a specific chapter in the history of arctic slaughtering. During his expedition of 1900, the Swedish zoologist Gustaf Kolthoff expressed the special value of hunting, when it was carried out for scientific tasks, as follows:

> [T]his musk-ox would no doubt be shot by the sealers, so I thought that we might shoot him just as well. [...] The knowledge of having killed the animal for a high purpose brings a feeling of satisfaction, and a high purpose it is when the meat of the animal contributes to one's livelihood – at the same time as the internal parts of the body, the hide and the skull are of value to science.[45]

Besides the game, there was another living feature in the Arctic: the indigenous population. From the moment the expedition had said farewell to home, the crew was surrounded by the uncivilized world. Many travel accounts depict the first encounter with the Northern folk at some length. At the same time, it was clear that the encounter was far from being equal. It was a meeting of a civilization and a savage tribe, the former seen as being superior to the latter. The Inuit, for example, were described as dirty, isolated, unwilling to co-operate and having no manners whatsoever. The attitudes towards the indigenous population were, however, not that simple. While the first expeditions did not became engaged with the indigenous people at all, the later explorers learned to take advantage of the Inuit's knowledge of the Arctic climate and landscape. Especially when the importance of dog teams in transportation became evident, explorers became more or less dependent on trading with the Inuit. Hunting, or more generally, surviving, was the other important skill that had to be transmitted from the Inuit to the explorers. Amundsen was eager to know how the Inuit people he met could manage to live in such a harsh climate. He did not see the Inuit as barbaric tribes but as people who had adjusted to the climate in such a way that Arctic expeditions could gain important information from them. Still, Amundsen was aware of his superior position – he was the master of the ship and, with help from his Inuit friends, he would soon be the master of the ice as well. In order to show their

---

44 Nansen, *Farthest North*, pp. 67–78, quotations from pp. 67 and 72.

45 Gustaf Kolthoff, as cited in Liljeqvist, *High Latitudes*, pp. 294 and 302. At the end of Kolthoff's expedition, in all 226 mammals and 1100 birds had been preserved.

power, he and his white crew blasted one of the Inuit igloos.[46] Many of the later reference books and articles still emphasize the scientific dimensions of these expeditions. Northern knowledge was 'learnt' from the Northern peoples in order to be able to carry out the scientific (and other) tasks of the expedition. From these accounts it is rather difficult to judge whether Nansen can really be called "one of the founders of neurology" or "a pioneer of modern oceanography".[47] For a small Nordic state like Norway, Nansen's achievements are, of course, worth highlighting, but it should be kept in mind that adopting a different point of view could lead to completely different conclusions. Many of the 'findings' of Western scientists, for instance, were already known by the Inuit and other Arctic peoples.

Finding It

> The Pole at last!!! The prize of three centuries, my dream and ambition for twenty-three years [...] What I see before me in all its splendid, sunlit savageness, is *mine*, mine by right of discovery, to be credited to me, and associated with my name, generations after I ceased to be [...] *Mine* at last ... The winning of the North Pole was a fight with nature [...], but due to [my] deathless ambition to know and to do, [I have] conquered. [I have] added to the sum of Earth's knowledge, and proven that the mind of man is boundless in its desire.[48]

Neither Nansen nor Amundsen could bring their country the *national* glory of being the first to reach the North Pole. For Robert Peary, the American explorer, it was a *personal* victory to stand on the North Pole as the first and only human, since he was the only white man of his team, the others being four Inuit men and a black servant. His 'team-mates' were mere spectators, witnesses to his heroic accomplishment. However, standing there was also *humanity*'s victory over the hostile nature as Frederick Cook put it, claiming to have reached the Pole before Peary.[49] Man had defeated the death-dealing Nature, as it was. Conquering, possessing and taming the last empty virgin lands meant dispossessing the cosmos of the Northern indigenous people and subjugating the Northern myths for a

---

[46] Roland Huntford, *Roald Amundsens oppdagelsesreiser i bilder* (Oslo: Grøndahl & Søn Forlag, 1988), p. 15.

[47] Roland Huntford in his introduction to Nansen's *Farthest North*, pp. xvii–xviii.

[48] Berton, *The Arctic Grail*, p. 563; Matthew A. Henson, *A Black Explorer at the North Pole* (Lincoln: University of Nebraska Press, 1989 [1912]), p. 142; Conefrey and Jordan, *Icemen*, p. 51. These are accounts of Robert Peary standing on the North Pole in April, 1909.

[49] Berton, *The Arctic Grail*, p. 592.

Western, scientific and masculine gaze. This dispossession was legitimized by measuring, mapping and naming the Arctic.

Certain measurements were needed to prove the *scientific* victory of really being there. To complete his work and to find out his precise location, Peary sledged ten miles on to make observations. After several similar short journeys and a series of observations, he was convinced that the Pole was somewhere within the small area he had covered. Similarly to Peary, Cook worked for hours trying to determine his location by measuring his shadow from fixed points at hourly intervals, trying to find where its length always remained the same.[50] At the same time, the other scientists were making their observations several miles away at the base camp. Ardent scientists were looking at nearly every situation and object from a scientific point of view. Nothing escaped them, be it a peculiar form of rock or plant, animal or human creature. This close encounter with nature was the thing Fridtjof Nansen had been searching for. He was against airship or balloon expeditions to the Poles, since they could never produce scientific results as important as expeditions carried out by ship or in sledges drawn by dogs. He was convinced that "[s]cientific results of importance in all branches of research can be attained only by persistent observations during a lengthened sojourn in these regions".[51]

Mapping formed an important activity during the journeys. The Northern regions were relatively unknown territories, and during every journey to these lands and seas, new enquiries were made for the purposes of mapmaking. This was, actually, one of the most important tasks of the geographers. Gerard de Geer, a Swedish geographer and geologist, defined the tasks of a geographer by emphasizing the cartographic techniques and knowledge in drawing maps and naming regions:

> Mere descriptions without maps or even eye-measure sketches by different kinds of tourists cannot claim any right to priority, while in such cases it may be left to the first real map surveyor to accept or omit names of that kind. Especially in regions which have become more generally visited by tourists it is easily understood how necessary is such a rule against trespass and poaching on the grounds of the genuine active geographers.[52]

---

[50] Henson, *A Black Explorer*, p. 157; Conefrey and Jordan, *Icemen*, pp. 51 and 59.

[51] Nansen, *Farthest North*, p. 11.

[52] Gerard De Geer, as cited in Urban Wråkberg, 'Minnets land: om den geografiske namngivningens historia i Arktis,' in Roger Sørheim and Leif Jonny Johannessen, eds., *Svalbard – fra ingenmannsland til del av Norge* (Trondheim: University of Trondheim, Centre for Environment and Development, 1995), pp. 121–42.

Moreover, the art of reading maps was complicated by the fact that sometimes the already existing maps were giving false information. This was explained by the actions of some early explorers, who had wanted to see new land so desperately that they had imagined having discovered it even where it actually did not exist. Hallucinations, sky emerging from the sea and odd features of nature made the Western eye see what it expected to see and the Western mind understand the sight according to the Western way of thinking.

Explaining the names given for different places in explorers' maps provides us an illustrative way to view the different processes and practices of the late 19th and early 20th century geographical imagination. It proves the power of geographical naming in the process of subordinating the Northern peoples and land. Launching an expedition which would spend several years in unknown territories without any connection to the civilized world was a financially insecure endeavour. Lectures given in geographical societies could be of some help for those who had managed to carry their previous journeys through successfully. Providing the expedition with all the necessary equipment was easiest if one knew the right type of people, those willing to open the doors to their personal treasuries. Looking at the polar regions on the map one notices some of the names of these 'patrons': for instance, the Don Pedro Christoffersen mountains in the Antarctica stand for the man behind Amundsen's rush for the South Pole and the Dickson Land in Svalbard is named after Oscar Dickson, the man who made Nordenskiöld's sailing through the Northeast Passage possible.

Memorialized were also the previous teachers or colleagues who had been supporting these journeys by legitimizing the explorers' plans from a scientific point of view. Of course, there might have been other, more egoistic reasons behind this, as Urban Wråkberg tellingly points out: by naming foreign mountains or bays after colleagues one might, perhaps, get a fjord or a peak carrying one's own name somewhere else.[53]

Finally, many of these regions are named after royalties. The Dronning Maud Land and the King Edward VII Land in the Antarctica, the King William Land and the Queen Elisabeth Islands in the Northern polar regions are there not only because these persons happened to be Kings and Queens at the time. There is a deeper connection between the royal houses, national parliaments and explorations. An efficient way of presenting new plans (and, thus, to receive financial support) was to present the expedition plan under the name of the nation. In England, a country with a long tradition of exploration and navigating to unknown regions in order to bring them under the powers of the crown, it was a national duty for the

---

[53] *Ibid.*

monarchy to be the central factor behind the expeditions.[54] For the Nordic countries, on the other hand, and especially for Norway, struggling in the Union with Sweden, journeys to the polar regions were a rite of passage in forming a national identity. "Would it be Norwegians who are showing the Way! Would it be the Norwegian Flag that would be streaming over our Pole!"[55] These words of Fridtjof Nansen to the Norwegian politicians deciding whether or not to financially support his journey towards the North Pole capture the moment when the goals of his expedition and the goals of the Norwegian nationalism became one.

Coming Home

> So we passed from town to town, from fête to fête, along the coast of Norway [...] Steamboats swarmed around, all black with people. There were flags high and low, salutes, hurrahs, waving of handkerchiefs and hats, radiant faces everywhere, the whole fjord one multitudinous welcome. There lay home, and the well-known strand before it, glittering and smiling in the sunshine, [...] and the sea rippling at my feet seemed to whisper, "Now you are at home".
>
> The ice and the long moonlit polar nights, with all their yearning, seemed like a far-off dream from another world – a dream that had come and passed away. But what would life be worth without its dream?[56]
>
> We surrender for this tragedy. The only thing there is left to do is to express our warmest gratitude for their [Andrée and his crew's] willingness to sacrifice themselves for the [Swedish] science.[57]

The late return of Salomon Andrée's hot air balloon expedition on 5 October 1930, exactly 33 years after his death in Vitön, and Fridtjof Nansen celebrating his return from the farthest North could be considered the extreme forms of failure and success. Still, there are many similarities among the attitudes towards these two expeditions. Both the explorers became national heroes, whom the whole nation seemed to be welcoming. Nearly twenty thousand schoolchildren were waving Norwegian flags to welcome Nansen when he arrived in Kristiania (Oslo). Markets for different products related to Nansen's expedition seemed unlimited: there was *Fram* tobacco, matches, tie pins, Nansen shirts and North Pole beer and wine (note the masculine target group) for sale. Numerous books and newspaper articles were

---

[54] See Richard Vaughan, *The Arctic. A History* (Phoenix Mill *et al.*: Alan Sutton, 1994), pp. 142-67.

[55] Fridtjof Nansen, as cited in Einar Arne Drivenes, 'Polarforskning – vitenskap eller politikk?' *Ottar*, No. 197 (1993), p. 8.

[56] Nansen, *Farthest North*, pp. 431-3.

[57] King Gustaf V of Sweden at Andrée's memorial service, as cited in Sverker Sörlin, 'Hemkomsten: De dödas färd från Vitön,' *Ymer*, Vol. 117 (1998), p. 55.

Plate 2   Unknown photographer, 'Lapon des
anciennes province Jämtland (partie sud)
ou du Härjedalen en costume d'hiver,' 1884
(with permission of Musée de l'Homme,
Paris, collection du prince Roland Bonaparte)

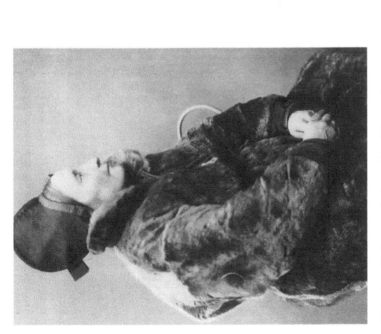

Plate 1   Unknown photographer, 'Femme lapone
des anciennes province de Jämtland ou du
Härjedalen,' 1884 (with permission of
Musée de l'Homme, Paris, collection du
prince Roland Bonaparte)

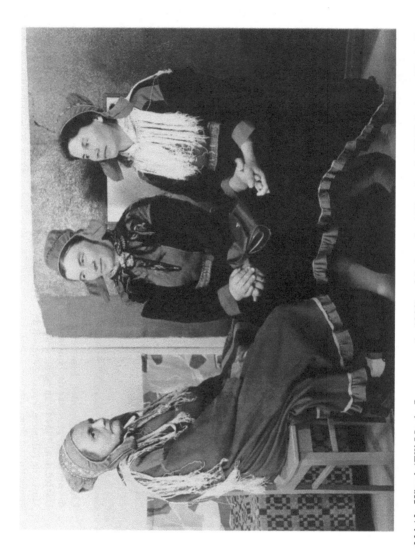

Plate 3   Erkki Ala-Könni, 'Elli Maria Jomppanen (s. 1900), sekä sisarukset Kaarina (s. 1923) ja Anna Maria Jomppanen; EMJ Menesjärveltä ja jälkimmäiset Lemmenjoelta,' 1961 (with permission of Tampereen yliopisto, kansanperinteen arkisto [Inari 13])

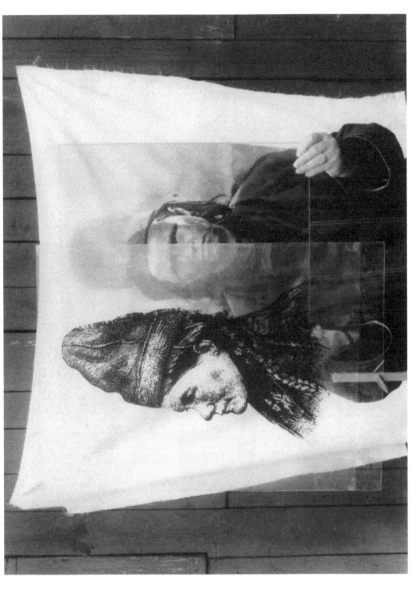

Plate 4 Jorma Puranen, 'untitled' from *Imaginary Homecoming*, 1994 (with permission of Jorma Puranen)

Plate 5   Jorma Puranen, 'untitled' from *Imaginary Homecoming*, 1991 (with permission of Jorma Puranen)

Plate 6   Jorma Puranen, 'untitled' from *Imaginary Homecoming*, 1991 (with permission of Jorma Puranen)

Plate 7 Jorma Puranen, 'untitled' from *Language Is a Foreign Country*, 1999 (with permission of Jorma Puranen)

Plate 8   Oscar Arnold Wergeland, 'Eidsvoll 1814' (with permission of Stortingsarkivet and Teigen Fotoatelier)

Plate 9 Carl Wahlbom, 'Gustaf Vasa vis Mora' (with permission of Uppsala universitetsbibliotek, Kart- & bildavdelningen)

published on both Nansen and his crew as well as on the tragic fate of Andrée, presenting the men as examples of scientific reasoning and persistence.[58] Moreover, both expeditions gained scientific knowledge "richer and deeper than any other humanitarian research had gained so far".[59] Nansen became the first Nordic professor in Oceanography. This fact may be important for the historiography of geosciences, but for us it is even more important to note how the explorers, both Nansen and Andrée, kept the myth of the masculine explorer alive. Although Andrée and so many other explorers never saw their homes again, they sacrificed their lives for the Science. Concerning those who finally managed to return home, it was all too often forgotten that the warm homecoming of polar explorers was made possible by the women staying at home. Similarly to other masculine accounts of the history of mankind, women, as well as feminine topics are absent from the history of polar exploration. Most of this history consists of completely gender blind reproductions of the masculine gaze. Explorer after explorer is put on a pedestal to hear the uncritical symphony of hegemonic glory. The heroic achievements of these men, it is argued, turned the 'invisible' polar expertise into a visible, world-wide realm.[60] What remains invisible is the gender specific production of the structures of gaining knowledge. When writing this article, we found but a few examples of feminist inquiry in this specific field.[61]

... Is Not to Go to New Places, But to Have Other Eyes?[62]

The historical dimensions of and the relationship between the Northern exploration and geography are still surprisingly narrowly covered themes in geographical

---

[58] Susan Barr, *Fram mot Nordpolen. En hundreårsbragd Fram-ferden 1893-1896* (Oslo: Schibsted, 1996).

[59] *Ibid.*, p. 123.

[60] See, for example, Berton, *The Arctic Grail*; Vilhjálmur Stefánsson, *Unsolved Mysteries of the Arctic* (Freepoint: Books for Libraries Press, 1972); Susan Barr, 'Norge inntar arenaen,' in Sigmund Nesset and Helge Salvesen, eds., *Ultima Thule* (Tromsø: Universitetsbiblioteket i Tromsø, 1996), pp. 202-12.

[61] See Bloom, *Gender on Ice*; Lena Eskilsson, *Masculinity and the Northern Space* (Umeå: The International Research Network on the History of Polar Science, 1996); Anka Ryall, 'Antarctica and the Question of Women,' Paper presented at Writing the Journey: a Conference on American, British and Anglophone Travel Writers and Writing, University of Pennsylvania, 10-13 June 1999.

[62] Marcel Proust, *Remembrance of Things Past, Vol. 3 The Captive*, as cited in *National Geographic*, Introduction to 'The Millennium series,' *National Geographic*, Vol. 193, No. 2 (February 1998), pp. 4-5.

research. The narrow scope of the existing research on the subject is in line with James Duncan and Derek Gregory's notion that, until recently, travel writing not conducted under the sign of the 'Science' was virtually ignored in geographical research, as being "read as an unproblematic record of heroism and triumphant discovery in which other cultures and other natures were shown to have surrendered their secrets before the powerful gaze of Western 'Reason'".[63] However, the traditions of exploration and narrating them had a major impact on the basic ideas of which geographical knowledge consists. In this chapter, we have covered this epistemological domain with the notion of *geographical imagination*, containing both what the geographers define as Geography and what is generally perceived as geography by people outside the borders of the discipline. This idea explains also our decision to treat Northern explorers in line with geographers as well as to cross the borders between Geography and other field sciences (especially Anthropology).

We have shown the dualistic nature of geographical imagination. Further, it has been emphasized that many of the opposite pairs are gendered and hierarchically constructed. It is more appreciated to act away from the familiar, to treat visual elements prior to the senses of, for instance, smell and taste, and to search for scientifically accurate explanations rather than let the objects of the study/gaze speak for themselves. Still, the world does not function in a black and white way. Nature is not an exclusively stabilized sphere of the feminine but rather a realm of different femininities against which the masculine gaze is produced. Our argumentation holds that the Northern nature has been feminized either by admiring its beauty and otherness (*i.e.* non-similarity) or by measuring and controlling it by means of scientific accuracy. Our reading of Gillian Rose's ideas on the aesthetic and scientific masculinity does not divide the world into two opposite parts but it does offer an analytic tool for further inquiries into the gender sensitive construction of knowledge.

In the context of Northern exploration, geographical imagination is present in the term *polar patriotism*. The term refers to the national as well as the personal (masculine) interests behind the travels into Northern regions. It also reminds us of the characteristics of a 'true polar explorer': he is not a daydreamer or a (fortune) hunter but a brave, sophisticated scientist who has the right gaze and is capable of making the right type of observations from a distance, but who is, if needed, also ready to encounter all the frightful beasts of the other side. In the eyes of the nation and its laymen, it was exclusively men with these characteristics who became national heroes and thus, as in the case of early 20th century Norway,

---

[63] James Duncan and Derek Gregory, 'Introduction,' in James Duncan and Derek Gregory, eds., *Writes of Passage: Reading Travel Writing* (London: Routledge, 1999), p. 2.

contributed to the rising need for a nationalistic mood and mental coherence. However, does this all only belong to the past? We are more than a hundred years away from the times when the Arctic was considered the unfriendliest place on the face of earth. Men such as Amundsen, Nansen and Peary made it, together with their crews, a place alike any other visited by members of the human race. They tamed 'the roaring inside her', they gave her a name and a civilized face. Today, even the last unexplored tracts of the world have been charted. So, why are there still men – and women – pushing on their way to the Arctic regions?

"Answering an urgent call to adventure," Ramón Larramendi decided to make a journey across the Arctic by using only traditional forms of transport.[64] Between the years 1990 and 1993 he turned his plans into reality, despite setbacks leaving "[no] part of the trip undone". His goal was not to set any records but "only to gain a better understanding of the Northern ways". From his travel account, published in *National Geographic*, one gets, however, a rather different image of his purposes. His story depicts the prevailing way the North is still dealt with today: trying to learn "the Northern way", Ramón *hires* two Inuit brothers to guide him across the ice. Although there are some sunny days, most attention is paid to the times when Ramón is scared to death because of the ice conditions, because he is lacking food and supplies, because he has to shoot several dogs and feed his favourites' "flesh to his ravenous sled mates". Having to battle with himself he finally finds the strength of making it until the end, taking the last steps and strokes to see his family. When Ramón notices being "a different person from the one who set out from Greenland", he is merely referring to the reflection in the mirror showing a strange face "with the scruffy beard and wild eyes". His travel account comes incredibly close to the ones published a hundred years ago.

Ramón's example of searching for a better understanding of the Northern ways shows how the ways to see the North and exploration are tightly linked with the heroic narratives of the past. The North is somehow frozen in time (following the vocabulary of Monica Tennberg's chapter), while the rest of the West takes the advantage of progress. By denying the North the possibility of change – by linking the Northern people exclusively with the traditional modes of transportation – the time-space gap between 'us here' and 'them there' is made even greater. The North becomes a nostalgic place against which the Western version of the Northern reality is constructed.

The central principle in feminist thinking is the trust in the ability to produce alternatively gendered 'realities'.[65] The preliminary step towards this 'reality' would be the rethinking of the notion of knowledge and the ways of producing it.

---

[64] Ramón Hernando de Larramendi, 'Perilous Journey. Three Years Across the Arctic,' *National Geographic*, Vol. 187, No. 1 (January 1995), pp. 120–38.

[65] See, for example, Nairn, 'Doing Feminist Fieldwork,' p. 146.

'Doing otherwise' in the context of polar exploration and research could be something Monica Kristensen Solås, the Norwegian glaciologist and expedition leader, is hinting at.[66] In the late 1980s, she took the same route to the South Pole as Roald Amundsen almost a hundred years ago. Just before reaching the Pole – Kristensen would have been the second human and first woman to travel this particular route – she decided to turn back. By doing this, she was breaking down and de-heroizing the masculine picture of a (male) polar explorer striving towards the ultimate goal.

This example from Kristensen is a suitable starting point for searching those "different eyes" as the quotation from Marcel Proust, included in the title of this concluding section and commonly repeated in relation to exploration, would suggest. But what could it mean? As the example of Ramón shows, it is almost the same, historically constructed gender blind eyes through which many explorers of the new generation still see the world. Hence, at least, this means continuing the privileging of sight. We argue, however, that geographical imagination should include more than sight. It should be open to appreciating sounds as well as the 'feminine' senses such as the smell and the taste. Not only seeing but also hearing the surroundings and the subjects of the study can be instructive. This would, however, require changes in the academic as well as the popular understanding of the constituents of geographical knowledge.

There is a call for the rethinking of 'fathers of geography' and the historical construction of the socio-cultural context which formed the background for our present actions and made the discipline predominantly colonial and masculine. Uncovering – or rather, critically rediscovering – those insufficiently one-sided and narrow-minded ways of seeing means questioning the position of the (Southern and anglophonic) geographer. We should move from *speaking of* the Northern people towards *speaking to* them and think about how, why, for whom and from what kind of power position we are constructing the knowledge through our representations.[67] We are familiar with explorers' accounts of their encounters with the Northern peoples but do we know enough about the Northern peoples' viewpoints on these same encounters?[68] Or could the alternative reality be

---

[66]   We would like to thank Bjørg Evjen for pointing to these recent developments in women's exploration.

[67]   Tzvetan Todorov, *The Conquest of America. The Question of the Other* (New York: Harper Perennial, 1992), p. 132. See also Audrey Kobayashi, 'Coloring the Field: Gender, "Race," and the Politics of Fieldwork,' *Professional Geographer*, Vol. 46, No. 1 (1994), pp. 73–80.

[68]   See, for example, Penny Petrone, ed., *Northern Voices: Inuit Writing in English* (Toronto: University of Toronto Press, 1998); Juri Rytchëu, *Die Suche nach der letzten Zahl* (Zurich: Unionsverlag, 2001).

produced by comparing the travel accounts written by men and women? Is the Northern nature produced differently in women's writing?[69] These are only some questions arising from the need to locate polar exploration and its cultural practices within the formations and inscriptions of power and to analyze travel writing not as an uncritical reprinting of polar patriotism, but as a scientifically legitimized collection of textual practices which have formed the characteristic gestures of the geographical imagination.

---

[69] The reading list could include, for example, Helen Thayer's *Polar Dream* (1993) where she recounts her North Pole expedition which made her the first woman to ski alone to the Pole. See also Ryall, 'Antarctica and the Question of Women'.

Chapter 7

The *Thing* as an Incarnation of Nordic Political Culture and Its Roots

Ulrich Albrecht[1]

Nordic political culture is credited with features such as being orientated towards compromise, based upon a mutual understanding of equalness, and a strong belief in related ground rules which regulate political discourse. The notion of non-violence enshrines the whole approach. The Swedish writer Per Olov Enquist speaks of "a world based upon the notion of *a society we are with*, not against."[2] Hans L. Zetterberg, once editor-in-chief of *Svenska Dagbladet*, states that "[i]ts key word is (the intranslatable) *lagom*, which means both 'reasonable' and 'middle-road.'"[3] Arne Ruth, the then head of the feuilleton of *Dagens Nyheter*, states in a more specific manner:

> Untranslatable in its exact sense, *lagom* means *just right*, as well as *in moderation*; *sufficient*, as well as *appropriate* and *suitable*. [...] The expression is said to derive from the drinking habits of ancient Sweden, where the jug of beer was passed around a group of men; 'making the rounds' is *gå laget om* in Swedish, and according to this etymology, *lag-om* is short for *laget om*. At any rate, the image of the drinking bout is a telling metaphor for the moral implications of the word.[4]

---

[1] This piece is dedicated to my beloved daughter Ruth who became a *Skandinaver* when she decided to continue her studies at Copenhagen University – and prompted her father to take Nordic studies more seriously.

[2] Per Olov Enquist, 'On the Art of Flying Backward with Dignity,' in Stephen R. Graubard, ed., *Norden – The Passion for Equality* (Oslo: Norwegian University Press, 1986), p. 70. From 1961 to 2001, Graubard was editor of the periodical *Daedalus. Journal of the American Academy of Arts and Sciences* where the articles collected in *Norden – The Passion for Equality* were published in 1984.

[3] Hans L. Zetterberg, 'The Rational Humanitarians,' in *ibid.*, p. 89.

[4] Arne Ruth, 'The Second New Nation: The Mythology of Modern Sweden,' in *ibid.*, p. 247. He also (p. 249) observes that "[a]lmost from the start, the green wave exploited *lagom*, since it proved to be ideally adaptable to the orthodoxies of ecological thinking."

The notion of a *round* and the reference to ancient history will be taken up in the main section of this chapter as a key tool to understand Scandinavian approaches to politics. There is obviously also a search for *jämliket*, "the egalitarian passion that has been almost as important as moral force", finds Ruth. He also refers to the "myths of paternal care and egalitarian national solidarity embodied in Per Albin Hansson's concept of the *folkhem*, the people's home".[5]

Enquist adds: "It's possible that there is no such thing as a 'national character.' There is, however, no doubt that patterns of acquired values-learned reactions exist, together with a great societal 'of course' that sooner or later leaves its imprint on all areas."[6] The prototypical incarnation of this basic attitude is reflected in the concept of *the middle way*, the title of one of the first foreign books about this Scandinavian speciality.[7] In the mid-1980s voices started to become audible arguing that the very features of Scandinavian political culture are bound to disappear. Among others Zetterberg assumes "that the emphasis on equality in Swedish politics will give way to an emphasis on freedom", Enquist writes about the Nordic countries as "an island group starting to crack" and Ruth observes that "Swedish political culture in the 1980s seems more fragile, less rooted in a firm conviction of its own values".[8]

Equality in the Scandinavian context has a firm material context, concretized in the Nordic welfare state. "In their welfare development, the Protestant areas secularized and materialized [Virgin Mary's] vision of a helpful hand at every turn of life, available to the fortunate and unfortunate, the articulate and inarticulate."[9] Nicolai Fredrik Severin Grundtvig, Danish theologian and poet, once coined the dictum:

Og da har i rigdom vi drevet det vidt,
når få har for meget, og færre for lidt.

And yet in wealth real far have we gone
when few have too much, and poor are none.[10]

---

[5]   *Ibid.*, pp. 243 and 250.

[6]   Enquist, 'On the Art of Flying Backward,' p. 74.

[7]   Marquis Childs, *Sweden – The Middle Way* (New Haven: Yale University Press, 1936). According to Gösta Rehn, the title of this book "became the slogan for all those who believed that there was indeed an ideal society on earth – or at least one that approximated such a society." See Gösta Rehn, 'The Wages of Success,' in *Norden*, p. 167.

[8]   Zetterberg, 'The Rational Humanitarians,' p. 85; Enquist, 'On the Art of Flying Backward,' p. 68; Ruth, 'The Second New Nation,' p. 241.

[9]   Zetterberg, 'The Rational Humanitarians,' p. 94.

[10] Quoted from Bent Rold Andersen, 'Rationality and Irrationality of the Nordic Welfare State,' in *Norden*, p. 113.

The Norwegian historian Hans Fredrik Dahl refers to the roots of what he calls the "Nordic equity syndrome": "Rural radicalism, then, seems to lie at the heart of the Nordic equity syndrome."[11] The Finnish sociologist Erik Allardt adds: "Even if the Nordic countries were not as free of feudalism as is sometimes imagined in standard, non-analytic historical presentations, the tendency to emphasize the role of the common land-owning peasantry in the development of the Nordic societies is wholly legitimate."[12]

A closer assessment of the ways and means of the Nordic culture of compromise deserves more than the attention of scholars interested in Scandinavian affairs. According to the German political scientist Martin Greiffenhagen one can observe the emergence of a worldwide "culture of compromise", as a new overarching paradigm in the conduct of political affairs. Compromise, characterized by mutual concessions, turns – in Greiffenhagen's view – into a general pattern. The basis is a thorough understanding of one's opponent, linked with empathy and the awareness of mutuality:

> We live in an era in which hierarchical structures come more and more under the pressure of public scrutiny. Networks, deregulations, horizontal modes of decision-making, new forms of negotiating, entwining politics, self organization and a political ethos of responsivity and acceptance in congruence with these new forms open up new perspectives for a new understanding of politics.[13]

Greiffenhagen sees strong opposition in the German preference for the decisionist maxim of 'either or', which he contrasts with compromise. For the political culture of democracy, he states, "compromise provides for a good formula: not any longer solely as a strategy of mutual concessions, but as an awareness of mutual openness and preparedness for consent encompassing all social processes."[14] Greiffenhagen prototypically cites the Nazi author Christoph Steding, a *Habilitationsschrift* of 1938, with polemics against Scandinavian political cultures:

---

[11] Hans Fredrik Dahl, 'Those Equal Folk,' in *Norden*, p. 104. Later (p. 110) he summarizes: "Traditionally, all [Nordic] countries have a basic five party system [...], all pursuing Nordic equity with energy and with the overwhelming support of their constituents."

[12] Erik Allardt, 'Representative Government in a Bureaucratic Age,' in *Norden*, p. 201.

[13] Martin Greiffenhagen, *Kulturen des Kompromisses* (Opladen: Leske + Budrich, 1999), p. 16 (translated by the author).

[14] *Ibid.*, p. 14.

It can be unambiguously stated for all three Scandinavian countries how neutralization and cultivation open up special chances for totally cultivated and neutralized Jewishness. In Denmark [...] one may fully unequivocal find the process of making everything Jewish (*Verjudung*) [...] It also made possible Kierkegaard, the highest possible embodiment of complete groundlessness [...] [This] led to intellectual depletion of the country, of the peasantry, which became completely dilapilated by the lures of the cities.[15]

Steding denounces empathy, the sense for ambiguity and compromise in the Scandinavian (and other) political cultures from a strictly decisionst position.

These features of creating public political space remain unique, and provoke questions about genesis and persistence. There developed a new kind of political public in Scandinavia which deserves general attention. The chief features of this kind of culture of compromise are visually depicted by the *round table* or earlier the Nordic *Thing*, a roundel in which participants met on an equal footing, seeking consensus in political affairs, in a period of decided non-violence (the peace of the Thing). This chapter is aiming to track the roots of this peculiar Nordic setting in a more general context, also in order to allow for an assessment of the wider implications (if Greiffenhagen is correct).

Round tables were, in addition, the key instrument in bringing about the peaceful or, as a more analytically oriented prefix states, negotiated revolutions in Eastern Central Europe, Scandinavia's Southern neighbours. Is there possibly a direct link between Nordic political culture and the demand by the electrician Lech Walesa and his followers in the Lenin shipyard at Gdansk, that Polish state authorities should meet them for far-reaching discussions at a round table? How did Walesa know of the concept? Perhaps he was inspired by the actors participating in the social resistance movement. Apart from writing, printing and distributing underground press, some of them "contributed to central opposition structures" such as the shadow government preparing the round table negotiations in 1989.[16]

The following is meant as a contribution to shed some light onto these questions. An idealization of Nordic approaches to politics is, however, not intended

---

[15] Christoph Steding, *Das Reich und die Krankheit der europäischen Kultur* (Hamburg 1938), as cited in *ibid.*, p. 23. Steding continues (p. 24): "Denmark has special reasons to fool herself, because Danish history has been marked by deconstruction into the absence of history. Sweden immediately follows suit to Denmark" (translated by the author).

[16] Justyne Balasinski, 'Culture and Politics in Transition Régimes: The Case of Theatre in Poland in the 1980s and 1990s,' Paper presented at the Politics and the Arts Group Symposium 'Politics and the Arts: Making Connections in Theory and Praxis,' Berlin, 23–25 May 2002, p. 6.

in what follows. Rather, concentration upon the Thing/round table concept should merely reveal a transcendent pattern of political style in the European tradition.

The Nordic *Thing* as a Concept of Finding Compromise

The heritage of the Thing, the meeting of the people in special spaces under the open sky, is clearly represented in the names of Nordic Parliaments: in Iceland one finds the *Althing*, founded in 930, allegedly, the oldest Parliament on the globe.[17] Again, the issue is not one of idealization: less than ten per cent of the population – adult male farm-owners – came to the *Althing* which met once a year and had no executive branch to carry out its decisions. Iceland's Golden Age, depicted in the "patriotic mystique of modern Iceland [...] as glorious democracy termed the Commonwealth or Free State", appears to have been 'golden' "only by comparison with the worse that was to follow."[18] In Norway there is the *Great Thing* or *Storting*,[19] and in Denmark the name is *People's Thing* or *Folketing*.[20] "The Swedish Parliament, in which Finland also had representatives, was founded in the fourteenth century; parliamentary tradition was fully institutionalized during the sixteenth and seventeenth centuries", reports Allardt.[21] The Nazis in Germany

---

[17] See Finnur Jónsson, *Det islandse Altings historie i omrids* (Copenhagen: Hest, 1922); Mathias Thordarson, *The Althing. Iceland's Thousand Year Parliament, 930–1930* (Reykjavik: Jonsson, 1930); Bjorn Thorsteinsson and Thorstein Josepsson, *Thingvellir: Birthplace of a Nation* (Reykjavik: Heimskringla, 1961).

[18] For a review of recent publications on Iceland's history, see Jared Diamond, 'Living on the Moon,' *The New York Review of Books*, Vol. XLIX, No. 9 (23 May 2002), pp. 59–60 (see p. 60 for the quotation).

[19] See Alf Kaartvedt, Rolf Danielsen and Tim Greve, *Det Norske storting gjennom 150 år* (Oslo: Gyldendal, 1964); Guttorm Hansen, *Der er det godt å sitte: hverdag på Løvebakken gjennom hundre års parlamentarisme* (Oslo: Aschehoug, 1984); Jo Benkow, *Folkevalgt* (Oslo: Gyldendal, 1988).

[20] See Svend Thorsen, *Danmarks Folketing: om dets hus og historie* (Copenhagen: Schultz, 1961); Torben K. Jensen, *Politik i praxis: aspekter af danske folketingsmedlemmers politiske kultur og lisverden* (Frederiksberg: Samfundslitteratur, 1993).

[21] Allardt, 'Representative Government,' p. 201. With regard to Swedish respective traditions, see Elis Wilhelm Hastadt, *The Parliament of Sweden* (London: Hansard Society, 1957); Eric Lindström, *Riksdag och regering: ansvarsfördeling, arbetsformer, beslutsprocesser* (Stockholm: Liber, 1981); Lars-Göran Stenelo and Magnus Jerneck, eds., *The Bargaining Democracy* (Lund: Lund University Press, 1996); Sören Holenberg and Peter Esaiasson, *De folkvalda: en bok om riksdagsstedamöterna och den representativa demokratin i Sverige* (Stockholm: Bonnier, 1998).

wanted to restore this old tradition by building some four hundred Thing places (*Thingplätze*) for the perverted purposes of "mass rallies, events of national dedication, choir games by the people."[22] A few of those sites have been completed, the first one being placed on top of the Brandberge near Halle in Saxony-Anhaltine.

During the ancient Things arms had to remain silent and peace prevailed (*Thingfriede*). Decisions ought to be based on unanimity, there was no majority rule. This enforced the search for consensus which till the present day appears to be characteristic for major decisions in the Nordic political culture. Approval was signalled by banging the arms, what entailed the commitment of the individual fellow Thing member (*Thinggenosse*) to contribute to enforcement of decisions if needed.

Things served to decide about all crucial legal matters, especially about war and peace. Nordic kings in the early Middle Ages were elected at the place, and they had to ride around after election to other lower-ranking Things to seek general recognition.

The locality for a Thing (*Thingstatt*) commonly was the top of a hill, a landmark, for obvious reasons a roundel, encircled by a hedge or similar minor markings. The Nazi architecture for *Thingplätze* clearly reflects this concept.[23] To boil the idea down to the core, the Thing was identical to what in present days is called a 'round table', which gained fame in the recent peaceful revolutions in Eastern Central Europe, beginning with Poland and the advent of *Solidarnosc*. The Thing was public space with special functions, following specific rules, which apparently shape political culture in Scandinavia till the present.

The circular space of the Thing is clearly visible in Nordic paintings of historic moments. A good case in point is represented by Oscar Arnold Wergeland's monumentous painting *Eidsvoll 1814* (the *Eidsvoll meeting* lies at the roots of the Norwegian national myth), which is shaping till the present day the background to the speaker's rostrum in Oslo's *Storting* (see plate 8). Gustaf Vasa, the Swedish national hero, is repeatedly portrayed in paintings as addressing a crowd from the centre of a circle. A case in point is Carl Wahlbom's engraving *Gustaf Vasa vid Mora* at Uppsala University Library (see plate 9).[24] And the various pictures showing the legendary Queen Thyra Danebod calling for the construction of the famed Danewerk fortifications clearly reflect the Thing situation which was

---

[22] *Der Große Brockhaus, 1934 ed.* (Leipzig: Brockhaus, 1934), p. 632 (translated by the author).

[23] Cf. the picture in *ibid*.

[24] See also the painting *Gustaf Vasa dalkarlarna i Mora* by Johan Gustaf Sandberg shown in Stockholm's Statens kunstmuseer.

traditionally used to cast such decisions.[25]

It is commonplace to trace the Thing back to religious rites, and worshipping of Germanic goddesses obviously lies at the roots of the development. The political circle, equality among participants, symbolized in the round table, is, however, much older. The key thesis of this chapter is that the North, in the process of christianization, incorporated features of these older traditions, adaptating them in a specific manner. This process then created the grand narratives which preform present political culture. During the Middle Ages, the Things lost in political significance and shrunk to instruments of local and regional jurisdiction.[26] In Denmark, Things served as courts until the middle of the nineteenth century. This strand of social development, the creation of a Nordic scheme of courts, will be omitted in the following.

In order to organize the argument, the following steps will be undertaken. The first one is to link the Thing to the legendary round table of King Arthur in Britain, a narrative which preoccupied much of the tales at courts in medieval Europe. The second step is to relate the notion of King Arthur's round table with its religious root, Lord's supper before crucification day, linking this notion with even earlier appraisals of the concept in the Old Testament.

First Recourse: King Arthur's Round Table

We know little about British King Arthur, living in the sixth century, or his famed "Table Ronde". The Benedictine monk Geoffrey of Monmouth (around 1100–1154) lists in his *Historia Regum Britanniae* Arthur as a powerful emperor. Older sources report in a much less favourable fashion about such a king who led his Kelts until he was killed in one of those numerous battles, the one near Camlan in 539. Father Geoffrey, however, depicted Arthur as the incarnation of the benevolent ruler whose royal table served as a grand example in medieval European court life, especially in France and Britain. From there the myth spilled over to Scandinavian courts.

According to the legend, the knights dined and wined at King Arthur's court at a round table, saving the host from having to solve the perennial riddle who of his distinguished guests deserved preferential treatment in the seating at the royal table because he was more distinguished than other fellow guests. Infights

---

[25] Cf. Lorens Frølich, *Thyra Danebod anlægger Dannevirke* in the Sønderburg Castle museum and Troels Trier's painting with the same title in the Vestbirk Højskole in Østbirk.

[26] One typical example is Volkert F. Faltings, *Die Dingprotokolle der Westerharde Föhr und Amrum 1658–1671, 2 volumes* (Neumünster: Wachholtz, 1990).

regarding ranking where commonplace among the daunting knights of the time. Thus everybody was pleased to find *him*self on an equal footing with everybody else seated around Arthur's famed table. In his playwright of 1941 of the *Chevaliers de la Table Ronde*, however, Jean Cocteau reminds us of the gender discrimination at King Arthur's court when he lets King Arthur speak to Guinevere: "I ask your pardon: You know when I ask you to quit right now, no women are entitled to participate in our council, according to the customs."[27] Yet, apart from gender discrimination an amiable, peaceful mood during the talks - or, as an old text put it, 'knightly fellowship dedicated to chivalry conduct' - was allowed.

Lech Walesa, the electrician and trade union leader who forced the Polish state into martial law and finally into compromise, resorted to this particular scheme - the round table - in order to open a new political space for dealing with the communist authorities. The dissident leaders were afraid of sanctions on the side of the authorities, which after all had imposed martial law upon the country. As the murder of the *Solidarnosc* Priest father Jerzy Popieluszko indicates, one must be aware of attempts against one's life. 'Chivalry conduct' should mean at least safe conduct for the participants in the new round table.

It is easy to state how Lech Walesa, who did not enjoy a higher education, arrived at the idea of a round table with the communist government. His home town Gdansk participated in the European admiration for the legendary king and his brave knights. There is an Arthur's Court near the Long Market, in the so-called King's gate in downtown Gdansk till the present day which is known to every child. The Gothic edifice originating in the late fifteenth century was once the meeting place of the Gdansk patricians who intended to copy the idea of fairness and equality in this Hansa town.[28] In Polish the notion of the round table has also a subtle secondary meaning: one has to negotiate without 'corners and Kanten', *i.e.* without non-declared diverging objectives and tricks.

---

[27] Jean Cocteau, *Dessins en Marge du Texte des Chevaliers de la Table Ronde* (Paris: Gallimard, 1941), as quoted here and translated by the author from the German version, *Die Ritter von der Tafelrunde* (Munich et al.: Kurt Desch, 1962), p. 32. Cocteau tends to ridicule the sacred round table when he lets Merlin, the great wizard, say, for example: "I am cross with you if you hide beneath the round table and if you mimic behind his back the king who is looking for you everywhere" (p. 13).

[28] The Gothic style edifice had been built by Abraham van den Block in the years 1476-1483. The main element is the huge meeting hall covering 450 square metres.

Second Recourse: Lord's Final Supper

Recourse to the last supper at which Jesus Christ convened his disciples as a round table will provoke resentment. Religious paintings, especially by the great masters, commonly depict the Lord in the midst of a long table, and not seated at a round table.

The long table was a late invention, made in order to allow for more convenient serving of dishes as well as for ceremonious purposes, to demonstrate from the high table to the lower ranks. Painters such as Leonardo da Vinci, by portraying Jesus and his disciples seated at a long table, pay tribute to several symbols of a new era: the Lord is depicted in an arrangement which reflects contemporary understandings of nobility, and the new sigificance credited to the individuum by seeking a possibility to deliver individual portraits of participants (difficult to accomplish if part of the party is painted with backs turned to the onlooker, as a picture of a round table would require).

The poor son of a carpenter from Bethlehem would never have seen a long table, but there are a only few paintings which show Jesus sitting in a circle, in the tradition of the Jewish lecturing hall which was molded by the circle enforcing the view towards the centre. Nomads in the Middle East till today seat themselves in circles around bowls of food for festive meals.

The Arthur legend reflects numerous hints towards the Lord's final supper. The number of twelve knights which the mythical King invites to his table resembles directly the twelve disciples of Jesus. Here and there is a most beloved follower: what John was to Jesus, Galahad was to Arthur. The Arthur legend pushes the mysticism with the number twelve into the extreme: the King is credited with twelve campaigns for Britain. In the final battle alone King Arthur is said to have slain twelve times eighty enemies.

The biblical tradition ("When the hour had come, he sat down with the twelve apostles", Luke 22, 14) focuses rather on the significance of the historic moment than on the seating arrangement. This creates problems among the disciples. Immediately "there arose also a contention among them, which of them was considered the greatest" (Luke 22, 24) – a problem well known to King Arthur and his knights. Jesus solves the conflict in a manner typical for him, providing thus for the eternal characteristics of the round table: "But one who is the greater among you, let him become as the younger, and one who is governing, as one who serves" (Luke 22, 26).

According to the bible there had been round tables long before Jesus lived. The *Exodus* contains the following instruction by Jachwe for the equipment in the holy tent in which he 1300 years earlier wanted to accompany the Israelites after the turn-over of the ten commandments on Mount Sinai: "You shall make a table of acacia wood [...] and make a gold molding around it. You shall make a golden molding at its rim around it. You shall make four rings of gold for it" (Exodus 25, 23 seq.).

But nobody was entitled to sit at this round table. "You shall set bread of the presence on the table before me always" (Exodus 25, 30), orders the God of the Old Testament. The round table with the breads of presence reappears 970 A.D. again with the construction of the Temple (Kings Book I, 7, 48). The revolutionary act by Jesus the Lord was to share the table with fellow human beings and to break the bread with them.

The myth of the round table, of openness for each one around it is continued in the *Relevation to John*: "If anyone hears my voice and opens the door, then I will come in to him, and will dine with him, and he with me. He who overcomes, I will give to him to sit down with me on my throne" (3, 20 seq.).

The topos is repeated in the same text (in Relevation 19, 9 the "marriage of the Lamb" is addressed in this vein. The revolutionary call for equity – and the rejection of hierarchy – is drastically apparent when the messenger of God "cried with a loud voice": "Come! Be gathered together for the great supper of God, that you may eat the flesh of kings, the flesh of captains, the flesh of mighty men" (19, 17 seq.).

The text suggests true equality: "Around the throne there were twenty-four thrones. On the thrones were twenty-four elders sitting, dressed in white garments, with crowns of gold on their heads" (4, 4). Whilst the final supper of Jesus offered a round seating for all disciples, the *Relevation* goes one step further and offers equality in governing, by seating the company of the Lord on thrones and capping them with golden crowns. The drastic invitation to "eat the flesh of kings and of captains" (*i.e.* of war) signals the final devotion to the end of hierarchy and war-making.

Provisional Findings

It should be noted that the round table of Jesus Christ as well as the one of King Arthur were not fully egalitarian. There was always a 'head', a figure in the lead, the undisputed chairperson: the Savior in the one instance, a King among knights in the other. Modern round tables also are led by extraordinary persons (in the case of the Central Eastern European revolutions of 1989, mostly higher-ranking clergy). There were members of the round who enjoyed preferred treatment by sitting next to the foreman, John respectively Galahad, again an indication of deliberative equality.

There remains more to add in scepticism to the claim of democratic perfection symbolized by the round table. Presumably the most sardonic comment has been written by the American Mark Twain in his 1889 satire *A Connecticut Yankee in King Arthur's Court*. Twain finds that "talks at the round table were mostly monologues", and that the usual hagglings of politics merely had been moved down the political ladder, and that their abandonment was a mere myth. All this is reason enough for Twain the satyric to dynamite King Arthur's court in the end.[29]

It deserves more research in order to substantiate the claim that the two recourses to older schemes of round tables indeed have patterned the Scandinavian notion of Thing and the ensuing political culture of equality and compromise. The present sketch is indeed not much more than a preliminary outline. Further investigation should reveal how much the legendary round table of King Arthur and the preceding biblical round tables influenced ancient Scandinavers in turning towards the Thing idea, very much in contrast to political styles elsewhere.

There is little doubt about the connectivity of Scandinavian political culture with European heritage. "The intellectual sources of our culture are to be found around the Mediterranean: Athens, Rome, and Jerusalem. The ideas and values in our culture spread northward from Rome, Florence, and Venice", wrote Hans L. Zetterberg.[30]

In recent times, the notion of round tables gained enormous currency in the peaceful revolutions in Central Eastern Europe. Lech Walesa, the *Solidarnosc* leader, whose movement in the end toppled the Communist Government in Poland, made the round table, as wry humour in this country puts it, to 'Poland's most exported commodity'. Polish political Catholicism will also have contributed to the idea of a round table. The promise of safe conduct in the Arthurian tradition certainly will have added to the attraction of this model for dissidents. Revolutions negotiated at round tables were copied from the Polish example everywhere in Communist Central Eastern Europe, for example in the GDR.[31] In the meantime world religious leaders also met at a special round table. Transnational Corporations have taken over the concept: DaimlerChrysler for instance has organized round tables with its suppliers of parts in order to arrive at better relationships in an era of flattened hierarchies, reportedly with success.

---

[29] Mark Twain, *A Connecticut Yankee in King Arthur's Court*, orig. 1889.

[30] Zetterberg, 'The Rational Humanitarians,' p. 86.

[31] Cf. Uwe Thaysen, ed., *Der Zentrale Runde Tisch der DDR. Wortprotokoll und Dokumente* (Opladen: Westdeutscher Verlag, 1999).

Chapter 8

The Northern Antipode to European Integration, or Why Everybody Expects Northern Norwegians to Be Sceptical of European Integration

Jochen Hille

Introduction

Within the academic mainstream and in political debate the inhabitants of Northern Europe, as well as the British, are usually described as the most reluctant Europeans. In this grading system the Norwegians get the worst grades and play the role of the 'nasty boys', who do not want to accept the 'external realities' of European integration.[1] Certainly, a closer look also reveals that European integration is not very popular in other parts of Europe. This can be seen in the French referendum on the Treaty of Maastricht in 1992 as well as in the debates on integration in Austria, Switzerland and even in the economically weak aspirants to EU membership.[2] Poor public support is one of the most central problems in legitimizing the political system of the EU. Therefore we find a widespread debate about the construction of a European identity which can supplement or substitute national or regional (e.g. northern) identities.[3]

---

[1] Martin Sæter, 'Norway and the European Union: Domestic Debate versus External Reality,' in Lee Miles, ed., *The European Union and the Nordic Countries* (London and New York: Routledge, 1996), pp. 133–49.

[2] See the webpage of the eurosceptics, at < http://www.eusceptic.org >.

[3] For an overview of concepts of European identity, see Kathrin Lorenz, *Europäische Identität? Theoretische Konzepte im Vergleich* (Berlin: Freie Universität Berlin, Arbeitsschwerpunkt Hauptstadt Berlin, 1999).

165

Encountering the North

Euroscepticism is not merely a phenomenon of Northern Europe in general and of the High North in particular. On the other hand there obviously is a basic rule in the three Nordic countries Finland, Norway and Sweden which can easily be detected from referendums on the EU issue and various surveys: the more northwards, the more eurosceptical.[4] In Norway there is a pattern in which the southern and densely populated areas are in favour of European integration, while the backbone of resistance can be found in the North.[5] For example, in the referendum of 1994 on membership of the European Union there was a clear-cut majority for joining the EU only in Oslo (66.6 per cent Yes-votes). In small communes (less than 2500 inhabitants) in Northern Norway, the opponents of European integration got 80.5 per cent of the votes.[6] We find the same pattern, although to a lesser extent in Finland and Sweden.

Northernness and euroscepticism are obviously correlated. But why is this the case? There are two basic ways to explain this correlation. The first is that accidentally eurosceptics tend to live in the North. The second explanation is that the collective (self-)definitions of the people in the North – this is what I mean by northernness – are shaping the image of European integration within the EU debate. In this chapter I will concentrate on this discursive role of the North.[7] In doing so, I mainly regard 'northernness' as a concept which contributes to the construction of Norwegian collective self-consciousness within the EU debate. Values and self-consciousness of a society, in turn, build the 'centre of legitimacy' which determines whether the population regards political decisions as right or wrong.[8]

---

[4] See Tor Bjørklund, 'The Three Nordic 1994 Referenda Concerning Membership in the EU,' *Cooperation and Conflict*, Vol. 31, No. 1 (March 1996), pp. 24–6; Anders T. Jenssen, Pertti Pesonen and Mikael Gilljam, eds., *To Join or Not to Join: Three Nordic Referendums on Membership of the European Union* (Oslo: Scandinavian University Press, 1998).

[5] Maps on the referendums of 1972 and 1994 clearly illustrate this basic regional structure in both referendums. They can be found on the webpage of the Norwegian Social Science Data Services, at < http://www.uib.no/nsd/diverse/eu-eng.htm >.

[6] See Tor Bjørklund, *Om folkeavstemninger: Norge og Norden 1905-1994* (Oslo: Universitetsforlaget, 1997), p. 179.

[7] For an understanding of discursive analysis as an approach to European integration studies, see Thomas Diez, 'Europe as a Discursive Battleground: Discourse Analysis and European Integration Studies,' *Cooperation and Conflict*, Vol. 36, No. 1 (March 2001), pp. 5–38.

[8] For culture and identity as the "core of legitimacy", see Jürgen Gebhardt, 'Politische Kulturforschung – ein Beitrag zur vergleichenden Analyse soziokultureller Ordnungszusammenhänge,' in Constantin von Barloeven and Kai Werhahn-Mees, eds., *Japan und der Westen, Band 3: Politik, Kultur, Gesellschaft* (Frankfurt: Fischer, 1986), pp. 60–77.

The first category of explanations stresses the socio-economic structure of the High North and the impact of political actors for the opposition towards EU integration. These approaches are valid and they can explain a good deal of euroscepticism in the High North.[9] Yet behind their heavy armour of empirical data and analysis of rational actors' choices, these explanations are vulnerable. Their Achilles heel is that they ignore the underlying conceptions and conditions which generate collective interests and give social movements a 'meaning'.[10] They fail in giving sufficient reasons for the fierceness, mobilization and continuity of northern Norwegian euroscepticism over decades.[11]

During the period of 22 years from the first referendum in 1972 to the second in 1994, the Norwegian economy was changed tremendously by the drilling of oil. Additionally, the social structure of Norway changed fundamentally due to urbanization, centralization, modernization, new generations and the impact of new values. Despite this, however, the geographical patterns of votes in both referendums on European integration were widely identical.[12] Furthermore, the impact of economic factors on the public debate is often overestimated. Surely in the campaign of the referendum of 1994, the supporters of integration were relatively more successful in winning votes by stating the 'economic necessity' of integration than the eurosceptics. But also the eurosceptics made some ground with economic arguments against integration (endangerment of the welfare system, the primary sector, unemployment caused by integration).[13]

---

[9] This is especially important in understanding why different occupational groups and interest organizations opt for or against EU membership. See Christine Ingebritsen, *The Nordic States and the European Unity* (Ithaca and London: Cornell University Press, 1998). On the level of individual *decisions*, economic expectations seem to be less obvious in their impact. See Anders T. Jenssen, 'Personal Economies and Economic Expectations,' in *To Join or Not to Join*, pp. 194–234.

[10] For the importance of identities within social movements, see Klaus Eder, *Kulturelle Identität zwischen Tradition und Utopie. Soziale Bewegungen als Ort gesellschaftlicher Lernprozesse* (Frankfurt/New York: Campus, 2000).

[11] For further reasons why explanations linked to identity are useful in explaining the Norwegian EU debate, see Jochen Hille, *Das norwegische EU-Referendum von 1994: Nationale Identität versus europäische Integration* (Berlin: Institut für Internationale Politik und Regionalstudien e.V. in Zusammenarbeit mit dem Otto-Suhr-Institut für Politikwissenschaft der Freien Universität Berlin, 2000), pp. 3–6.

[12] See the Norwegian Social Science Data Services, at <http://www.uib.no/nsd/diverse/endring.gif>.

[13] See Maria Oskarson and Kirsten Ringdal, 'The Arguments,' in *To Join or Not to Join*, pp. 149–67.

Additionally, the *de facto* EU integration of Norway in matters of economy and visa regime (membership in the European Economic Area, Schengen) has bolstered negative economic effects of non-integration. Norwegian law is widely adapted to the *acquis communautaire*. Therefore, the most persuasive argument of the supporters of integration is that only full EU membership will give Norway the ability to influence decisions made in Brussels which Norway has to adapt to anyway.[14] From this perspective, further integration and Norwegian full EU membership is a relatively small step for the whole Norwegian economy and society (with some exceptions such as agriculture and fishery).

The question whether Norway will be an EU member-state in the future is often asked but all in all it is of minor importance. Because it is connected with the legitimacy of the whole process of European integration, the crucial question is: Why is euroscepticism widespread in the North? For this reason, I will investigate how the concepts of northernness and European integration relate to one another. I will describe the popular clichés about the North and the way they are consciously or unconsciously used in the EU debate. To do so I will describe how concepts of integration theory (*core Europe*, functionalistic integration theory) shape the view of the North as peripheral. Then I will focus on the traditional images of northernness related to (continental) Europe and the way the Norwegian eurosceptics are projecting these ideas.

Northern Norway – Stronghold of the Eurosceptics

In this section, I will outline the importance of the resistance against European integration in the sparsely populated areas of Northern Norway.[15] The referendum-campaigns on European integration in Norway were major political struggles. For example, in the campaign of 1994 the dominant movement against integration, 'Nei til EU' ('No to EU') had 140,000 members and the 'Europe

---

[14]   Svein S. Andersen, *Norway: Insider AND Outsider* (Oslo: Advanced Research on the Europeanization of the Nation-State, 2000), at < http://www.arena.uio.no/publications/ wp00_4.htm >. For the importance of this arguments within the debate of 1994, see Kirsten Ringdal, 'Velgernes argumenter,' in Anders T. Jenssen and Henry Valen, eds., *Brussel midt imot Folkeavstemningen om EU* (Oslo: Gyldendal, 1996), pp. 45–66.

[15]   In the three counties usually referred to as Northern Norway (Nordland, Troms and Finnmark), the density of the population is very low: seven, seven and two inhabitants per square kilometre, respectively. The Norwegian average is fifteen inhabitants. See Official Statistics of Norway, *Statistical Yearbook of Norway 2000. 119th Issue* (Oslo/Kongsvinger: Gyldendal, 2000), p. 69.

Movement' 35,000 members.[16] In a country with around 4.4 million inhabitants this degree of political mobilization is hard to achieve even by large social-economic interest groups such as single unions. 'Nei til EU', by and large, represents the majority of the eurosceptics in Norway.[17] For this reason I will use eurosceptics and 'Nei til EU' as synonyms. Other important social and political actors, like the Centreparty, the interest groups of the primary sector and the Socialist Left Party, support 'Nei til EU' and are closely linked with this organization.[18]

The Norwegian eurosceptics won both referendums on European integration with just a small percentage of the votes (in 1972 with 53.5 per cent, 1994 with 52.3 per cent). Therefore, the small number of voters in Northern Norway and in other peripheral areas of Norway plays a key role for the outcome of decisions on European integration. In this strategic setting, the few people in Northern Norway play a relatively important role in shaping the policy toward the EU of the whole country. According to Per Pettersen, Anders Jenssen and Ola Listhaug, the higher mobilization of the periphery in the referendum of 1994 partly compensated for the decline in population in the peripheries since the first referendum in 1972.[19]

Additionally, the interests groups which are defending the subsidies for the rural and most outlying areas of Norway have the organizational and financial basis to organize the resistance against EU integration.[20] Of course, the special economic structure and the enormous wealth of Norway must be mentioned in this context. Without the income from natural resources (oil, gas and waterpower) Norway could, according to Christine Ingebritsen, simply not afford the subsidies for the peripheries and Northern Norway, which is an important argument for

---

[16] See the website of 'Nei til EU', at <http://www.neitileu.no> and Anders T. Jenssen, 'Ouverturen,' in *Brussel midt imot Folkeavstemningen om EU*, pp. 18–9.

[17] Only the 'Social Democrats against EU' are also of some importance. Furthermore, there are some other groups, mainly with fundamentalist Christian or racist motives, but their impact on the debate is very small.

[18] For the socio-economic background of the members of all important social movements concerned with EU integration, see Anders T. Jenssen *et al.*, *Medlemsundersøkelsen: holdninger og sosial bakgrunn blant medlemmer i Nei til EU: Europabevegelsen og Sosialdemokrater mot EU: dokumentasjon* (Trondheim: Universitet i Trondheim, Institutt for sosiologi og statsvitenskap, 1994).

[19] See Ola Listhaug, Anders T. Jenssen and Per A. Pettersen, 'The EU Referendum in Norway: Continuity and Change,' *Scandinavian Political Studies*, Vol. 19, No. 3 (September 1996), p. 279.

[20] See Christine Ingebritsen, 'Norwegian Political Economy and European Integration: Agricultural Power, Policy Legacies and EU Membership,' *Cooperation and Conflict*, Vol. 30, No. 4 (December 1995), pp. 349–63.

opposing integration. The Norwegian farmers enjoy the highest subsidies of all agricultural producers in Europe. The Norwegian regional policies are aimed at enabling people to keep on living in the peripheries. This policy has not stopped the centralization and urbanization during the last decades, but compared with Sweden and Finland these trends have been slowed down. So the regional policy has kept the peripheries populated and dependent on subsidies from Oslo. Both are reasons for the relatively strong standing of the peripheries within the EU debate and the widespread euroscepticism.[21] From this point of view we must see the reluctance of northern Norwegians, and people in the periphery in general, as a powerplay against the political and economic centre of the Oslo area.[22]

Yet the importance of the northern Norwegian resistance against integration lies not only in this strategic and interest group based setting. More than this I regard the symbolic role of Northern Norway as crucial for the rejection of European integration in Norway. Euroscepticism is much about belonging to the nation and to the North. These feelings of belonging are communicated through symbols like the North as the edge of Europe. Therefore, the question how concepts of 'the North' correspond to European integration is important.[23] With this interpretation I go far beyond the purely arithmetic impact of the no-votes in Northern Norway as it is already described by scholars. We must use the term Northern Norway in a much broader way because symbols are not restricted by the administrative borders of the three counties of Troms, Nordland and Finnmark. It therefore does not make much sense to consider Northern Norway as only the northern Norwegian districts. Of course, these districts are connected with the concept of northernness. It is, however, more useful to regard Northern Norway mainly as a set of ideas which are only loosely linked to a precisely defined territory. To say it more accurately, it should be seen as a set of clichés and the interaction of these clichés in the debate on European integration.

In this debate the High North is being constructed as an antipode of an integrated Europe, as an alternative society to the 'central European society'. The real outcomes of the referendums have served to consolidate this concept of Northern Norway. At the same time the euroscepticism in Northern Norway gives rise to the idea that the debate on European integration is a struggle between North and South. Finally, this conceptualization of a political struggle, as a struggle between geographical spaces, is crucial for the understanding of the EU debate in Norway because even the name of the country (Norway means: 'way to the North') implies the definition of belonging to the 'North'.

---

[21] See *ibid.*, pp. 350 and 356–8.

[22] Hille, *Das norwegische EU-Referendum*, pp. 34–44.

[23] See Diez, 'Europe as a Discursive Battleground.'

Collective Identities, Space and European Integration

Before I go further in analyzing the way northernness is correlated with integration, I have to outline my basic theoretical viewpoint. It is necessary to make clear how concepts of space, collective identities and European integration are linked. Twenty years ago constructivist approaches were often regarded as 'soft' explanations for political decisions and as merely useful when other explanations failed.[24] Today, political science still tends to favour materialistic approaches as more 'realistic' and sound. Yet because of the convoluted problems in defining reality and interests, constructivism is winning ground.

The main reason for this is that even when one assumes that human beings are dedicated to materialistic goals, one has to deal with the problems of defining these goals and the collective group itself. The definition of collective interest is dependent on assumptions about collective identities which describe the political subject. Therefore, interests and identities are always shaping each other; they are interdependent. To divide between these two factors is only possible by analytical means.[25]

How do the constructivist and materialistic approaches relate to European integration? While one can experience integration in small groups (sports club, friends, congregation etc.) personally, the integration of states, societies, people and 'Europe' are merely "imagined communities".[26] To decide whether we support integration or not, we need to have some idea with whom our 'people' is going to be integrated. Therefore, the question of integration is simply a question of unification with other (imagined) communities. This process of integration can only be thought of in some abstract way, by imagining new collective groups like the 'Europeans'. We need some kind of metaphor to grasp 'integration'. It makes a big difference whether we understand European integration as a network of the élite, a concept of core and periphery, a 'reunification of the community of Christians' and so on.

A further argument for regarding constructivism as a basic explanation pattern is that we are used to understanding communities and 'people' as linked to some

---

[24] See David J. Elkins and Richard Simeon, 'A Cause in Search of Its Effect, or What Does Political Culture Explain,' *Comparative Politics*, Vol. 11, No. 2 (January 1979), pp. 127–45.

[25] For constructed identities, see Shmuel Eisenstadt, 'Die Konstruktion kollektiver Identität im modernen Nationalstaat,' in Bernd Henningsen and Claudia Beindorf, eds., *Gemeinschaft. Eine zivile Imagination* (Baden-Baden: Nomos, 1999), pp. 197–211.

[26] See Benedict Anderson, *Imagined Communities: Reflections on the Origin and Spread of Nationalism* (London: Verso, 1993).

kind of territory – often the nation-state.[27] Concerning this article, beyond the nation-states, the most important geographical categories are antipodes like North – South and East – West. Geographical concepts are closely linked to essentialist concepts of cultural units and the ideas of borders and conflicts between those units.[28] The popularity of Samuel Huntington's estimation according to which the major political conflicts of the future will mirror cultural cleavages between the major 'civilizations' may not just be a self-fulfilling prophecy, but also shows that essentialist approaches to culture and space (West against the rest) are still in vogue.[29]

Of course, 'the High North of Norway' is not in danger of clashing with the West. Norway is defined as a part of the western world, even though in the EU debate of 1994 it was sometimes argued that Norway could be isolated in matters of security policies between the American pillar of NATO and a Western European armed force.[30] However, Norway is defined in a special way within the Western world and in the context of European integration. The main message of this definition is that Northern Norway (and the North of Europe as a whole) lies on the periphery of Europe. This mind-mapping of the High North coincides with models of European integration and the traditional reception of 'continental Europe' in Norway and Northern Europe. The High North is always seen as a border region – a peripheral area. Now, I will draw attention to the way the North and the peripheries are pictured by neofunctionalist integrational theory and by the concept of 'core Europe'.

Integration in a Neofunctionalist View – a Model of Gravitation

The most powerful model to explain integration in Europe is functionalist. Neofunctionalism is grounded in the idea that, for functional reasons, integration is an ongoing process which is leading to political integration by 'spill-over' from economic functions to political functions. The major importance of a functionalist

---

[27]   Gertjan Dijkink, *National Identity and Geopolitical Visions: Maps of Pride and Pain* (London and New York: Routledge, 1996).

[28]   For the construction of space, identity and political units, see *ibid.* and John Agnew, *Geopolitics: Re-visioning World Politics* (London and New York: Routledge, 1998).

[29]   Samuel P. Huntington, 'The Clash of Civilizations?' *Foreign Affairs*, Vol. 72, No. 3 (Summer 1993), pp. 22–49.

[30]   For this reason, it has been argued that Norway should join the EU. See Espen B. Eide, 'Adjustment Strategy of a Non-Member: Norwegian Foreign and Security Policy in the Shadow of the European Union,' *Cooperation and Conflict*, Vol. 31, No. 1 (March 1996), pp. 69–104.

approach lies in its practical impact for political decision-makers. Neofunctionalism is used as a blueprint for how to build up a United Europe. The idea that European integration is necessary for functional reasons is a major way to legitimate European integration.[31] This can be seen in the Norwegian EU debate, where the pro-integration movement is most successful when it stresses the economic benefits of integration.

Neofunctionalism draws a picture of integration as necessity. Here, political integration is the outcome of economic integration. This expected spill-over effect is a major reason for the implementation of the common market and the Euro. The European economic and political élites believe that these steps towards integration make further integration necessary. In this way neofunctionalism appears as an apolitical theory without any normative objectives. But, of course, theories are never apolitical. They always shape our view in some special direction.

By stating that integration is something 'natural', neofunctionalism determines a direction towards an ever deeper integrated European Community. Whether or not this really happens, and how close a union would be established, does not just depend on the diverse political aims of political actors. It is also an open question whether an integrated Europe is really a better way of organising society. Big does not have to be beautiful. Nobody knows if the Euro, by binding the national economies together in too inflexible a way, will bring more disadvantages than benefits. It is even harder to tell whether it is possible for 15 (or 27 in a few years) nation-states with different standards of living, languages, cultural backgrounds, welfare systems etc. to find solutions for their problems by building up some kind of 'common sense'.

To predict whether integration will fail or not is combined with all the problems of predicting the future of complex systems. A closer look at forecasts on such things as global warming, the stock markets, rates of future economic growth or the development of political systems reveals that they have much in common with prophesy: they are either too imprecise to be used or they are simply wrong. Therefore, the way in which a theory on integration describes integration is much more interesting than the forecasts it gives. For this reason I want to stress the core metaphor of neofunctionalism which is based on a simple analogy to the way the 'real' world works. This analogy is *a model of gravity*. It is based on a model of

---

[31]   For a criticism of neofunctionalism, see Andrew Moravcsik, 'Preferences and Power in the European Community: A Liberal Intergovernmentalist Approach,' in Simon Bulmer and Andrew Scott, eds., *Economist and Political Integration in Europe: Internal Dynamics and Global Context* (Oxford: Blackwell, 1994), pp. 29–80.

space and an understanding of gravity of mass. The concept of *Kerneuropa*[32] (core Europe) – suggested once again in a speech by the German Minister for Foreign Affairs, Joschka Fischer, in May 2001 – also includes this gravitational model. According to this idea, member-states of the EU which are more able and willing to deepen integration are creating a gravitational pull for integration.[33] The idea of *Kerneuropa* is not the same political concept as the neofunctional model. Yet, in the way it builds up an understanding of the 'gravitational' (integrational) relation between centre and periphery, it uses the same metaphor of 'gravity'.

Of even greater interest is why these models are appealing to political decision-makers and the public picture of the process of European integration. Why do these concepts appear sound? The reason is that these models are strongly connected with the belief in 'soundness' which we connote with physical models of gravity. Here, we can combine the ideas of neofunctionalist integration theory with the spatial conceptualization of Northern Norway. This model does not just have a high degree of suggestive power, it also fits perfectly into the way the élites and many researchers like to view European integration. In this model the more densely populated core of Europe, which is often seen as consisting of Germany, France and the Benelux countries, is moving forward towards deeper integration, while the peripheral areas are following this 'development' more slowly. This model draws its suggestive power from the analogy to a generally accepted model in physics which is most famous for describing the relation between the sun and the planets. A model, by the way, which in itself is a synonym for the victory of the Enlightenment a few centuries ago.

Of course, one can argue against this statement that the arguments and empirical evidence which support neofunctionalist views are much more complex, and that it is not the basic metaphor of a 'model of gravitation' which backs the theory, but more the evidence. In this view the metaphor is just used to make a complex theory understandable for common people and 'common decision-makers'. Scientists, too, need some metaphor in order to grasp and believe an idea. Even if we suggest that social scientists are able to keep some critical distance from the suggestive power of these constructs, complex political and economic processes can only be made intelligible by such simple metaphors.

---

[32] The term is based on a concept of the German Christian Democratic Party. See Karl Lamers and Wolfgang Schäuble, 'Überlegungen zur europäischen Politik. Positionspapier der CDU/CSU-Bundestagsfraktion vom 1. September 1994,' *Blätter für deutsche und internationale Politik*, No. 10 (October), 1994, pp. 1271–80.

[33] Joschka Fischer, 'Vom Staatenbund zur Föderation – Gedanken über die Finalität der europäischen Integration,' speech delivered at Humboldt University Berlin, 12 May 2000, at < http://www.auswaertiges-amt.de/6_archiv/2/r/r000512a.htm >.

In this model, a strengthening of the core (*Kerneuropa*) or a higher density of the core will attract the surrounding peripheries more and more. For example, Richard Münch argues that "all of Europe sees the EU as the core of the idea of the good life".[34] From this standpoint, European integration is propagated as a synonym for the future of Europe and the peripheries are defined as areas which are orbiting around the centre. So Norway is outside the core of Europe – like a planet orbiting around the sun. Within this picture Oslo is on a near orbit but Northern Norway on the outermost orbit where the maps begin to be imprecise. Below I will make clear that this position of Northern Europe in the 'planet-like system' of European integration is not just a concept of non-Northerners. It is also deeply rooted in the way the North defines itself as a periphery.

Here we must acknowledge that, concerning their relation to the 'core of Europe', different peripheries are defined in diverse ways. For example, Turkey is imagined to be on a distant orbit for reasons of economic potential, religious not-belonging, insufficient human rights policies etc., while Northern Europe as it is claimed by the concept of the 'Northern Dimension for the European Union' could easily define itself as a core of Europe – or at least as the northern part of the core. Yet, in a political struggle centrality and peripherality are only partly bounded by a real geographical setting. Even Switzerland – the most central country in Western Europe in terms of geography, Christianity, democratic tradition, standard of living and functioning administration – is regarded as peripheral within the debate. There is a great deal of literature about the backward Swiss people who do not want to accept the EU as the future of Europe.[35] For example, an advertisement in Germany for some sweets, imitating the Euro coins, states that: "When the people of Switzerland will have calculated long enough, they will join the Euro, too."[36] Here again, European integration is described as the natural path for the future. It is obvious that Münch is thinking about the poor peripheries in southern and eastern Europe – the 'backyards' of Europe. But this view totally ignores the effect which the distinction between peripheries and centres has in the periphery. It simply offends the people in the peripheries because it

---

[34] Richard Münch, *Globale Dynamik, lokale Lebenswelten. Der schwierige Weg in die Weltgesellschaft* (Frankfurt: Suhrkamp, 1998), p. 274 (translated by the author).

[35] See Wolf Lindner, Prisca Lanfranchi and Ewald R. Weibel, eds., *Schweizer Eigenart – eigenartige Schweiz: Der Kleinstaat im Kräftefeld der europäischen Integration* (Bern, Stuttgart, Vienna: Akademische Kommission der Universität Bern, 1996); Benedikt Loderer, 'Zwei Schweizen,' *Ästhetik und Kommunikation*, Vol. 30, No. 107 (December 1999), pp. 45–50.

[36] "Und wenn die Schweizer lange genug gerechnet haben, kommen sie auch." Poster quoting a politician of the Christian Democratic Party of Germany, Fürstenberger Strasse, Berlin, 10 April 2001.

declassifies them as remote and primitive. As a result of their weak economies and the corresponding lack of self-confidence, people in the southern and eastern peripheries are far more forced to accept being devalued, although we can also see that they try to define themselves as 'central European' by pointing to their common history with 'core Europe'.

Certainly, models are only relevant for political mobilization when they reach political actors and are believed and accepted by the people. In fact it does not matter for a Norwegian EU referendum if the heads of scientists in central Europe are full of ideas of 'core Europe' and remote, backward-oriented areas 'circling' around it. So do people really believe that they are in the periphery and do they admire the core of Europe as a synonym for 'the good life'?

While the superiority of core Europe over peripheral areas might be accepted in those weaker southern and eastern peripheries, the economically strong and politically stable peripheries of Northern Europe do not accept the connotation with remoteness and backwardness. The suggestion that the peripheral society has to look with a feeling of admiration at the core of the EU as 'the epitome of the good life' is fiercely rejected by the people in the northern peripheries. Backed by the rich economic natural resources and by the strong networks of a corporate organized civil society, eurosceptics in remote areas of Northern Europe are self-conscious enough to draw the periphery as the locus of 'the good life'. In their view, it is 'central European society' which should look to the periphery as an example of the good life.[37]

The Pictures Fit Together – Norway as the European Periphery, Brussels as the Core and Northern Norway as a Satellite of Both of Them

To win a political struggle is mainly about projecting a positive image connected with one's own opinion and producing negative images of the opponents and their political aims. Yet to produce a totally new image of political belonging, space and good society is very hard for a political actor. On the other hand it is quite easy to gain support when a political actor just paints an existing, deeply rooted picture in a new way.

Of course, there are many ideas concerning northernness, continental Europe, the relation between centres and peripheries, the locus of the good life, spatial

---

[37]   The argumentation of the EU opponents is mainly based on confidence that Norwegian society and the Norwegian state is superior to the EU. See the basic program of 'Nei til EU', at <http://www.neitileu.no/man/organisasjon/serv/bjelker.html>. Certainly political movements generally need some ideological goal. In the case of 'Nei til EU' it is striking that strong emphasis is on non-materialistic goals.

ideas of political, ethnic and cultural belonging. The idea of northernness includes myths from the 19th centuries' national-romantic period, like the Nordic sagas, and ethnical belonging to Germanic tribes as well as social democratic universalistic welfare states regimes[38] and the image that the Nordic states and societies are ideal democracies. Some of these ideas can be related to the EU debate while other facets of this idea of northernness appear to be rather apolitical.

There is already a lot of literature discussing the question of how northernness is connected with European integration. Most of this literature is a part of the political struggle about EU integration. The pressure for integration has forced political actors and the northern European societies to define their collective identities and interests in the context of European integration. Heavily connected with this process of collective self-definition of the North there is a smaller branch of literature which detects the traditions of political thought and ideologies in the North related to 'core Europe'.[39]

Broadly speaking, the results of these discourses are that the idea of northernness tends to be an antipode against European integration, or at least incompatible with it, but we should never forget that this is not a 'natural' result. It is partly a product of the EU debate itself. As the Finnish concept of the Northern Dimension for the European Union illustrates, northernness can also be defined as a driving force of European integration. Especially in Norway, however, this is not the dominant way northernness is interpreted, and it is not the common way people believe in it.

Some of these discourses are linked with categories of geographical space more or less directly while others are not. A sharp distinction between lines of argumentation concerning the content of northernness is not possible. For example, the refrain of the 'Independence song' of the EU opponents is relying on the community of the 'people of the North' as the political subject: "Yes, we are many here in the North, who can go better ways [than the EU] and we do not want to be ruled from Brussels."[40] Yet, at the same time the EU scepticism expressed in

---

38 See Gøsta Esping-Andersen, *The Three Worlds of Welfare Capitalism* (Princeton: Princeton University Press, 1990).

39 Hille, *Das norwegische EU-Referendum*, p. 13. For the discourse on the ostensible superiority of the Nordic welfare system, see Bo Stråth, *Folkhemmet mot Europa. Ett historisk perspektiv på 90-talet* (Stockholm: Tiden, 1992) and Ole Wæver, 'Nordic Nostalgia: Northern Europe after the Cold War,' *International Affairs*, Vol. 68, No. 1 (January 1992), pp. 77–102. For the impact of Lutheran thought on integration, see Teija Tiilikainen, *Europe and Finland: Defining the Political Identity of Finland in Western Europe* (Aldershot et al.: Ashgate, 1998).

40 The "Independence Song" of 'Nei til EU' (Selvstendighetsvisa) is included in the Compact Disc *Nei '94*, 1994. The translation is by the author.

the song is based on totally other categories and lines of argumentation such as the EU's lack of democracy and the labour market policies of the EU. So diverse rational motives for resistance against European integration are linked with the geographical setting. As in this example, connotations of the North (and/or the national community) are often used as residual categories. They produce a feeling of common identities and values which are tied to the national or 'northern' community. Here we can detect a clever double strategy of the eurosceptics. On one hand, they give plenty of economic and political reasons why joining the EU would be disadvantageous to Norway.[41] By this, and by ruling out racists from their organization, they respond to any accusation of being backward-minded nationalists in the periphery of Europe. On the other hand, using election posters with Norwegian flags and slogans like "The country is ours! Shout no!"[42] the European Union is pictured as incompatible with the North and the nation. In this way they give their rational and often heavily special interest-based resistance the necessary emotional underpinning. All political movements need some kind of common identity and a meaningful goal to mobilize people. The crucial question is whether or not they succeed in doing this. The enormous success in mobilizing the people reveals that the double strategy of 'Nei til EU' won the hearts and minds of the majority of Norwegians and especially of those in the High North of the country.[43]

To explain why this worked I will sum up the statements of the very diffuse discourses which are connected with the North. Afterwards I will concentrate on the more direct implications of the feeling of 'geographical distance from Europe' which is connected with ideas of a conflict between centre and periphery. According to Bo Stråth, the Scandinavian welfare model is an ideological antipode against integration. For large parts of social democracy in Northern Europe as well as in other parts of the world Northern Europe is a synonym for a more just, peaceful and equal society. According to Ole Wæver, Nordic in this sense stands for 'being better than Europe', but with the end of the Cold War this connotation lost much of its suggestive power.[44] 'Norden' is somehow linked with these ideals and is a symbol of modernity. This is still true, even though one sometimes gets the impression that today the content of this view of modernity has partly shifted

---

[41] 'Nei til EU', *Norge og EU: virkninger av medlemskap i Den europeiske union* (Trondheim: Aktietrykkeriet i Trondhjem, 1994).

[42] In Norwegian: "Landet er vårt! Røyst nei!" This slogan is printed as the central message on many posters with different additional slogans concerning other political content.

[43] We can see this by the extremely high turnout of 89 per cent in the referendum of 1994 and in the high degree of membership in 'Nei til EU'.

[44] Stråth, *Folkhemmet mot Europa*; Wæver, 'Nordic Nostalgia'.

from the image of the most 'equal societies' to that of the most advanced tele-communication sector.

For large parts of the electorate the 'welfare system' has lost much of its brightness under the siege of neo-liberal thoughts in the 1980s and 1990s. Yet of greater importance is that the welfare system is thought of as universal and not necessarily bounded by the special geographical setting of the North. Concerning the EU debate in Norway and in Northern Norway, however, it is obvious that the eurosceptics used the image of being better than the EU. This is expressed in their main slogan: "Solidarity or Union."[45] Solidarity was used in a very broad way which partly shifted to a more geographical meaning. By solidarity the eurosceptics mean not just that Norwegians should defend their welfare system but also that they should keep the peripheries alive. So in the case of Northern Norway 'solidarity' means to uphold the status quo with high subsidies.

The Primary Sector's Interest and National Identity

Strongly connected with this solidarity with the periphery of Norway is the discussion about farmers and fishermen opposing European integration. Without any doubt the primary sector in Norway is the backbone of resistance against European integration. 94 per cent of the voters working in the primary sector voted 'No' in 1994.[46] The former farmers party – the Centreparty – and the interest groups of the primary sector and of the Norwegian peripheries in general used their political resources to campaign against EU membership. Yet, we should not forget that this was compensated for by parts of the political élite and interest groups campaigning for integration.[47] As in all developed countries, the small percentage of the population occupied in the primary sector are of low impact at the ballot box. Thus, the importance of the agricultural interest groups in many OECD-countries does not directly follow from their numerical strength in elections. The power basis of these interest groups is based on strategic ideas such as the self-sufficient food production of a country in times of war and instability. It is also based on the lack of powerful, organized interest groups within the political

---

[45] *"Folkestyre eller Union / Miljøvern eller Union / Solidaritet eller Union"* are according to 'Nei til EU' the central slogans of their campaign of 1994. They emphasize that democracy is their main goal. See 'Nei til EU', *Kampanjeplan 1994. Tactical Plan for the Campaign for the Referendum of 1994*, p. 5 (first introduced to the public in June 1994).

[46] Bjørklund, *Om folkeavstemninger*, p. 193.

[47] The Employers' Association and the Conservative Party supported membership. For the impact of political actors on the EU debate of 1994, see Detlef Jahn *et al.*, 'The Acteurs and the Campaign,' in *To Join or Not to Join*, pp. 61–81.

system which could counterbalance the agricultural interests.

Last but not least, it is based on the symbolic value of farmers who cultivate the land and provide the people with food. This symbolic function is also backed up by the idea that modern societies are rooted in agricultural societies. Therefore, peasant culture is often construed as the basic tradition and culture of the nation. To reassure the people of the meaningfulness of their nation, it is quite common to stress the rural 'roots of the nation'. Think, for example, about Soviet war memorials from World War II with martialistic iconographies of farmers and soldiers defending Russian soil. The whole genre of Hollywood western movies idealizes farmers who are on the trail to the west to cultivate the land. Swiss farmers defending their country against 'foreign knights' is also a myth about farmers as the core of the nation. To illustrate how deeply rural life is still connected with the idea of the nation, just compare it with another socio-economic group which in modern society is larger and much more important than the farmers: the civil service – would you ever connect them with the nation?

This comparison illustrates how deeply the idea of the nation is connected with the primary sector. Interest groups of the primary sector are therefore able to use this connection to influence the definition of national interest in public political debates like the EU debate. Here, Norway is not an exceptional case. Farmers are often regarded as the core of national identity, but Norway is an extreme case of strong idealization of the farmers and fishermen as a symbol of the nation. One reason for this is that in the 19th century when Norwegian nationalism arose other socio-economic groups on which national identity could be projected were simply missing. The civil servants were Danish influenced, the King was Swedish (Union with Sweden from 1814–1905). The urban class of citizens and the cities were weak. There was no significant class of noblemen or warriors. So the inventors of the nation (mainly intellectuals) took what they could use: the Vikings – to produce an image of a glorious history. They idealized farmers and fishermen in the peripheries of the country as the core of Norwegianness.[48]

Those socio-economic groups which were already there, for example Sámi communities, did not seem to be a suitable vehicle for Norwegian nationalism in the 19th century. 'Nei til EU' sometimes claims that the Norwegian nation-state is better in safeguarding the interests of the Sámi communities than the EU.[49] This argument is surely not central to euroscepticism in Norway (like democracy or welfare) but it is targeted at the votes of the Sámi people and on the solidarity with

---

[48]   For the development of Norwegian identity in the 19th century, see Øystein Sørensen, ed., *Jakten på det norske* (Oslo: Gyldendal, 1998). For the relation of primary sector and national identity, see Hille, *Das norwegische EU-Referendum*, pp. 39–42, and Ingebritsen, 'Norwegian Political Economy.'

[49]   Swein Lund and Carlotte Persen, 'Norske samer og EU,' in *Norge og EU*, pp. 247–53.

the 'natives' within the whole population. Furthermore, stressing the responsibility of all Norwegians for the Sámi people seems to be aimed at the guilty conscience felt in some sections of Norwegian society for the discrimination of the Sámi during the past centuries.

It is not surprising that the Vikings play a minor role in today's EU debate, even though some pictures on publications of 'Nei til EU' obviously hark back to the Viking heritage.[50] But as pointed out before, the primary sector was heavily involved in the EU debate for economic reasons. By using their status as a national symbol, they enabled the No-side to get an advantage in defining national interest and identity as incompatible with European integration. For example, the small (quite typical northern Norwegian) fishing-boats are one of the most popular motifs in publications and election posters of 'Nei til EU'. They do not really indicate the actual number of people working on the boats and being opposed to European integration. Rather, the fishermen are used as a symbol for the national interest in general. Here northernness, peripheral interests and identities merge with ideas of national identity. This combination clearly emphasizes the interpretation that the North and the European Union do not fit together well.

'Europe Movement' or Integration as a Necessity

The supporters of European integration are very conscious of the connection between the nation and the primary sector. They know that they are at a disadvantage in the struggle to win the votes and hearts of the electorate. To "understand Norwegian traditional scepticism about Europe", they do not tire of emphasizing that the connection between farmers and the nation is a construction of the 19th century. For this reason euroscepticism appears as an expression of irrationality and backwardness. According to them much of this national heritage has always been a part of the European heritage.[51] They are keen on deconstructing this farmer ideology as an obstacle on the path to the future which is seen by 'Europe Movement' as defined by a more or less integrated Europe.

Supporters of integration stress that Norway must join the EU in order to influence it and to strengthen national sovereignty within the EU. Empirical data clearly supports this view. When positive respondents in Norway were asked for their arguments for EU integration, 57 per cent mentioned the "necessity of joining the EU to defend independence" and 47 per cent stated economic reasons. Yet, less than one per cent mentioned the "belonging, or the wish to belong to Western

---

[50] See, for example, 'Nei til EU', *Ja til Folkestyre* (1994).

[51] See Europeisk Ungdom, *Norge og Unionen Overnasjonalitet i det nye Europa* (Flekkefjord: Hegland trykkeri, 1992), pp. 123–5 (translated by the author).

Europe" as an argument for integration.[52] So the emotional support for European integration is weak, even among the supporters of EU membership. This observation is backed up by the numbers of people mobilized in the large social movements of the 1994 campaign. Here 'Nei til EU' with, at its peak, 140,000 members clearly outnumbered the 'Europe movement' with around 35,000 members. The 'Europe Movement' convinced people mainly by rational, functionalist arguments rather than by 'heart', and even supporters regard EU integration as a necessary tactical move to uphold Norwegian interests within the EU. This view was more often shared in the urban, southern areas of Norway – mainly in the capital – than in the northern periphery.

So people in the south of Norway are not eager supporters of European integration, but they are more likely to accept integration as a kind of 'unavoidable fate'. They also have a limited trust in the EU system and the 'core of Europe'. Otherwise, they would not believe in the possibility of influencing the EU by entering it. This predisposition for supporting an 'insider-strategy' mirrors a weaker feeling of distance from the 'core of Europe' than the widespread fundamental opposition towards integration in the peripheries.

European Integration and the Distance to Power

The above argument is strengthened by the belief that Norway is more democratic than the EU system. The argument that EU membership will have a negative effect on democracy is the most prominent argument of the Norwegian eurosceptics. It is also the argument which is most important to the voters who voted 'no' in the referendum of 1994.[53] As I pointed out earlier the supporters of European integration argue the other way around, namely that Norway is losing democracy because it has to adapt to EU laws anyway.

Here it does not matter which interpretation is right or wrong. To compare different levels of democratization is a highly normative task. What matters in the present context is what the majority of the electorate believed: without any doubt, regarding arguments concerning democracy the no-side was superior.[54] Because of a tautology, this is not that surprising: the Norwegian democratic system is better suited to the Norwegian collective understanding of democracy than other democratic systems, and regarding the EU, it was a discussion about a 'would-be polity'. Today nobody knows what the final form of the EU will be. Therefore, the

---

[52]   The respondents were allowed to mention up to three arguments. The average was 2.19 answers. Therefore the total number of entries add up to more than 100. See Oskarson and Ringdal, 'The arguments,' p. 152.

[53]   See Ringdal, 'Velgernes argumenter,' pp. 45–66.

[54]   See Oskarson and Ringdal, 'The arguments,' p. 152.

possibilities to interpret the future of the EU system are extremely wide. The eurosceptics imagine the final form of the EU as a European super-state with a distant and mighty centre of power. This is what they are opposed to. This view includes the idea that space and distance are important to democratic legitimation. Somehow this seems to be natural because in many languages (e.g. German, Norwegian, English) one uses the same word, distance, both for space and for hierarchical, non-transparent structures of governance. In ancient times, the distance to the power centre was really of some importance to the ability to influence decisions, but with modern means of transport and information technology 'distance to power' is merely a metaphor for lacking power.

In Norway's EU debate, 'distance to power' appeared very often. Norway was characterized as 'close-to-the-people democracy'[55] while Brussels was described as the distant and non-transparent centre. According to the basic program of 'Nei til EU', "Der skal være makt der folk bor" ("power should be there, where the people live").[56] This strong connection between space and power must be seen in the framework of traditional patterns of political mobilization and political thinking in Norway. I suppose that it is not accidental that Stein Rokkan, who emphasized centre-periphery relations as an important cleavage, was a Norwegian.[57] The real distances and remoteness of large parts of the country have some suggestive power for viewing space as a category of major importance. Yet, more striking for stressing distance and centre-periphery relations as crucial is Norwegian history which is characterized by the struggle between countryside and centres for economic resources, political power and the definition of the leading culture. This fits together with the observation of Norwegian scholars that the EU debate reanimated traditional cleavage-patterns between urban centre and rural periphery which originated in the patterns of mobilization of the last century.[58]

However, the patterns of centre-periphery relations in Norway have always displayed a second dimension, namely one about the setting of Norway in relation to 'continental Europe'. Here it is often argued that "Norwegian self-consciousness

---

[55] This aspect of democracy is very important for the EU debate. The term 'close democracy' here mainly describes the strong relation between society and the widely unitarian Norwegian state. It is mainly based on the idea that the 'democratic state and the society' ('samfunn') are a close unit which is endangered by EU integration. See Hille, *Das norwegische EU-Referendum*, pp. 68–82.

[56] See the basic program of 'Nei til EU', at < http://www.neitileu.no/man/organisasjon/serv/bjelker.html >.

[57] See Stein Rokkan and Derek W. Urwin, *Economy, Territory, Identity: Politics of West European Peripheries* (London, Beverly Hills, New Delhi: Sage, 1983), pp. 1–18.

[58] See Anders T. Jenssen, *Hva gjorde EU-striden med det ideologiske landskapet i Norge?* (Trondheim: Universitet i Trondheim, Institutt for sosiologi og statsvitenskap, 1995).

as a peripheral state of the European mainland mirrors continental clichés about
Norway".[59] Therefore, Norwegians are suspected of lacking a feeling of belonging to continental Europe. Empirically, this view is backed up by a large quan-
titative, comparative study concerning values in western countries. Here, Beate
Huseby and Ola Listhaug come to the conclusion that Norwegians (and Icelanders)
have the weakest feeling of belonging to Europe.[60] Theoretically, this feeling of
not-belonging to Europe is explained by a wide range of arguments. Some lines of
argumentation stress the geographical distance to the High North. Some are based
on a cleavage between Catholic influence in continental Europe and Lutheran
traditions in northern Europe. It is hard to say how valid these theories are in
explaining Norwegian feelings about Europe. But they all paint the same picture
of Norway – it appears as a distant satellite of continental Europe. Europe is
defined as the centre and Norway, and especially the peripheral parts of Northern
Norway, as being at the outermost limits of the continent.

   These two levels of centre-periphery relations or self-definitions are strongly
intermingled. This can be seen in the popular slogan from the EU debates of 1972
and 1994: "Langt til Oslo lenger til Brussel" ("It is a long way to Oslo, but even
further to Brussels"). Here we see that the political conflict over EU integration
and the definitions of political space and belonging appear as the same.

Conclusion

Northernness and euroscepticism are correlated: the further north the more
eurosceptical is the electorate. Partly, this is accidental in that rural and sparsely
populated areas can be found more often in the High North, and sparsely populated
areas are traditionally more eurosceptical. The special interests of rational actors
who want to defend their subsidies enforce this resistance. These materialistic
reasons for the strong euroscepticism in Northern Norway lead to the interpretation
that the High North is the 'natural' stronghold of euroscepticism.

   At the same time, as I have shown in this chapter, a network of corresponding
ideas concerning the High North and the North supports this view. The major
content of these ideas is that northernness and European integration do not fit
together. They are antipodes. This idea is partly based on traditional views in the
High North concerning continental Europe. It is also based on the Norwegian

---

[59]   Stein Kuhnle, 'Norwegen,' in *Aus Politik und Zeitgeschichte*, B 43 (1992), p. 19
       (translation by the author).

[60]   See Beate Huseby and Ola Listhaug, 'Identifications of Norwegians with Europe: The
       Impact of Values and Centre-Periphery Factors,' in Ruud de Moor, ed., *Values in
       Western Societies* (Tilburg: Tilburg University Press, 1995), pp. 137–62.

centre-periphery conflict which projects the conflict between Oslo and the periphery onto the relationship between Norway and EU integration. The political actors in the Norwegian EU debate are part of this 'story-telling' and re-invention of the North as an antipode to European integration, they follow these traditional patterns of thinking. This is partly a calculated strategy to fulfil their political goals and partly unconscious.

It is not only the Norwegian and northern view on European integration that finds an antipode between northernness and integration. Prominent concepts of European integration like 'core Europe' and the 'gravitation model of neo-functionalism' support the view of northern Europeans being pulled into a gravitation field. With all the existing connotations of 'being peripheral' this leads to strong resentment against integration. All the corresponding ideas concerning the High North and 'central Europe' see the High North as the 'natural antipode' of European integration. In this context it is not surprising that everybody expects the people high up in the North to be the most reluctant Europeans and vice versa that Brussels – the distant centre – is regarded as threatening. This is obviously true for the strong movement of the eurosceptics.

What is less regarded is that even the supporters of European integration follow these ideas. They rely heavily on functional arguments for joining the EU. Integration is seen as an ongoing process within which Norway is forced to integrate. Therefore the main difference between supporters and opponents of integration in Norway is a strategic difference. The opponents regard non-integration as the best way to uphold national sovereignty while the supporters argue that only full membership will give Norway the ability to influence political decisions made in Brussels. The people in southern, urbanized areas of Norway are more likely to follow this line of argumentation. One reason for this is that the distance towards the core of Europe is believed to be shorter than in the peripheral North.

To get a different view I would propose turning the maps of Europe around or drawing them using other factors to determine the centre and periphery like, for example, the GDP per head, the density of civil society, the degree of education, the efficiency of public administration etc. On these maps the 'core of Europe' would encompass the most eurosceptical societies like Norway, Sweden, Iceland, Greenland, Great Britain, and the core of Europe would shift to the North. In this scenario, the protagonists of European integration could not easily deny that euroscepticism is a phenomenon of the most central European areas and not the absurd standpoint of a few 'backwoodsmen' high up in the peripheral North. Perhaps decision-makers would then take the well-founded euroscepticism concerning the lack of democracy and the endangerment of the welfare states more seriously.

Chapter 9

Spaces of Change Frozen in Time: Global Climate Change in the Arctic

Monica Tennberg

This chapter studies practices of representing climate change and its impact in the Arctic. The first part comprises a short theoretical discussion on globalization focusing on the relationships between space, time and environment. This is followed by a short historical review of climate discourses since the early 19th century. Next, the spatiality and temporality of representations of Arctic climate change in scientific and political discourses are discussed. The chapter suggests that in many of these discourses the temporal dimension of the representation is frozen. One can see the past dominate the current scientific discourses. The future of climate policy on the national, international and regional levels is constrained and limited by that which is possible at the present time. Possible future options remain unavailable. The frozen temporal aspects of current Arctic climate discourse in science and politics constitute a dilemma, since climate change and its impact on the region are above all a problem of the future.

Space, Time and the Environment in the Globalized World

The globe has shrunk in terms of the extent and intensity of environmental problems and their impacts on ecological, economic and social processes around the world. For example, the Arctic will be one of the regions most affected by future climate change. The largest temperature increases are expected for winters in the northern part of the northern hemisphere. The Arctic as a whole is expected to warm more than the global mean, although within the Arctic there will be both warming and cooling in different regions.[1]

---

[1] Arctic Monitoring and Assessment Program (AMAP), *Arctic Pollution Issues: A State of the Arctic Environment Report* (Oslo: AMAP, 1997), p. 160.

Due to large-scale changes such as these, globalization theorists argue that our understanding of space and time relations is changing fundamentally. Transformation of the objective qualities of time and space forces us to change the way we represent the world to ourselves. According to David Harvey, globalization is a compression of relations of time and space. These relations are constructed as discourses, representations and practices, with experience, perception and imagination contributing to them.[2]

Globalization may be defined as the circulation of images on a novel, global scale, including images of the entire globe. Our image is one of the global environment and global concerns. The environment in the globalist environmentalist discourse can be understood as one place, as the whole earth, or as spaceship earth. The view of the globe and its climate is a particular kind of construction; it is produced by computer models, alternative scenarios for future developments and remote monitoring by satellites that provides a view of the entire globe. In representing the environment as different kinds of spatial constructions, the question arises of how different realities, representations and imaginations relate to each other. Nature and society are intertwined in the modern globalized world: representations of nature cannot be distinguished from representations of society. When we talk about changes in the environment, we talk about economic, cultural, societal and political changes at the same time. According to this understanding, the environment is an intersubjectively constructed set of meanings associated with the human-environment relationship. The process of defining the relationship is a conflictual process among different actors locally, nationally, regionally and internationally. Therefore it is a politically relevant area of study.[3]

Discourses of global and Arctic climate change can be found in summaries of scientific reports for policy-makers and in national reports on efforts to tackle the problem through international climate cooperation. In these texts, the Arctic and its future are produced and reproduced through different practices of representing the relations of space and time. Henri Lefebvre divides practices of space into three categories: 1) concrete practices of space that transform the environment into buildings, streets and cities, and changes in the landscape; 2) practices of space belonging mainly to planners and officials of state that produce different representations of space in the form of maps, statistics and graphics; and 3) spatiality found in these representations of spaces.[4] The strength of this approach is that materiality, representation and imagination are not separated. Indeed, different human realities are based on the interaction of the three. In the ideal case,

---

[2] David Harvey, *The Condition of Postmodernity* (Oxford: Blackwell, 1989), p. 240.

[3] See, for example, Steven Yearley, *Sociology, Environmentalism, Globalization* (London: Sage, 1996).

[4] Henri Lefebvre, *The Production of Space* (Oxford: Blackwell, 1991).

following Lefebvre, the directly experienced spaces of representation and the conceptual representations of space relate to the lived contextuality of spatial practices.

To date, the discussion on globalization seems to have been dominated by the analysis of the spatial dimension. Space has conquered time. Globalization theorists often assume that the path towards the globalized world is one way. The future will be based on current trends of development. One way to unlock the one-dimensionality of the debate on globalization is to take the temporal dimension seriously. Time can be understood as a continuum, that is, an extension from the past to the future, or as repetitive cycles of seasons and changes. It can have several layers of rapid everyday and political events, or conjectures regarding to social and economic structures, including changes in these structures through different decades and long-term, very slow changes, which *Annales school* historians call 'longue duree'. The human-environment relationship is usually understood as 'longue duree'. The multitude of time conceptions calls our everyday understanding of the uni-directional flow of time into question. Our modern understanding of time has transformed into periods following each other in a linear fashion. The periods are fixed, with clear lines of division. The multitude of changes and layers of events, conjectures and long-term developments are restricted to the pre-modern, modern and postmodern periods. Time has also become scarce: the complexity of society limits the time available for co-ordination and calculation of future actions. The fear of ecological catastrophe on a large, even global, scale has limited our view of the future. Time is no longer an open horizon with endless opportunities: it is constrained or limited by the choices made today.[5]

A Short History of Climate Discourses

Climate has been used as a source of explanation for human behaviour, the organization of government and the fate of societies throughout history. We are still far away from the wish presented by James Croll in 1885 in his *Discussions on Climate and Cosmology* (1885): "It is hoped that the day is not far distant when the climate controversy can be concluded."[6] Early thinkers on the relationships between climate, society and human beings were concerned with the impact of climate on human civilizations, societies and individuals. There was strong doubt whether societies and civilizations could flourish in the northern regions. The

---

[5] Helga Nowotny, *Time. The Modern and Postmodern Experience* (Cambridge: Polity Press, 1994).

[6] James Croll, *Discussions on Climate and Cosmology* (Edinburgh: Adam and Charles Black, 1885), pp. 14–15.

northern climate was seen as detrimental to humans, leading to wild and violent behaviour. The end of the 18th century saw a change in the climate discourse, with the thinkers of the time calling for a more scientific approach to the issue. The thinkers of the early Enlightenment considered that 1) cultures are determined or at least strongly shaped by climate; 2) the climate of Europe had become more moderate since ancient times; 3) new settlements would change the American climate through agriculture; and 4) the amelioration of the American climate would make it more fit for European types of civilization and less suitable for primitive native cultures. If people had any impact on climate, it was one of improvement. Agriculture was supposed to have a positive impact on the climate. However, new regions could become dangerous for settlers because of the climatic influences of water, air and respiration.[7]

The time also saw an increased scientific interest in the Arctic, especially the behaviour of glaciers, prompted by the stories of early explorers and mariners. The mystery of ice ages generated great interest among researchers, especially in the role of ice in transforming the landscape. In 1837, the Swiss scientist Louis Agassiz suggested that at a distant time in the past ice sheets had covered an area from the North Pole to the shores of the Mediterranean and Caspian Seas. The ice had threatened the life existing on the Earth at the time. He argued that granite boulders in the Swiss Jura Mountains must have been transported to their locations by glaciers extending beyond their present limits. He called it the *Eiszeit* – the Ice Age. Soon it was realized that the Ice Age had not been a single period of glaciation. While some researchers were identifying the succession of ice ages and locating them on the geological timetable, others sought explanations for the behaviour of glaciers. The most widely held view, one proposed by Charles Lyell, was that glaciation and subsequent warming must have resulted from movements of the earth's crust – continuous risings and fallings that turned lands into seas and seas into lands with accompanying changes in snowfall and the formation of ice sheets. Another explanation, suggested by Louis Adhemar, was that the Earth's climate was influenced by its orbital path and by variations in the angle of its axis in relation to the Sun; the changing position of the Earth in relation to the Sun explained the changing climatic conditions. Following Adhemar's theory, James Croll concluded that ice ages were caused by changes in the Earth's orbit. An ice age was most likely to occur when the Earth's angle of tilt was smallest, for, it is then that the polar regions receive the smallest amount of heat.[8]

The discovery of the greenhouse effect, and especially the human impact in exacerbating it, is connected in climate history with two names in particular:

---

[7]   James Rodger Fleming, *Historical Perspectives on Climate Change* (Oxford: Oxford University Press, 1998), p. 18.

[8]   See Windsor Chorlton, *Ice Ages* (Alexandria: Time, 1983), pp. 83–5 and 100–105.

Joseph Fourier (1827) and Svante Arrhenius (1896). In a recent book on climate history, James Rodger Fleming finds singling out a couple of names in the history of discovering the greenhouse effect problematic. Fourier was not really interested in the effect of changes in the atmosphere on the climate but was more convinced of the impact of solar radiation as an explanation for the greenhouse effect. Arrhenius, on the other hand, did not consider climate change caused by human impact, especially that due to the increase of CO_2 in the atmosphere, to be a problem. He saw it as a positive impact and an improvement in the climate. He thought that the Arctic would witness a considerable warming in the future due to the human impact on its climate. In 1904, Arrhenius made the connection between the increase in carbon dioxide and industrialization. This meant that the focus of climate discourse changed considerably, from the impact of agriculture to the effects of industrialization.[9]

By the 1950s, climate change had found its way onto the public agenda for the first time. Scientific information available at the time suggested that a warming of the earth's atmosphere had begun in the late 19th century and continued to the 1940s. Scientific interest in climate change in the polar regions started to bloom internationally around the time of the International Geophysical Year (1957–1958). A network of stations and observations such as that which had been centred in the Arctic during the first and second International Polar Years was now developed in the Antarctic. Moreover, new means of studying glaciers and the atmosphere were implemented, such as satellites, rockets and ice core drilling.[10] In this era, there were many theories of climate change available. The main theories looked for explanations in changes in the Earth's orbit; changes in solar radiation or sunspots or cosmic dust; in lunar-sonar tidal influences; in changes of land masses and mountains; changes in atmospheric circulations or oceanic circulation; changes in atmospheric composition due to volcanic dust; or changes due to polar magnetism and continental drift. Scientists working in the Arctic, such as Hans W:sson Ahlman of Sweden and Alfred Wegener of Germany, contributed to the development of theories on climatic change.[11] Their theoretical work was based on empirical work done on Arctic expeditions.

By the early 1970s, the human impact on the climate was defined as a critical environmental issue in two studies carried out at the Massachusetts Institute of

---

[9] Fleming, *Historical Perspectives on Climate Change*, pp. 79 and 82.

[10] See J. Tyzo Wilson, *I.G.Y. The Year of the Moon* (London: Michael Joseph, 1961).

[11] Martin Schartzbach, *Alfred Wegener. The Father of Continental Drift* (Berlin: Springer-Verlag, 1986); Sverker Sörlin, *Hans W:sson Ahlman. Arctic Research and Polar Warming. From a National to an International Scientific Agenda 1929–1952* (Umeå: Umeå University, Centre for Regional Studies, 1997).

Technology (MIT).[12] At the end of the 1970s the understanding of the human-environment relationship based on the lessons in climatic changes in historical times was that 1) the climate is not fixed; 2) the climate tends to change rapidly rather than gradually; 3) cultural changes usually accompany climate changes; 4) what we think of as the normal climate at present is not normal in the longer perspective of recent history and 5) cool periods of the Earth's history are periods of greater than normal climatic instability.[13] At a conference on the Arctic climate in 1973, W.W. Kellogg pointed out that:

> climatic research is bringing the polar regions into a new perspective, because they are not only sensitive indicators of climatic change, with temperature fluctuations three to five times larger than the temperature fluctuations of the hemisphere as a whole, but because they also have large masses of ice in them that respond to temperature changes in a special way.[14]

The special interest in the poles concerned so-called 'flip-flops' in the climate circuit. In the climatic history of polar regions, either the poles have been frozen or they have not, with seemingly no stable condition in between. When one pole freezes, the other pole freezes shortly thereafter. For most of the history of the earth, however, the poles have been unfrozen.[15]

The human dimension has developed rather slowly and late in the scientific discourses on climate change. At a conference in 1994 on the internationalization of the Arctic, Marianne Stenbaek called for the study of the human dimensions of global change in the Arctic and pointed out the lack of a human dimension in the large-scale scientific projects in the region.[16] Nowadays, archaeological excavations around the Arctic and collections of oral histories of indigenous

---

[12] Massachusetts Institute of Technology (MIT), *Man's Impact on the Global Environment. Assessment and Recommendations for Action. Report of the Study of Critical Environmental Problems* (Cambridge: MIT, 1970); MIT, *Inadvertent Climate Modification. Report of the Study of Man's Impact on Climate* (Cambridge: MIT, 1971).

[13] John Gribbin and H.H. Lamb, 'Climatic Changes in Historical Times,' in John Gribbin and H.H. Lamb, eds., *Climatic Changes* (Cambridge: Cambridge University Press, 1978), p. 81.

[14] W.W. Kellogg, 'The Poles: A Key to Climate Change,' in Gunter Weller and Sue Ann Bowling, eds., *Climate of the Arctic* (Fairbanks: Geophysical Institute, 1975), p. vi.

[15] *Ibid.*

[16] Marianne Stenbaek, 'Human Dimensions of Arctic Global Change and Trends in the Human Dimension of the Arctic,' in Peter A. Friis, ed., *The Internationalization Process and the Arctic. Proceedings from Arctic Research Forum Symposium* (Roskilde: Roskilde University, 1994), p. 22.

peoples as well as documentaries on observed changes provide accounts of the human experiences of climate change and its impact in the Arctic.[17]

The Past in the Present in the Arctic

In the current summaries of climate research communicated to decision-makers, the Arctic is seen as a polar region and changes there as comparable to those in the Antarctic. The tradition of scientific work seems to strongly dictate this position and the comparison between the two polar regions. For current climate researchers, the Arctic is seen as part of the global climate system. There is also a very long tradition of regarding nature as a machine, and this tradition continues today. The report by the Arctic Monitoring and Assessment Program (AMAP) describes the Arctic as a refrigerator in the equator-to-pole transport of energy. The heat exchange between the Arctic and areas farther south is important for understanding contaminant global climate change.[18] Indeed, the importance of the Arctic for the global climate is apparent, for while the global climate clearly affects the region, the region – as the AMAP report points out- has an impact of its own on the global climate. These reciprocal effects are based on the view of the ecosystem as a whole encompassing the atmosphere, soil, ocean, clouds and their interactions.[19]

The Intergovernmental Panel on Climate Change (IPCC) report from 1997 notes that the polar regions are "very different in character". The Arctic is defined in the report as the area lying within the Arctic Circle; the Antarctic includes the area within the Antarctic Convergence, including the Antarctic continent, the Southern Ocean and the sub-Antarctic islands. The Arctic is described as "a frozen ocean surrounded by land" and the Antarctic as "a frozen continent surrounded by ocean".[20] Defining the area in the northern hemisphere in such a circumpolar global way is a problem for researchers using different criteria of definition. The most common definition of the circumpolar north is based on the Arctic Circle and encompasses the area north of the Arctic Circle (66°32'N). This is the area of the

---

[17] See Robert McGhee, *Ancient People of the Arctic* (Vancouver: UNB Press, 1996); Michael LeTourneau, 'Salmon in Sachs. Film Depicts Temperature Change,' *Northern News Service* (27 November 2000); Julie Cruikshank, 'Glaciers and Climate Change: Perspectives from Oral Tradition,' *Arctic*, Vol. 54, No. 4 (December 2001), pp. 377–93.

[18] AMAP, *Arctic Pollution Issues*, p. 14.

[19] *Ibid.*, p. 161.

[20] Intergovernmental Panel on Climate Change (IPCC), *The Regional Impacts of Climate Change: An Assessment of Vulnerability. IPCC Special Report. Summary for Policymakers*, at < http://www.grida.no/climate/ipcc/regional/508.htm > .

midnight sun. The Arctic is also often defined as the area north of the 10°C July isotherm. Another criterion used as the boundary is the treeline, which is the border between southern forests and northern tundra. This changes with the climatic conditions. Also climatically relevant is the definition based on permafrost. The presence of discontinuous permafrost is sometimes used to define the subarctic, in contrast to the Arctic, where permafrost is continuous. On balance, the Arctic includes many climates.[21]

According to the IPCC the Arctic is special compared to the Antarctic: "the Arctic is more vulnerable than the Antarctic because of its sensitive and fragile ecosystems and the impacts on traditional lifestyles of indigenous peoples and because climate changes are expected to be greater."[22] It is true that some four million people, including several indigenous peoples, live within the Arctic region. In the Antarctic, the population consists of scientific groups visiting the region. Indigenous peoples' groups are minorities in most Arctic regions. In northern Norway there are 380,000 inhabitants, in northern Sweden 270,000 and in northern Finland 200,000. According to the AMAP estimates, there are about 85,000 Sámi, of whom 50,000 live in the Arctic region. In Iceland there are 270,000 habitants without any indigenous peoples' groups. Only in Greenland does the majority of the population (55,000) belong to an indigenous people, the Inuit. In the Russian North, there are about 2 million people, of whom 67,000 belong to different indigenous peoples' groups. In northern Canada, there are 47,000 indigenous people, representing about half of the total population of the region. Alaska has half a million habitants, of whom 15 per cent belong to different indigenous peoples' groups.[23]

The Arctic regions are described mostly as non-modern, very traditional societies. According to the IPCC, the impacts of climate change seem positive for human beings in general. Agriculture will become easier. Where fisheries are concerned, the IPCC expects that marine ecological productivity should rise although it has some reservations about the negative impact of ultraviolet-B (UV-B) on fish productivity. The warming of the climate will create new opportunities for shipping, the oil industry, transport, mining, tourism and migration. Sea ice changes projected for the Arctic have major strategic implications for trade, especially between Asia and Europe. All of the negative consequences of climate change seem to affect the indigenous peoples in the region. Human communities in the Arctic will be "substantially affected" by the projected physical and

---

21 AMAP, *Arctic Pollution Issues*, pp. 6–7 and 16–19.

22 IPCC, *The Regional Impacts of Climate Change*.

23 AMAP, *Arctic Pollution Issues*, pp. 52, 56 and 59–65.

ecological changes.[24] The effects will be particularly important for indigenous peoples who lead traditional lifestyles. The report does not take into consideration that there are both indigenous and non-indigenous people living in the Arctic who do not live traditional lifestyles. The distinction between indigenous/traditional and local/modern life styles is a simplification of the situation.

Tradition is very strong in the descriptions of the indigenous cultures in North America to be found in the AMAP report of 1997. The main sources of income and livelihood in the Arctic are hunting, fishing and gathering. They are important as well in northern Canada, but indigenous peoples there also participate in the wage economy when possible. The non-indigenous population in the region hunt and fish as well but not to the extent that indigenous peoples do.[25] The traditional livelihoods are also a relevant factor in the case of the Russian Arctic. As indigenous peoples in Russia have had limited opportunities for higher education available to them, they have found it difficult to participate in the industrial economy, with the result that the unemployment rate among the population is high. Most indigenous peoples make their living in reindeer herding, trapping, hunting, fishing and making handicrafts.[26]

In the Nordic countries, modernization is more evident. Greenland, Iceland and Faroe Islands depend on fisheries and the fishing industry. In Greenland, about one-fifth of the population is dependent directly or indirectly on hunting activities.[27] In Sweden and in Finland, the Sámi engage in reindeer herding although many of them are in occupations similar to those held by the rest of the population. In Norway, the majority of the Sámi combine farming with fishing and reindeer husbandry. The local population in northern Sweden are employed in mining, electricity, water services, forestry or public services. The same is true in the case of Finland. The service and tourist industries are rapidly growing in Finnish Lapland.[28]

When we talk about climate change and its potential impacts in the Arctic, we are discussing the future of the region. The focus is on developments that will take place in next 50 to 100 years. Climate change is generally defined as an additional stress factor to other changes taking place in the environment in the near future. However, where humans and societies are concerned, the perspective seems somehow frozen despite the fact that we are talking about the future of the region. The forces of globalization and modernization will change the region, its

---

[24] IPCC, *The Regional Impacts of Climate Change*.

[25] AMAP, *Arctic Pollution Issues*, p. 56.

[26] *Ibid.*, p. 65.

[27] *Ibid.*, p. 60–62.

[28] *Ibid.*, p. 62–4.

economies, societies and cultures in the future. Yet, one cannot see these changes taken into account credibly in the evaluation of impacts of climate change. The efforts to tackle the problem are very much related to constructions of the problem for political decision-makers at the local and regional level.[29] The report of the planning meeting of Arctic Climate Change Impact Assessment (ACIA) in spring 2000 concludes that ministers want to know what climate change and its impacts will mean for the people.[30]

The Extended Present of Climate Politics

Scientific debates concerning climate change tend to focus on the impact of this change on the Arctic as if there is nothing to be done about the situation except to prepare for the consequences. Political argumentation among local and regional decision-makers, although using science as part of the argumentation, is more concerned with the real social and economic consequences of the climate change. Thus, although scientific and political debates have some common topics, they are also different from one another. Climate change caught the attention of the 10th anniversary ministerial meeting of Arctic environmental cooperation (The Arctic Council), held in Rovaniemi, Finland, in June 2001. The Arctic Council brings together political decision-makers from Nordic countries, Canada, the United States and Russia to discuss common concerns. A number of indigenous peoples' organizations from the Arctic participate in the cooperation as permanent participants.

The concern among the decision-makers is the impact of global climate change in the Arctic: "According to an overwhelming majority of scientists and long-term observations by local people climate change is taking place with strong, variable and largely unpredictable effects on nature and communities in the Arctic."[31] Impacts on the human communities are the main concern. Changes in ocean currents are possible and carry the risk of severe effects on living conditions in some parts of the Arctic. The implications at the regional and global level could

---

[29]   Thomas J. Willbanks and Robert W. Kates, 'Global Change in Local Places: How Scale Matters,' _Climatic Change_, Vol. 43, No. 3 (1999), pp. 601–29.

[30]   Arctic Climate Change Impact Assessment (ACIA), _Report of a Meeting and Workshop to Plan a Study of the Impacts of Climate Change on Arctic Regions_, 28 February – 1 March 2000, Washington, DC, USA, at < http://www.acia.uaf.edu/pages/background.html#scoping >.

[31]   Ministry for Foreign Affairs of Finland, 'Conclusions of Chair,' in _10 Years of Arctic Environmental Cooperation_ (Helsinki: Ministry for Foreign Affairs of Finland, 2001), p. 10.

create "a vicious circle": thawing permafrost poses dangers to communities and industries in many parts of the Arctic and could, with increasing emissions of greenhouse gases, accelerate climate change.[32] The potentially positive impacts, the transport of natural resources using new transport routes, were discussed by the Council as well. The Arctic oil and gas resources are part of world markets and many expect these to be in growing demand by the world markets.[33]

There is an obvious need for more detailed information on the impacts of global climate change in the Arctic: "The Sámi want to know how the global changes will affect us? How will our fisheries be impacted? How will our reindeer herds survive? Are they in danger?"[34] For the Inuit, it will be "serious, unavoidable and certainly a key issue". The question for them is whether they will be able to adapt to and manage the projected impacts.[35]

The Finnish minister of the environment called upon those countries producing the highest amounts of greenhouse gases per capita to take the first steps in reducing greenhouse gas emissions.[36] This call is directed to all the Arctic states since their indicators are high compared to those of other states in the OECD region. Among the OECD countries, the average CO_2 emissions per capita are 10.92 tonnes. Canada, Finland and United States have higher CO_2 emissions per capita than the OECD average, *i.e.*, 15.75, 11.59 and 20.10 tonnes respectively. Denmark is just below the OECD average with 10.81 tonnes. The Russian average is 9.64 followed by Norway (7.77) and Iceland (7.69). The lowest average is in Sweden (6.05). However, these numbers are higher than any indicators elsewhere in the world: the non-OECD average is 4.83 and the Middle East's average is 5.78. The Arctic states' indicators are considerably higher than the averages in China (2.32), Latin America (2.15), Asia (1.03) and Africa (0.96).[37] The northern areas within the Arctic states are not major greenhouse gas polluters.[38]

---

[32] *Ibid.*

[33] Paavo Lipponen, '10 Years of Arctic Environmental Cooperation,' in *ibid.*, p. 16.

[34] Anne Nuorgam, '10 Years of Arctic Environmental Cooperation,' in *ibid.*, p. 42.

[35] Aqqaluk Lynge, 'From Environmental Protection to Sustainable Development: An Inuit Perspective,' in *ibid.*, p. 53.

[36] Satu Hassi, 'Opening Remarks,' in *ibid.*, p. 14.

[37] International Energy Agency (IEA), *Selected Energy Indicators for 1998* and IEA, *Selected Energy Statistics for 1998*, both at < http://www.iea.org/statist/keyworld/ >.

[38] See Danish Energy Agency, *Climate 2012 – Status and Perspectives for Denmark's Climate Policy*, at < http://www.ens.dk/publikationer/2000/climate2012 >, p. 16; United Nations Framework Convention on Climate Change (UNFCCC), *Canada's National Report on Climate Change*, p. 95. All UNFCCC documents referred to in this chapter can be found at < http://unfccc.int/resource/natcom/nctable.html#a1 >.

The message to the World Summit on Sustainable Development formulated at the 10th anniversary meeting of Arctic environmental cooperation concludes that "[t]he fate of the Arctic is largely dependent on progress in global efforts to adjust human economic activities to the capacity of nature. Global action, with the circumpolar North as an active partner, is essential for the future of the Arctic."[39] However, when looking at the international negotiations, this connection between regional and global concerns is not highlighted, not even in the reports of the Arctic states themselves. In political assessments, the concern over the future climatic changes in the Arctic does not get much space either. The nature of vulnerability is also evaluated in some of the political statements. In Norway, most attention has to date been focused on effects on ecosystems and their vulnerability to climate change. Because of its geography and long coastline, Norway may, however, be more vulnerable to changes in the frequency of weather patterns and extreme events such as storms, flood and spring tides than to climate change caused by increases in mean temperature.[40] The work done in the United States concludes that it is in part sensitive, but not vulnerable, to the impacts of climate change.[41]

In some of the assessments, such as those carried out in Canada, Iceland, Sweden and Russia, evaluation of the impacts in the Arctic have been mentioned. In Canada, Arctic concerns are evident since over 40 per cent of Canada is situated north of 60° north latitude. The vulnerability of ecosystems is the main concern.[42] In the Icelandic evaluation, warming is expected to have positive effects in most respects on the island itself, although its effects on the fishing banks are less certain. Of great concern to Iceland is the possible effect of climate change on ocean currents, which could cause drastic change in climate and fisheries.[43] In Sweden, sub-arctic ecosystems are seen as being sensitive to climate change owing to long generation times, slow growth and irregular reproduction. Climate change is expected to affect the ecology of higher mountain areas, which have a limited adaptability. The fate of the forests is also uncertain: spruce forests are expected to be the most vulnerable to rapid changes in the climate.[44] In the Russian

---

[39] Ministry for Foreign Affairs of Finland, 'Conclusions of Chair,' p. 12.

[40] UNFCCC, *Norway's Second National Communication under the Framework Convention of Climate Change*, p. 14.

[41] UNFCCC, *U.S. Climate Action Report*, pp. 19–20.

[42] UNFCCC, *Canada's National Report on Climate Change*, pp. 20–21.

[43] UNFCCC, *Status Report for Iceland Pursuant to the United Nations Framework Convention of Climate Change*, p. 50.

[44] UNFCCC, *Sweden's National Report under the United Nations Framework Convention of Climate Change*, p. 10.

Federation, a substantial shift to the north of the permafrost zone, which currently occupies 58 per cent of the national territory, could take place. This shift and decomposition of the permafrost will primarily influence human settlements with their infrastructure, roads, airports and energy facilities (oil and gas pipelines, power stations and so on).[45]

Some of the Nordic countries are more concerned about the impacts of climate change around the world than in the Arctic regions of these countries as such. In Denmark's evaluation, the vulnerability of Greenland to climate change is noted. The Danish assessment states that "Denmark only accounts for about 1/1000 of the World population and covers even less of the global land surface. Climatic impacts in Denmark are therefore insignificant in economic evaluations of the seriousness of the total global impact, and in discussions of which degree of intervention is justified."[46]

A concern for international developments due to climate change can be found in the Finnish report: "The greatest problems in Finland would therefore be the indirect effects of changes elsewhere in the world." These could cause problems of refugees and of food production, and lead to a decline in the world economy.[47]

Time seems to be running out on international climate cooperation. The first efforts of climate cooperation started in late 1980s. All Arctic countries signed and ratified the United Nations Framework Convention on Climate Change in Rio de Janeiro at the United Nations' Conference on Environment and Development (UNCED) in June 1992, which aims at the "stabilization of greenhouse gas concentrations in the atmosphere at a level that would prevent dangerous anthropogenic (man-made) interference with the climate system." The Convention did not contain any timetables or targets for emissions cuts. It was soon realized that the Convention was not enough and further actions were needed. According to the 1997 Kyoto Protocol, industrialized countries have a legally binding commitment to reduce their collective greenhouses gas emissions by at least 5 per cent compared to 1990 levels by the period 2008-2012.[48] All Arctic countries belong to the group of industrialized countries (Annex 1 countries) in the Kyoto Protocol. It has taken five years since 1997 to finally agree on the details of the Protocol. International climate cooperation now covers the first commitment period until 2008-2012. The so-called Kyoto mechanisms, joint implementation and

---

[45] UNFCCC, *First National Communication of the Russian Federation*, p. 52.

[46] UNFCCC, *Denmark's Second National Communication on Climate Change*, p. 12 and 64.

[47] UNFCCC, *Finland's National Communication under the United Nations Framework Convention on Climate Change*, p. 101.

[48] Michael Grubb, *The Kyoto Protocol. A Guide and Assessment* (London: Earthscan and Royal Institute of International Affairs, 1999), p. 37.

emission trade, will not take effect until later this decade, with little expected impact during the first commitment period. The short-term thinking in international climate cooperation and long-term decisions required to advance sustainability in climate and energy policy are not easily balanced. Decisions between different sources of energy and the development of transportation and network systems are long-term structural decisions with a perspective of 20–30 years into the future. At the same time, the idea of sustainable development needs to be operationalized in concrete cooperation on issues related to energy and climate concerns. Short-term thinking seems to dominate international climate cooperation.

Most of the Arctic countries will have problems in achieving the aim of stabilizing emissions at the 1990 level. In Canada, United States, Sweden, Norway, Iceland and Denmark emissions have grown during the 1990s. Both Finland and Russia managed to stabilize their emissions at the 1990 level by the end of the decade.[49] The explanation is economic. Both countries suffered a severe economic decline, which cut the emissions. A common concern for the Arctic states is the increase in the quantities of emissions in the future and the problems of meeting the commitments during the Kyoto period. The Kyoto collective target for industrialized countries in 2012 will be achieved through cuts of 8 per cent by the European Union and of 6 per cent by Canada. The EU will meet its target by distributing different rates to its member states. The common target for EU countries commits Denmark to a reduction of 21 per cent. This is a challenge for Denmark since measures put in place have not yet resulted in the expected reduction of emissions. As a member of the EU, Finland is expected not to increase its emissions by 2008–2012. This has created a political debate in Finland over whether an increase in the use of nuclear power might be a solution.[50] Agreement between EU countries allows an increase of 4 per cent for Sweden. However, it seems that Sweden will not be able to adhere to that limit.[51]

---

[49] UNFCCC, *Report on In-depth Review of the Second National Communication of Canada*, pp. 19–21; *Report on In-depth Review of the Second National Communication of the United States of America*, pp. 19 and 26; *Report on In-depth Review of the Second National Communication of Sweden*, p. 23; *Report on In-depth Review of the Second National Communication of Norway*, p. 25; *Report on In-depth Review of the Second National Communication of Iceland*, p. 16; *Report on In-depth Review of the Second National Communication of Denmark*, pp. 22 and 28; *Report on In-depth Review of the Second National Communication of the Russian Federation*, p. 8; Tilastokeskus, *Energiaennakko 2000*, at < http://www.tilastokeskus.fi/tk/yr/ye_energiaenn.html >.

[50] UNFCCC, *Finland's Second National Communication on Climate Change*, pp. 6–7.

[51] UNFCCC, *Report on In-depth Review of the Second National Communication of Sweden*, p. 18.

Outside the European Union, Norway is allowed to increase emissions by up to 1 per cent and Iceland up to 10 per cent. In Norway, however, the expected rise of emissions of greenhouse gases is considerably higher (16 per cent) compared to the 1990 level. The Kyoto Protocol allows Iceland to increase its emissions by 10 per cent during the commitment period. This will not be easy since projections indicate a considerable increase by 2010 because of HFCs (hydrofluorocarbons). Russia for its part is to stabilize its emissions. While the Russian Federation's emissions in 2000 should be lower than the 1990 level by at least 24 per cent, the country may have problems with long-term implementation of stabilization. For Canada, the Kyoto target of reducing greenhouse gas emissions to 6 per cent less than the 1990 levels by 2008–2012 is an "enormous challenge". The U.S. withdrew from international climate cooperation in spring 2001. The projection for U.S. emissions predicts that in 2010 greenhouse gas emissions will be 26 per cent above the 1990 level.[52]

The End of an Era? Debates Stuck in the Past and in the Present

Globalization of the environment is above all a new, qualitatively different understanding of the human-environment relationship. Only some of the themes in the discussions on the globalization of the environment have reached the Arctic. Mostly the Arctic is being globalized as part of the scientific debate on climate change and its impacts. However, the details of expected changes and their impacts are very much unclear at the local level. Other aspects of globalization relevant to the Arctic are outside the current scientific and political debates, namely the role of the Arctic states in energy consumption and production, emission trends and the potential of regional political efforts to tackle the issue. In terms of political and social changes, the Arctic societies are presented as remote, traditional societies which seem to be mostly outside the forces of globalization.

In the Arctic, there is a sense in much of the climate literature that the end of an era has come, including the threat that something that we know as the Arctic, something unique and special, will disappear from the world as the warming occurs

---

[52] UNFCCC, *Report on In-depth Review of the Second National Communication of the Russian Federation*, pp. 12 and 23; UNFCCC, *Report on In-depth Review of the Second National Communication of Norway*, p. 25; UNFCCC, *Report on In-depth Review of the Second National Communication of Iceland*, p. 16; UNFCCC, *Report on In-depth Review of the Second National Communication of Canada*, pp. 20–21; UNFCCC, *Report on In-depth Review of the Second National Communication of the United States of America*, pp. 19 and 26.

and can never be replaced.[53] This is the way the Arctic is presented for consumption in images on globalization.[54] It is based more on the past than on expectations for the future. Of course, as such, these images based on the past, nature and authenticity are very strong, and they can be used politically. However, at the same time, these views of the Arctic make the region marginalized in the area of climate politics. Most disturbing is that the perceived and abstract representations of the changing human-environment relationship do not relate to everyday life and experiences in the region. The Arctic seems to remain frozen in many of the representations of climate change in the region.

---

[53]   Claes Bernes, *The Nordic Arctic Environment – Unspoilt, Exploited, Polluted?* (Copenhagen: The Nordic Council of Ministers, 1996), p. 219.

[54]   See Scott Lash and John Urry, *Economies of Signs and Space* (London: Sage, 1994).

Chapter 10

Competing Industries and Contested Nature in Finnish Lapland after World War II

Leena Suopajärvi

Much of Finnish Lapland was burnt to the ground in the last phases of World War II in 1945. Retreating German troops burnt almost half of all the buildings in the region, and in some municipalities the destruction was almost complete. For example, in the city of Rovaniemi, the administrative center of the province, about 90 per cent of the buildings were ruined. The railroad, roads, bridges and ferries were bombed and telephone and telegraph connections destroyed. Laplanders coming back home from the war and evacuation had to start their lives and reconstruction empty-handed.[1]

World War II was a turning point in the environmental history of Lapland. Finland needed all of its resources to recover from the war, and the natural resources in the North were harnessed for that effort. Immediately after the war, the state began an extensive program for the utilization of the natural resources in Lapland, the focus of which was economic growth and industrial development. The Finnish environmental sociologist Ilmo Massa has described this policy as *ecological colonialism*; development of the local traditional sources of livelihood or protection of nature were not on the agenda in the first decades after the war.[2]

It was not until the 1970s that the large-scale exploitation of natural resources in Lapland became a subject of critical public debate. Since that time, environmental issues have had a place in the regional discussion.

---

[1] Martti Ursin, *Pohjois-Suomen tuhot ja jälleenrakennus saksalaissodan 1944–1945 jälkeen* (Oulu: Pohjoinen, 1980), pp. 32–49.

[2] Ilmo Massa, *Pohjoinen luonnonvalloitus: suunnistus ympäristöhistoriaan Lapissa ja Suomessa* (Helsinki: Gaudeamus, 1994), pp. 204–205.

Contested Nature

In this chapter, I will discuss why nature is such a contested issue in Finnish Lapland. I argue that the reason for the constant dispute in environmental questions is that nature is an important resource base for different branches of industry. According to a survey made in the mid-1990s, the livelihood of almost half (47 per cent) of the people in Lapland is based on nature, the occupations concerned being found in agriculture, forestry or tourism.[3] My argument is that the ways in which nature is understood and used in these different industries conflict with each other, which causes problems for combining alternative sources of livelihood in the region.

The traditional way of life, for example reindeer herding, is still part of life in Finland's northernmost province. Although the population is now concentrated in small cities and villages, nature-based hobbies such as berry-picking, hunting and fishing are part of the northern way of life. Different uses of nature are part of the everyday practices learned in childhood and carried on into adult years. This idea has been captured in the concept of *traditional ecological knowledge (TEK)*, which is defined as "a cumulative body of knowledge and beliefs, handed down through generations by cultural transmission, about the relationship of living beings (including humans) with one another and with their environment. Further, TEK is an attribute of societies with historical continuity in resource use practices; by and large, these are non-industrial or less technology advanced societies, many of them indigenous or tribal."[4] Traditional ecological knowledge is cultural capital in its embodied state, a result of life-long learning acquired by doing.[5]

Modernization began in Lapland after World War II. Modernization is usually understood as industrialization and urbanization, but in Lapland the development was different; the region has always been – and still is – a sparsely populated area with a low industrialization rate. The core of modernization was the mechanization of the old ways of working and large-scale utilization of the region's natural resources. The mechanization of logging sites and construction work, for example in hydropower projects, gave rise to a working class of professional lumberjacks or builders and destroyed the traditional combination of small-scale farming,

---

[3]   Liisa Kajala, *Lappilaisten näkemyksiä metsien hoidosta ja käytöstä* (Helsinki: Metsäntutkimuslaitos, 1997), p. 8.

[4]   Fikret Berkes, 'Traditional Ecological Knowledge in Perspective,' in Julian T. Inglis, ed., *Traditional Ecological Knowledge* (Ottawa: International Development Center, 1993), p. 3.

[5]   Pierre Bourdieu, 'The Forms of Capital,' in John G. Richardson, ed., *Handbook of Theory and Research for the Sociology of Education* (New York: Greenwood Press, 1986), p. 243.

subsistence economy and seasonal forest work. More efficient, large-scale utilization of forests, ores and hydropower developed after the war time in the modern spirit: nature was valuable only to the extent that it was productive and furthered economic prosperity.[6] This idea is still alive. Lapland produces raw material for the paper industry and hydropower for the national grid, and is promising ground for mining interests in the new millennium.

Nowadays, an important branch of industry in Lapland is tourism, which has developed rapidly since the 1980s. Tourism is a service-based sector whose markets are global, and, accordingly, can be understood as a phenomenon of the postmodern or late modern era. Lapland's main tourist attraction is its relatively clean and peaceful nature, which tourists from the South expect to be an authentic and pure experience.

These three different industries have deeply varying interests and values with regard to nature in the North. I depict these tensions in the figure below.

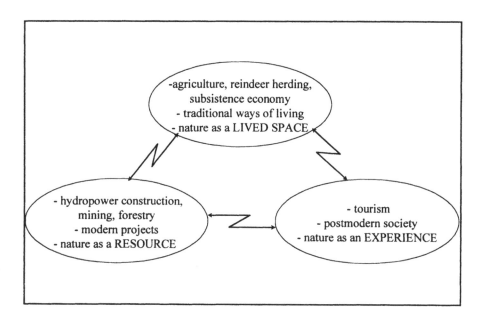

Figure 10.1 Different Industries and Their Relationship to Nature

6   Ari Haavio, 'Approaches to Man's Relation to Nature,' in Anna-Mari Konttinen, ed., *Green Moves, Political Stalemates. Sociological Perspectives on the Environment* (Turku: University of Turku, 1996), pp. 11–13.

The value of nature in the North has also been recognized in the area of nature conservation. Nature has become a contested resource in Lapland: there have been disputes – even lawsuits – between reindeer herders and logging companies, lumberjacks and conservationists, tourism entrepreneurs and reindeer herders. In this chapter, I will discuss competing industries and contested nature in Finnish Lapland and take an empirical look at the time since World War II. I base my arguments on data which I have collected on two environmental conflicts situated in Lapland. The first set of data is drawn from my doctoral thesis, which dealt with disputes about harnessing the Ounasjoki River and building a reservoir, Vuotos, for hydropower production.[7] As part of my study, I interviewed 21 local activists who were engaged in the disputes. Nine of the activists supported the hydropower construction, twelve opposed it. The interviewees were the core activists in the disputes, mainly men (18), middle-aged or older, and living in rural Lapland. In the interviews we discussed not only the conflicts but also the human-nature relationship on a more general level.

The second source of data for the present chapter consists of statistics collected by the regional authorities during the years studied, which give a picture of the significance of the different industries in the period after World War II.

Nature as a Space for Traditional Livelihoods

Traditionally, life in rural Lapland has been based on various uses of natural resources, such as small-scale farming, reindeer-herding, fishing, hunting and berry-picking. After World War II, over 400,000 people had to be settled in Finland and accommodated in Finnish society from areas lost to the Soviet Union. For example, in Lapland almost 2,200 farms were established on the basis of a land acquisition Act issued in 1945. The Finnish state supported agriculture by setting up new farms and giving land to the farmers in a policy that continued until the 1960s. In 1950, there were about 13,500 farms in Lapland and 56 per cent of the workforce was employed in agriculture. In 1959 there were about 4,850 farms more, making a total of 18,350. As a matter of fact, Finland is the only Western European country in which the number of farms grew after World War II.[8]

---

[7]   Leena Suopajärvi, *Vuotos ja Ounasjoki-kamppailujen kentät ja merkitykset Lapissa* (Rovaniemi: University of Lapland, 2001).

[8]   Uuno Varjo, *Lapin maatalous 1950–1959* (Rovaniemi: Lapin seutusuunnittelun kuntainliitto, 1967), p. 26; Lapin Seutukaavaliitto, *Väestö ja työvoima Lapissa 1960–2030* (Rovaniemi: Lapin Seutukaavaliitto, 1988), p. 68; Massa, *Pohjoinen luonnonvalloitus*, pp. 205–207.

In hindsight the policy of resettlement can be criticized. New farms were set up in remote marshlands, where the conditions for farming were harsh. The policy had been based on a scientific optimism regarding the development in cultivation techniques and in plant improvement but also on the fact that the state owned most of the forest areas in Lapland, where settlement was easier than on privately owned fertile banks of the rivers.[9]

The new farms were small and life was harsh; agriculture was not a sufficient source of livelihood. For example, according to research done in eastern Lapland in the early 1970s, most of the households in the area received income from seasonal forest work in addition to small-scale farming. The subsistence economy was also an important source of livelihood: hunting, fishing and berry-picking accounted for about seven per cent of the households' income. At the time, researcher Timo Järvikoski was of the opinion that the share of additional income from natural products was even bigger but that households underestimated the proportion in their responses.[10]

The regional development between northern and southern Finland was totally different in the decades after World War II. While agriculture was still growing in the North, the development of industry and the service sector were in full swing in the South. Opportunities for employment grew more rapidly in southern Finland in the period 1950–1975.[11] In Lapland, the combination of farming, subsistence economy and seasonal forest work finally broke down at the end of the 1960s. When the postwar settlers got old and retired in the last decades of the 1900s, their children usually refused to continue farming or did not have the financial possibilities for it. The result was rapid migration to the South, which was later called the "Great Move" because in the period 1967–1969, for example, the net migration from Lapland was almost 14,000 people. In the space of a few years, Lapland lost one tenth of its population because there were no jobs available in industry or the services and primary production had lost its importance rapidly.[12]

The decline of agriculture in rural Lapland has continued until the present time. In 1975 over 19 per cent of the workforce was working in that sector; five years later, the figure had fallen to below 16 per cent. According to the statistics of the Regional Council of Lapland at the end of the last millennium, there were about 2,000 farms left and primary production accounted for 6.4 per cent of the

---

9   *Ibid.*, pp. 207–208.

10   Timo Järvikoski, *Kemihaaran allasalueen väestötutkimus* (Helsinki: Vesihallitus, 1975), p. 44.

11   Tapani Valkonen, 'Alueelliset erot,' in Tapani Valkonen *et al.*, *Suomalaiset. Yhteiskunnan rakenne teollistumisen aikana* (Juva: WSOY, 1985), p. 228.

12   Lapin lääninhallitus, *Lapin läänin kehityksestä 1970- ja 1980-lukujen vaihteessa* (Rovaniemi: Lapin lääninhallitus, 1982); Lapin Seutukaavaliitto, *Väestö ja työvoima*.

jobs in the province.[13]

A specific feature of Lapland's primary production is reindeer herding, which is still an important way of life, especially in the remote areas in the northern and eastern parts of the province. In the mid 1990s, about 1,000 families earned their basic income from reindeer herding and another 800 obtained an important part of their living from it.[14]

Although most Laplanders make their living in the services nowadays, nature continues to have an important role in northern life. Nature-based hobbies are popular in Lapland. According to a study done in 1991, almost every fifth adult in Lapland had been hunting during the previous year. Every second Laplander had been fishing in the previous year and almost all households were using and even selling berries that they had picked themselves.[15]

The Finnish sociologist Pertti Rannikko has discussed the differences between the rural and urban human-nature relationship. In his view, in rural areas people's relationship to nature is functionally organized and based on activity, whereas in towns people see nature largely in terms of aesthetic experiences and recreation.[16]

In my own interviews as well, local people stressed the practical dimension when talking about nature. Childhood memories and experiences of fishing and berry-picking trips form the basis of the local cultural meanings attached to nature. When local rural people talk about the environment, they are speaking about their own environment, not nature on a general level. For them nature is their home area – the surroundings they have experienced, where all their memories are present and the work of earlier generations lives on. Local people have learned during their life to collect and use the products of nature, and these practices continue. In the following excerpt, one of the activists who opposes Vuotos tells about the meanings she attaches to the local nature:

> Our skiing tracks have been there in Kuikkavaara; we have been skiing there; and I have skied there always. And then there's fishing and swimming and picking cloudberries. My aunt and I have been fishing; the two of us were fishing with a hook and line. And you could see when the fish bite; the water is that clear. And everything. Of course there was also work planting trees further away in the Vuotos area. You have seen how important it has been to Father; he always talks about it.

---

[13] Lapin lääninhallitus, *Lapin läänin kehityksestä*; Regional Council of Lapland, *Statistics*, at < http://www.lapinliitto.fi > .

[14] Kajala, *Lappilaisten näkemyksiä metsien hoidosta ja käytöstä*, pp. 46–7.

[15] *Ibid.*, pp. 49–54.

[16] Pertti Rannikko, 'Ympäristötietoisuus ja ympäristöristiriidat,' in Timo Järvikoski, Pekka Jokinen and Pertti Rannikko, *Näkökulmia ympäristösosiologiaan* (Turku: Turun täydennyskoulutuskeskus, 1995), pp. 78–9.

The Laplanders whom I interviewed also stressed that knowledge about living in the northern nature is a skill that cannot be learnt from books. This view is strongly expressed, for example, in the discussion of nature conservation in Lapland. Local people argue that southern conservationists do not really understand life in the North. For example, here is the viewpoint of a man who supports hydropower construction on the Ounasjoki River:

> And if we start nature conservation, it will mean that we don't accept that human beings live in the area. This is the difference. And, and my basic idea is that ... I am such a conservative that I demand, and think it is right, that people who live here, who are born here, can still continue to live here and work here. And, and ... now that right is becoming unclear, disappearing, if we accept this kind of one-sided nature conservation. And it is certain that Laplanders, most of the Laplanders, want to protect nature. They don't want to exploit this nature. We are losing if we, for example, cut down the trees in our backyards. That's obvious.

In conclusion, for the traditional livelihoods, for small-scale farmers and locals engaged in a subsistence economy, nature is defined by the practices learned during one's lifetime. Nature is a place to live.

Nature as a Resource for Modern Utilization

After World War II, the Finnish state began to utilize Lapland's natural resources extensively. The nation-state recovering from the war needed its northern resources as well if it was to get the wheels of the metal and paper industry turning. The powerful prime minister Urho Kekkonen, who would later become president, wrote a program for the development of Northern Finland in 1952. In his view, the state had to take the leading role in the utilization of the natural resources in the North, because there were not enough private funds for the work.[17] This is in fact what happened, and state-owned companies took the leading role in harnessing the northern rivers for hydropower production and in forestry and in mining as well.

Finland had lost about one-third of its hydropower production to the Soviet Union, and the resulting lack of energy was hindering industrial development and modernization, for example, electrification of rural areas. The wood-processing industries and the state wanted to get domestic energy so that Finland would not be dependent on expensive foreign fuel and on international trade cycles. With the southern rivers already in hydropower production, the harnessing of the Kemijoki River, the longest river in Finland, began right after the war. The state made a

---

[17]   Urho Kekkonen, *Onko maallamme malttia vaurastua?* (Helsinki: Otava, 1952).

contract with the power company Pohjolan Voima Ltd. providing that the company would get permission to build a dam and hydroelectric plant at the mouth of the Kemijoki if it agreed to build bridges over the river. The bridges had been destroyed in the last phases of the war, and the reconstruction of Lapland was delayed because building materials were hard to transport across the wide river.[18] The Isohaara power plant was taken on line in 1948 and marked the starting point of the hydropower construction project along the Kemijoki. To date, 19 power plants have been built on the river and its tributaries along with two large reservoirs on the upper course of the river. Construction continues, and the harnessing of the Kemijoki has become one of the largest hydropower building projects in Europe.

The northern forests were also taken into production after World War II because Finland had lost one-tenth of its forest resources and one-fourth of its paper industry in the war. New pulp mills were established throughout the country, including Lapland, where improved roads and timber floating had made large-scale logging feasible. The state has always been the biggest land-owner in Finnish Lapland. After the war, it owned almost 80 per cent of the land in the province and thus there was not even any need to bargain over wood prices with farmers. In the 1960s, over seven million cubic meters of wood, especially pine, were cut every year. In recent years, the amount has been about 4.7–5.2 million cubic metres per year. Environmentalists have argued that logging has exceeded the limit of sustainable growth in the region. On the other hand, experts in forestry economics have argued that because of intensive forestry, which includes afforestation, the forest resources in Lapland are growing faster than ever.[19]

In the mining sector, gold has been the most important metal found in the region. In 1868, there was a gold-rush to the northern parts of Lapland, and prospecting continues to this day. A few mines have been opened – and closed – during the decades after the war but, according to Ilmo Massa, until the 1980s experts had estimated that Lapland was not notably rich in ore.[20] At the time of writing, there is only one active metal ore mine (Cr) in Lapland, in Elijärvi, but deposits have been discovered that support the prospect of opening new mines in the region. The reason for the revival of mining is that the research and knowledge on the richness of the region's soil has increased since the Geological Survey of Finland opened its regional office and began systematic study of the area. For example, gold, copper and nickel deposits have been identified. In addition, there

---

[18]   Kustaa Vilkuna, *Lohi. Kemijoen ja sen lähialueen lohenkalastuksen historia* (Helsinki: Otava, 1975), p. 387.

[19]   Varjo, *Lapin maatalous*, p. 24; Lapin lääninhallitus, *Lapin läänin kehityksestä*, p. 15; Massa, *Pohjoinen luonnonvalloitus*, pp. 225–8.

[20]   *Ibid.*, p. 248.

is a plan to open a new mine in central Lapland, where ores belonging to the platinum metals have been discovered.

Although the environmental history of Lapland since World War II is a story about deriving maximum profit from natural resources, the state has justified its policy in terms of regional development as well. For example, the then prime minister Urho Kekkonen argued strongly that the goal was also to industrialize and modernize the country's northern province.[21] In his view, utilization of the natural resources, large construction projects and logging sites were in fact necessary for development and economic well-being in the area. For example, there were 2,500 employees working at the site of the Petäjäskoski power plant at the height of construction in the 1950s. In addition, the services and trade received a boost from the large job sites, and municipalities got additional tax revenues.

Industrialization in Lapland never succeeded as the state had hoped; there was neither private capital enough in the region for competing in the market nor an industrial tradition in the modern sense. In fact, Lapland remained a sparsely populated area. Of course, modern technological development did reach Lapland, with tractors, excavators and chainsaws increasingly replacing manpower, horses, spades and saws on the job sites. Technological development required professional workers, and hence the unskilled workforce, mainly small farmers, began to lose seasonal job opportunities. The rapid development of modern equipment was one reason for the high unemployment in Lapland in the late 1960s and early 1970s, when tens of thousands of young Laplanders – mainly children of smallholders – had to migrate to southern Finland and to Sweden to seek employment. There were not enough jobs for the baby-boom generation born right after the war.

Despite the change from manpower to mechanized work in the utilization of natural resources, there are still many people in rural Lapland who think that intensive use of natural resources, for example harnessing the northern rivers for hydropower production, is the main solution to keeping the remote areas alive. Here is the opinion of one of my interviewees, who supports the construction of hydropower as a solution for employment:

In this unemployment situation and misery that we are experiencing, domestic energy production should of course be built and we would have jobs here and energy, too.

Although the utilization of natural resources has been essential for economic development in Lapland, there have also been critical voices raised against this policy. Intensive forestry has had serious effects on the northern nature. The forest areas have been treated merely as wood production sites, and biodiversity has been diminishing. Pine has been the main forest product, with other species losing

---

21 Kekkonen, *Onko maallamme malttia vaurastua?*

ground. Peatlands have been drained, and clear-cut areas have become extensive indeed after years of intensive forestry. Laplanders have lost their rich hunting areas and bountiful berry places to the interests of large-scale forestry. Yet there are Laplanders who have lost their home areas in more concrete terms. In the late 1960s, two large reservoirs were built at the headwaters of the Kemijoki River in order to maximize hydropower production in the power plants on the lower course of the river. The construction of the reservoirs, Lokka and Porttipahta, resulted in the inundation of an area 630 thousand hectares in size. Almost 600 people, mostly farmers, reindeer herders and fishermen, had to move away.[22]

The construction of hydropower on the Kemijoki River also destroyed the traditional fishing culture throughout the river basin. When the Isohaara dam and power plant were built at the mouth of the river at the end of 1940s, salmon could no longer get upstream. The local population tried to get compensation from the state for this loss and, after 30 years of struggle, occupied the Ministry of Justice in 1979. Compensation was paid after this, but the Laplanders felt they had received no more than pocket money, which was hardly enough to compensate for the loss of the salmon and the traditional fishing culture.[23]

Laplanders felt helpless against the alliance of the state and large-scale industry, which justified the exploitation of the region's natural resources in the name of the national interest. Local criticism against the so called "public interest" emerged in the last years of the 1970s, when there arose local movements against hydropower construction. The members of these movements shared the idea that most of the benefit derived from Lapland's natural resources had mainly profited southern industry, "big money" and the state, whereas all of the disadvantages were suffered by the local people. As one put it:

> It was not until then that we began to ask, for example, among the people who were against the building of Vuotos, if it really is in the public interest and what that public interest is that tramples on the local peoples' possibilities to earn their livelihood, and even hinders the local decision-making. The decisions are made elsewhere, not locally by local people ... And things are promoted that have not been proven to be useful for the local people.

In eastern Lapland, where Vuotos was to be built and which was also an area of intensive logging, the local people who opposed intensive utilization of the natural resources thought that the local structure of industry and employment was

---

[22] Massa, *Pohjoinen luonnonvalloitus*, pp. 230–32 and 242–3.

[23] Timo Järvikoski and Juha Kylämäki, *Isohaaran padosta Kemijoen karvalakkilähetystöön. Tutkimus Kemijoen kalakorvauskiistasta* (Turku: University of Turku, 1981).

becoming one-sided as a result of ecological colonialism. As one resident of the area observes:

[The Vuotos area] is a part of the way of life that Laplanders have relied on. And if you take such a large part away, it will destroy the sources of livelihood which have been available in the area. And although the reservoir would not drown houses, it will hurt our sources of livelihoods badly. Which, to tell the truth, are already threatened by other activities that the state is carrying on. In any case, it is speeding up the development that is detaching, for example, the reindeer herders from their tradition.

From the local point of view, one problem is that the degree of refinement of Lapland's natural resources is still low. Although natural resources are heavily utilized, refining industries have not developed as well as in the other parts of the country. In 1999, in the remote areas of Lapland, *i.e.* in half of its municipalities, the rate of refining was under 15 per cent, whereas the average for the entire province was about 22 per cent and for Finland as a whole about 27 per cent.[24] In conclusion, after World War II Lapland's natural resources were harnessed for the needs of a growing industry and nation-state, and nature was seen as a resource for economic development. Environmental considerations and critical opinions of the local people were neglected until the end of the 1970s, when local action against hydropower construction emerged in the region with the support of a rising national environmental movement.

Nature as an Attraction for the Tourism Industry

Before World War II, Lapland was an attraction in small-scale tourism providing outdoor activities for the Finnish upper class. The war destroyed practically the whole infrastructure of tourism, and, for example, important tourism sites such as the Petsamo region and Sallatunturi Fell were lost to the Soviet Union. Although reconstruction of the skiing centers begun after the war, large-scale construction in the tourist industry did not start until the 1960s, when the building of the eight main tourist centers in different parts of Lapland's fell area began. The rising standard of living, increased leisure time and private cars had put Lapland within reach of the middle class.[25] The real boom in tourism in Lapland began in the 1980s. A period of economic growth and the growing popularity of downhill skiing prompted the construction

---

[24] Regional Council of Lapland, *Statistics*.

[25] Harri Hautajärvi, *Lapin läänin matkailuarkkitehtuurin historia* (Oulu: University of Oulu, 1995).

industry to invest in holiday building in Lapland, and tourist centers grew rapidly. In the 1980s, tourists' overnight stays in Lapland grew by eight per cent annually. This figure does not include, for example, tourists who had bought their own holiday cottage in Lapland.[26]

According to statistics collected by Statistics Finland, in 1990 there were about 1.3 million registered overnight stays in Lapland and in 2000 as many as 1.6 million. Nevertheless, although the volume of tourism has grown, employment in the industry has not developed correspondingly. In 1990, there were about 6,200 man-years in the business, whereas nowadays the estimate is about 4,000. This decline is due to efforts in the industry to achieve optimal efficiency. For example, tourism frequently offers only seasonal employment, with employees being laid off during the low seasons.[27]

Nature is the main attraction for tourists coming to Lapland. Both Finnish and foreign tourists say that the clean, beautiful and peaceful nature is the most attractive thing in the region. In the whole industry, nature-based tourism, such as skiing, hiking and fishing, is growing rapidly.[28]

In regional plans and political programs, tourism is seen as a cornerstone for prosperity in Lapland in the future, but tourism also alters the northern nature. Skiing centres in Lapland are nowadays like small cities, with hundreds of cottages and hotel rooms, restaurants, shops and slalom slopes. Hannu Hautajärvi, who has studied the architecture of Lapland's tourism-related construction, has criticized the way in which the skiing centres have been built. He asks who would benefit from the leisure accommodation that has been built in the North in recent years. Luxurious units, often built as investments, stand empty – yet heated – most of the year and are wholly contrary to the principle of sustainable development.[29] Littering and erosion are also some of the marks left on nature by the growing tourist business.[30]

From the local point of view, large-scale tourism entails other disadvantages as well. Tourism is expanding its area in nature; local people sometimes feel that they are being pushed out and that they will have to give up their traditional usufructuary right. For example, reindeer herders have questioned why they and

---

[26] Massa, *Pohjoinen luonnonvalloitus*, p. 219; Hautajärvi, *Lapin läänin matkailu-arkkitehtuurin historia*, pp. 140–43.

[27] Massa, *Pohjoinen luonnonvalloitus*, p. 219; Liisa Kajala, ed., *Lapin metsästrategia* (Helsinki: Maa- ja metsätalousministeriö, 1996), pp. 42–4; Regional Council of Lapland, *Statistics*.

[28] Kajala, *Lapin metsästrategia*, p. 44.

[29] Hautajärvi, *Lapin läänin matkailuarkkitehtuurin historia*, p. 142.

[30] Massa, *Pohjoinen luonnonvalloitus*, p. 219.

their source of livelihood have to make room for tourists who are hiking in the wilds 'just for fun'.[31] Another important question is the distribution of the profits from tourism. Some of my interviewees took the view that tourism centers do not increase prosperity in the surrounding area. As one commented:

> Do they count that there will be so many tourists that we will all make our living serving them? I don't believe it. I don't even know if we could, if our mentality is obliging enough. We are sulky sorts. And it wouldn't benefit every village. They will die out, those remote villages. There might be one or two tourist centers in a municipality and some small ones, but they will not put bread on everyone's table.

Despite some critical voices, tourism is an important industry for economic development in Lapland. The relatively clean and peaceful nature is the main reason for tourists' coming to Lapland. Nature is an attraction, but there is also a dilemma involved: how can we increase the number of tourists but, at the same time, keep the damage to nature limited.

Nature as a Conservation Area

Local public action against the exploitation of natural resources in Lapland emerged in the end of the 1970s. In 1979 and 1980, people in the Ounasjoki River valley and eastern Lapland, where the Vuotos reservoir had been planned, organized themselves in local committees and began an active campaign against hydropower building. Laplanders wanted to defend their home areas and traditional sources of livelihood. For years local people had seen changes happening in their surroundings, but without having any real possibility to influence the decisions made.

Authoritarian planning and decision-making by the state and powerful companies had embittered the minds in Lapland. One example of the hydropower builders' lack of concern for the consequences in the area is a story about the maps used in the planning process. The power company was presenting the plans for the construction of the Ounasjoki River, which included maps describing the changes in the river valley. But the scale of the maps was so large that entire villages were covered when the planner drew the new waterline. Local people did not even get detailed maps showing the anticipated loss of land, nor did they receive information about changes in the natural conditions.

---

[31] Piia Varanka, *Lappi matkailun näyttämöllä. Saamelaiskulttuuri ja luonto matkailun kulisseina* (Rovaniemi: University of Lapland, 2001), p. 95.

Both movements were successful and reached their goal of protection of the river nature in the early 1980s. In 1982, the Finnish government decided that the Vuotos reservoir would not be built.[32] The Ounasjoki River was protected by a more official proceeding, with the Finnish Parliament passing a special Act to that end in 1983.

At the regional level, the success of the local movements can be attributed to the rise of a regional consciousness at the end of the 1970s. Local people began to criticize the bureaucratic state and a national discourse about a unanimous Finnish nation that had prevailed since the war.[33] For example, the regional campaign for salmon compensation was covered constantly in the regional newspapers, and demands for the rights of the local people received a great deal of support. Examples of people who had lost their homes because of the building of the reservoirs Lokka and Porttipahta also embittered the people:

> People had to obey, because they were that war-time generation who had to do what they were told. But then, when we heard examples of what had happened there, we here [in the Ounasjoki valley] did not obey any more. We had learned that you have to fight for yourself, and they'll take everything if you don't.

At the national level, an awareness of environmental issues also grew among the general public because the Finnish media were reporting more and more about environmental problems and pollution. Among the nationally important background forces for the Laplanders' demands were several environmental protests around Finland, in which local people rose up, for example, against aerial herbicide sprayings of willow sprouts or the draining of the lakes. In addition, the Finnish Association for Nature Conservation was strengthening its organization at the regional and local levels. Environmentalists and local people found themselves on the same side in those days, and also in the cases of the Vuotos reservoir and the Ounasjoki River, local committees for the protection of the areas received support

---

[32] The building of the Vuotos reservoir again came to the fore in the political debate in the end of the 1980s. The Finnish government gave permission to build the reservoir in 1992. The case went to appeal and the Supreme Administrative Court is currently (2002) reviewing the case.

[33] Anssi Paasi, *Neljä maakuntaa: maantieteellinen tutkimus aluetietoisuuden kehittymisestä* (Joensuu: University of Joensuu, 1986), p. 206; Esa Konttinen, 'Uusien liikkeiden tuleminen subjektiviteetin puolustamisen kulttuuri-ilmastossa,' in Kaj Ilmonen and Martti Siisiäinen, eds., *Uudet ja vanhat liikkeet* (Tampere: Vastapaino, 1998), pp. 198–9.

from the national environmental movement.[34] Environmental issues became part of the political agenda in Finland at that time. The Ministry of Environment was founded in 1983 and the Greens had representatives elected to Parliament in the same year. The first nature conservation areas in Finland were founded in the early 1900s, but it is especially since the beginning of the 1980s that the legislation for conservation of specific nature types has increased rapidly. Because the state has owned most of the area in Lapland, it has been quite easy to reserve areas for conservation. Nowadays, about 90 per cent of the conservation areas in Finland are situated in Lapland. The European Union's Natura 2000 conservation program now encompasses more than 2.9 million hectares of conservation areas in Lapland, about one-third of the area of the province. The areas proposed for inclusion in the Natura 2000 Network mainly consist of existing conservation areas, wilderness areas, and sites covered by already existing protection programs.[35] There are different levels of protection: for example, there are some areas where the public are not allowed to go, but conventional protected areas, such as national parks, are open to all. Indeed, they are an important tourist attraction in Lapland.[36]

Although nature protection has advanced rapidly in the northern parts of Finland, environmentalists have criticized logging in the old forests in Lapland. In the late 1980s and early 1990s, environmentalists' criticism was translated into episodes of protest in Lapland's forests, when some of them tried to prevent the felling by non-violent action. The parties in these confrontations were environmentalists, local lumberjacks or logging machine operators, the police and the national media, which reported widely on what was a new kind of action in Finnish environmentalism.

An important change can be seen in environmental protests in Finland when comparing the late 1970s and early 1980s with the early 1990s. In the first phase, local people and environmentalists usually shared the concern about the protection of nature and hence were on the same side. In the conflicts in the early 1990s, according to the media coverage, the environmental activists and the local people were on opposite sides in the confrontations. Pertti Rannikko describes the situation as follows:

---

[34] *Ibid.*; Pertti Rannikko, 'Ympäristökamppailujen aallot,' in Ari Lehtinen and Pertti Rannikko, eds., *Pasilasta Vuotokselle. Ympäristökamppailujen uusi aalto* (Helsinki: Gaudeamus, 1994), p. 13.

[35] Lapland Regional Environment Centre, *Networks of Conservation Areas*, at <http://www.vyh.fi/eng/environ/naturcon/natureco.htm>.

[36] Kajala, *Lapin metsästrategia*, pp. 10 and 27–9.

In order to understand this we have to take into account the world of the young educated urban reporter, where a certain 'anti-bumpkin discourse' has played a key role in the recent years. This world sees rural people as strong, politically and occupationally well-organized group which has been better able to protect its own interests than other groups have.[37]

According to my own study, there were three gaps separating the local people and environmentalists in the environmental conflicts. The first was a generation gap. In the disputes in the early 1990s, activists in the environmental groups were usually young people, whereas the local people involved were usually in late middle age. The local, older generation was born during or right after the war and had experienced the total deprivation that followed the war in the region. Also, because of the constant unemployment in the region, finding work was (and still is) a very important goal for Laplanders. Hence, it was difficult for older people to understand why youngsters, of the same age as their own children, tried to prevent them from doing their work. The younger generations involved in these confrontations were the first generation in Finland who had grown up in a society where information about environmental problems, environmental activism and environmental politics was a constant topic of public discussion.

Secondly, a gap was also evident between rural and urban residents. Rural people questioned the right of urban people to interfere in their life-style. In their view, urban people themselves live the "easy life" with all the modern technology and services readily available to them. As a matter of fact, environmentalists' practices in urban life appear to be quite anti-environmental when looked at from the rural perspective. On the other hand, as already argued, rural people are often described in the national media as an army of lumberjacks or farmers who are only interested in their own economic well-being and financial gain.[38] This is true in the sense that local people want to make their living in their home area, but there are not many job opportunities to choose from in rural Lapland. Then again, an image of greedy rural people has arisen because the media tends to build antagonism in news coverage. It is important to keep in mind that the local people are not a homogeneous unit; rather, there are also different opinions in the countryside about nature conservation. For example, children and women are seldom heard in the media when nature protection is being discussed.

---

[37] Pertti Rannikko, 'Local Environmental Conflicts and Change in Environmental Consciousness,' *Acta Sociologica*, Vol. 39, No. 1 (1996), p. 67.

[38] *Ibid.*, p. 64; Kirsi-Marja Poikela, *Leipää vai luonnonsuojelua: sanomalehtikirjoittelun näkökulma Pohjois-Suomen vanhojen metsien suojeluohjelmaan* (Rovaniemi: University of Lapland, 1998), p. 79.

Thirdly, environmentalists who wanted to protect the old forests in the North in the confrontations in the early 1990s were usually from southern Finland. From the local point of view, they represented a continuation of regional inequality: first, the state and southern industry exploited northern natural resources and then, with the wave of environmentalism, southern people wanted to protect the northern nature. The surroundings of the local people were constructed as "wilderness", *i.e.* pure and untouched nature, despite the fact that people had lived in the area for centuries.

According to the surveys, Laplanders are more critical towards nature conservation than people living in southern Finland. More than half (55 per cent) of the people in the province think that the level of nature protection is already sufficient. Laplanders are also more often worried about the effects that nature conservation could have on employment and the standard of living in the area. However, this is not the whole truth: about one-fourth (27 per cent) of the people in the region support more conservation.[39]

My own data indicates that criticism has basically been directed against the ways in which nature conservation has been carried out in Lapland. Local people have felt that they have not been heard in the planning processes. On the other hand, they think that there should be compensation for the restrictions and losses which might affect the local people following the establishment of new conservation areas.

Conclusions

Finnish Lapland has gone through a rapid socio-economic change during the five decades since World War II. In the beginning of the 1950s, Laplanders mainly lived from a combination of small-scale farming, seasonal forest or construction work and a subsistence economy. At the beginning of the new millennium, the main industry is services, comprising a combination of public services and tourism.[40] Despite the rapid changes in society locally, the northern nature has maintained its central role in life in Lapland. Almost half of the Laplanders still make their living in occupations which are related to nature.

As I have argued, different industries have different kinds of interests in the northern nature. For the local people, who practice the traditional sources of

---

[39] Kajala, *Lappilaisten näkemyksiä metsien hoidosta ja käytöstä*, pp.19–20; see also Rauno Sairinen, *Suomalaiset ja ympäristöpolitiikka* (Helsinki: Tilastokeskus, 1996), pp. 136–9.

[40] The share of the service sector is over 80 per cent of the jobs in some municipalities, when the average in Lapland is about 67 per cent and in the whole country about 65 per cent. See Regional Council of Lapland, *Statistics*.

livelihood, nature is a place to live in, whereas for the forestry, mining and energy businesses it is merely a resource. The tourism industry in Lapland bases its marketing on experiences of and in the northern nature. It is difficult to combine the interests of a subsistence economy, industrial utilization of natural resources, and tourism, given the conflicting relationships to nature of these competing perspectives. From the local residents' point of view, the problem has been that the economic capital and national interests have had hegemonic positions in the decision-making concerning the northern nature. Nevertheless, the role of the local people has become stronger in the last few years, with changes in the legislation and planning processes. For example, residents are now among the parties heard in the environmental impact assessment procedure. In all, the environmental history of Lapland is an example of the fact that environmental issues are not situated at the margins of the society. Environmental questions are questions of power, equality and the future, *i.e.* questions of the basis on which the future will be built.

Chapter 11

Can Europe Be Told from the North? Tapping into the EU's Northern Dimension

Pertti Joenniemi[1]

The Challenge

The European Union has been furnished with a Northern Dimension (ND). The initiative, taken originally by Finland in 1997, has landed on the Union's agenda yielding policy documents, high-level conferences and some projects pertaining to Europe's North. It outlines, in terms of the spatial markers used, a sphere that reaches far beyond the northernmost North. The initiative aims, in one of its aspects, at turning northernness into a representational frame and regime that nurtures communality and influences the relations between the Union, its northern member states, some accession countries and Russia as well as Norway as non-applicants. The neo-North embedded in the move offers a joint arena for those already 'in', actors on their way 'in' and the ones that remain 'out'. In essence, it mediates in their relations and contributes to what Christiansen, Petito and Tonra have called the "fuzziness" of the European Union by blurring established divisions.[2]

This chapter probes, using the initiative as a starting-point, into the question whether Europe can be told from the North. It seeks to explore the constitutive aspects of the discourse waged around the Northern Dimension, including what remains in the shadows and what is obscured from sight. The North is hence not approached as a marker with a given content and unproblematic status. Instead, the aim is to expose its open, contingent and unstable nature and, in that context, view

---

[1] I would like to thank the European Security (EUR) group at Copenhagen Peace Research Institute for comments on a draft version of this contribution.

[2] Thomas Christiansen, Fabio Petito and Ben Tonra, 'Fuzzy Politics Around Fuzzy Borders: The European Union's Near Abroad,' *Cooperation and Conflict*, Vol. 35, No. 4 (December 2000), pp. 417-31.

critically the ontological and methodological orientations that have pertained to research focusing on the initiative.

In its final part the chapter moves on to explore the vistas of the EU – and Europe more broadly – against the backdrop of changes in the representational frames underlying the European configuration. In particular, the question is pursued how these vistas resonate with a marker such as the North. It is assumed that some of them are more open to northernness informing policy responses whereas others reject such a marker out of hand as a non-European departure. The catapulting of northernness into a legitimate departure in the context of the Union's representational politics should – against this background – be quite indicative as to the balance between the different configurations and the way their relationship is unfolding.

Success or Failure?

One of the obvious issues surrounding much of the ND-related debate pertains to whether the initiative has been successful or if it should be predominantly regarded as a failure. Those with a positive attitude towards the initiative claim that it has clearly advanced since having been presented in 1997.[3] The way the initiative has been presented has stayed within the limits of what is acceptable in the discourse on the EU and it has become legitimate to 'speak' Europe from the point of the North. It is stressed that the initiative has been recognized and provided with some status. Consequently, the Northern Dimension no longer stands out as a proposal advanced by a single member state. Instead, it figures as an item common for the Union as a whole constituting a theme on the Union's agenda to be tackled and followed up in a variety of ways.

But these achievements notwithstanding, the question remains whether the initiative has been a successful one. Making it in to the sphere of intra-Union diplomacy does not seem to satisfy the critics.[4] For example, there are no mechanisms of negotiation that would allow for a constant exchange of views, and hence progress remains very much dependent on the will and the diverse interests

---

[3]   See Lassi Heininen, 'Ideas and Outcomes: Finding a Concrete Form for the Northern Dimension Initiative,' in Hanna Ojanen, ed., *The Northern Dimension: Fuel for the EU?* (Helsinki/Bonn: The Finnish Institute of International Affairs/Institut für Europäische Politik, 2001), pp. 20–53.

[4]   See various contributions in Gianni Bonvicini, Tapani Vaahtoranta and Wolfgang Wessels, eds., *The Northern EU: National Views on the Emerging Security Dimension* (Helsinki/Bonn: The Finnish Institute of International Affairs/Institut für Europäische Politik, 2000).

of the countries in charge of EU-presidencies. David Arter remarks that often the reception of the initiative has been "lukewarm". At worst, "it was regarded with cynicism and suspicion."[5]

The proponents have mostly been on the defensive in lacking broadly plausible arguments (installing a new spatial marker into the discourse is taken to be too abstract) that would work in favour of the ND whereas it has been relatively easy to gain support for the opposite view. The initiative, seen as a routine move by one of the member states, thus remains quite contested. The dominant view seems to be that for it really to count, the initiative has to yield quite tangible results in terms of institutions or specific processes. The media, for one, has been rather suspicious about the move. The landing of the initiative on the EU's agenda is not viewed as constituting proof of its significance and, in fact, the opposite is held to be true. The Northern Dimension is quite often seen as an expression of provincialism and regarded as a kind of 'noise' from the fringes with the rule being that the EU is spoken into existence at the core. More specifically, the initiative is comprehended in terms of a conduit that quite harmlessly channels some worries specific to the EU's northern borderlands.

The conspicuous pessimism as well as the predominance of rather critical appraisals in the ND-related discourse calls, no doubt, for interpretation. Moves of closure tend to prevail, and they do so particularly in the public debate. But what explains the negative mood that seems to dominate much of the commentary? More precisely, what are the discursive codes that go against 'speaking' North in a broader European context in the first place? Why is the Northern Dimension not given the benefit of doubt instead of jumping to conclusions already at an early stage of its lifecycle? At least two different reasons appear to account for this state of affairs.

It seems that the debate has been rather shallow in historical and fixed in ontological as well as epistemological terms. The North's history has been largely forgotten, *i.e.* there is no drawing upon the past as a discursive resource. There has been no reminding of that the marker constituted, up to the Napoleonic wars, a core pole in the construction of Europe. It has a long history of constituting an internal – and not external – marker of Europeanness instead of being seen as a pristine, God-given and autonomous frontier to be continually pushed back towards the fringes. Yet only scant efforts have been launched to bolster the credibility of

---

5   David Arter, 'Small State Influence within the EU: The Case of Finland's "Northern Dimension Initiative",' *Journal of Common Market Studies*, Vol. 38, No. 5 (December 2000), pp. 677–97.

the North as a relevant representation by the utilization of its considerable temporal heritage.[6]

More generally, the marker remains embedded in perceptions of immobility and permanency. It is depicted, similarly to the other main markers on the compass, as being frozen, fixed and pre-set as a natural geographic marker. Given its position as part of the 'deep structures', northernness is so firmly naturalized and sedimented that it is difficult to comprehend that in the end also the North forms a discursive construct with changing boundaries and meaning. It is conducive to change over time and may therefore also be opened to influence as a geopolitically informed marker of power and space. It bends to performative discourses. Having been northernized, *i.e.* pushed further towards the edges, and emptied of its previously rather rich political, social and cultural content, it may, at least in principle, also be re-furnished with such qualities.

The inclusion of a genealogical perspective – one providing the history of the present in terms of the past – and breaking with the prevailing essentialist and naturalizing tendencies obviously leads to more general questions about centrality and peripherality – or, for that matter, the core departures in the construction of political space. If centrality and peripherality are not given and geographically fixed properties but qualities that change over time, what makes for a centre and how did a marker such as the North turn into a sign of utmost peripherality? How was it deprived of its previous political and social content and what are the dynamics and changes detectable over time? Moreover, could the current post-Cold War situation stand for a 'formative moment' and invite for a new 'regime of truth' – to employ some postconstructivist terminology – allowing the North to re-conquer some of its former positions in a Europe with region-building as a major constitutive principle? Questions pertaining to genealogy seem important, but so far they have been largely absent from the discourse, including the scholarly efforts of placing the Northern Dimension and the 'neo-North' in perspective.

This is to say that there seems to be some friction governing interpretations as to the significance of the initiative. This, then, appears to account for the somewhat shallow nature of the debate and the lack of historically informed insight. Although it is readily admitted (with all the talk concerning the new situation that has emerged with the fall of the walls) that Europe now harbours an exceptional openness and might experience a change of historical proportions, there has hitherto been little confidence in that some of the answers could come from the periphery and pertain to markers such as the North. This is nothing surprising as

---

[6]   For an exception, see Marko Lehti, 'Competing or Complementary Images: The North and the Baltic World from the Historical Perspective,' in Hiski Haukkala, ed., *Dynamic Aspects of the Northern Dimension* (Turku: Turku University, Jean Monnet Unit, 1999), pp. 21–45.

such taking into account that as to the choice of representational frames, it has almost without exception been the core that has defined the edges.[7] The self-evidence of sovereignty has been beyond doubt and there has been very little space for regionality as a contending principle. The core preserves its sovereignty inasmuch as it copes with the administering of existence and is able to use its right of defining order, including the usage and meaning given to the cardinal geographic markers. This has been the normal state of affairs and a revision would constitute, it appears, a radical break with far-reaching consequences.

There is, consequently, a need to explain away and downgrade the ND. The North constitutes – according to a dominant reading – a site imbued with passivity, escapism, dependency and a considerable degree of helplessness. It is the land of the shadows and death, or more concretely, a source of troubles rather than an asset and a resource to be drawn upon in coping with radical change and in the construction of a new, post-Cold War Europe. The idea of a broad North co-constituting what Europe and the EU is about, lacks in credibility. It hence tends to be foreclosed that such a marker – signalling the effort of bringing about a European mega-region – could suddenly provide the ground for some autonomous initiative.

It is hence to be expected that the debate has predominantly focused on the specific processes of promoting the ND and has been rather factual, statist and outcome-oriented. The stress has been on the instrumental rather than the ideational. This also applies to the scholarly discourse where the North has been viewed and tested, it appears, against a factual background rather than regarding it as a representation and a kind of lens or prism through which the world is being viewed. What escapes scrutiny, therewith, is that the installing of a new and different lens may, as such, have a rather profound impact. It reshuffles the comprehension of basic issues, and does this quite apart from anything factual. Switching over to a new lens makes Europe look different: Some things are magnified and become more visible whereas others become shaded and slip out of sight.

In fact, the lens metaphor appears to be very apt here; the initiative has provided – one may claim – the North with an improved standing as a coinage through which the environment is viewed. The marker works in terms of framing. Northernness has advanced – in this perspective – by gaining an accepted standing within the confines of the EU as a site of power. This, then, appears rather significant as such and has to be accounted for. It has to be tackled as a generative move in the sphere of speaking power and space, and not only evaluated in view of the consequent, and often more specific processes set in motion or the 'hard',

---

[7]   The theme has been elaborated by Ronny Ambjörnsson, *Öst och Väst. Tankar om ett Europa mellan Asien och Amerika* (Stockholm: Natur och kultur, 1994).

empirically measurable outcomes. A move pertaining to codes impacting the very construction of the EU has been made by one member country advocating the introduction of a different lens, and the discursive rules governing what can be legitimately purported as a basic departure have proved to be flexible enough for the Union to approve of northernness as a representational frame and ordering devise.

Postmodern Voices

Scholarly analysis surely deviates from the ND-coverage and debate in the media. A fuller and more nuanced picture has been provided, although also the research focusing on the ND seems to call for a decoding as well as an exploring of its shades and shadows.

What seems apparent is that the many postmodernist IR-scholars actively engaged in the debate about the Baltic Sea[8] and Barents co-operation[9] have been less observant regarding northernness and the Northern Dimension. They clearly departed from that the Baltic and Barents configurations are not natural but rather arbitrary in essence. They are premised on regionality and yet constructions worth supporting. However, in the case of the ND, the choice has been one of remaining at the sidelines. The features of multiperspectivism, the aims of transcending borders and paving space for non-state actors, or a number of other unmistakable features of postmodernism embedded in the initiative have not tempted analytical curiosity. The efforts of breaking out of the confines of the traditional geopolitical and realist understandings that characterized the Cold War period have not been extended further, and there has actually been scant interest in the Northern Dimension among those that have previously engaged themselves in combining postmodern scholarship, regionalism and politics. The 'school' – as it has been called – tends to view region building as a practice instead of regarding regions and related markers simply as (re-)discovered entities. They have contributed to the regionalist 'revolt' that has taken place in Northern Europe over the past years and

---

[8]   See, among others, Ole Wæver, 'Regionalization in Europe – and in the Baltic Sea Area,' in *Co-operation in the Baltic Sea Area. Report from the Second Parliamentary Conference at the Storting, Oslo, 22–24 April 1992* (Stockholm: Nordic Council, 1992), pp. 16–21; Ole Wæver, 'Culture and Identity in the Baltic Sea Region,' in Pertti Joenniemi, ed., *Cooperation in the Baltic Sea Region* (New York: Taylor & Francis, 1993), pp. 23–48; Iver B. Neumann, 'A Region-building Approach to Northern Europe,' *Review of International Studies*, Vol. 20, No. 1 (January 1994), pp. 53–74.

[9]   See Olav Schram Stokke and Ola Tunander, eds., *The Barents Region: Cooperation in Arctic Europe* (London, Thousand Oaks, New Delhi: Sage, 1994).

feel, in general, at home with cascading identities as well as marginal, obscure, anti-heroic and openly constructivist themes such as the North.[10] They have shown themselves to be quite fast in sensing new options and chances open to spectacle, but have nonetheless abstained from embracing northernness, the ND and the role of regionality in the context of the EU. As testified by Iver Neumann:

> A tightly knit epistemic community of 'Nordic' foreign policy intellectuals played a conspicuous role in producing the knowledge that was used to prop up these several ideas. The battle-cry was that under the prevailing postmodern conditions, state sovereignty is relativized in favour of a new European medievalism where different political issues are settled on different political levels.[11]

The question thus emerges what might explain this relative neglect among scholars that have been less constrained by rules that in general apply to IR and foreign policy research. The omission is worth exploring at some length as it might help to bring into the open some of the limitations embedded in the discourse on the EU's Northern Dimension.

Obviously, several explanations are on offer, although interpreting the silence of the 'school' is a speculative endeavour as such. There might, despite the fact that the ND appears in general to be a theme preferred by the postmodern regionalists, be the feeling that the groundwork has largely been done and the novelty of constructing regionalist regimes may simply have worn off. There is no need for repetition or projection into an even larger sphere. The analysis already carried out may be taken to be equally applicable to the ND as the Baltic and Barents projects (although it used to be assumed that the path of broadening would be one of the Nordic region extending into a Baltic direction).[12] No extra effort is called for – despite the fact that the Baltic and Barents projects focus on carving out specific regions whereas the Northern Dimension is more about the constitutive principles in the context of the EU and Europe at large as such. This is so as the train was set in motion already some time ago. The postmodern scholarship restrained itself, it seems, to the more short-term issues and regionalist vehicles with concrete visibility instead of dwelling upon a broader, more abstract and sedimented signifier such as the North.

---

[10] See Ole Wæver, 'The Baltic Sea: A Region after Post-Modernity?' in Pertti Joenniemi, ed., *Neo-Nationalism or Regionality: The Restructuring of Political Space around the Baltic Rim* (Stockholm: NordREFO, 1997), pp. 293–342.

[11] Neumann, 'A Region-building Approach,' p. 67.

[12] See, for example, Wæver, 'The Baltic Sea' and Lehti, 'Competing or Complementary Images'.

It could also be of relevance that Finland, as a state, succeeded in making a move prior to the scholarly community having developed ideas regarding the North as a more general vision and policy frame that reaches beyond the already existing regional arrangements. This, then, left scholars in a somewhat inconvenient position. They had either to stay aloof or assume a position of backing an idea and a proposal put forward by a particular statist actor. And what may have further contributed to the unease consists of the initiative being set up in a manner that immediately pushes things over for the EU to settle. It reduces, one could suspect, the influence of the local actors with societal concerns replacing statist ones and, instead, invites the Union to increasingly influence, steer and strengthen its grip on the Baltic, Barents, Nordic and Arctic constellations.

The emergence of the neo-North could also be seen as something unwarranted in inviting the EU to pursue politics towards homogeneity and uniformity. It would hence work by narrowing down the space available for the regionalist 'revolt' – one usually celebrated by the postmodernists – to unfold. The ND could be viewed, if this is the way the initiative functions, as a disciplinary kind of move, one aiming at rolling back the recent "fuzziness" rather than contributing to it. The initiative aims, as concluded according to one analysis at "the construction of a unified international actor, a modern subject with eternal essence." It is, as such, admitted that the move of bolstering northernness might contribute to the erasure of the reified East-West border and invite for a vision of 'Europe' understood as increasingly open to diversity. However, the verdict comes in the end down on the side of seeing the ND as "being highjacked and subsumed within a more traditional modern discourse."[13] The initiative is, along the lines of such critique, too much championed for its tendencies of harmonization and control: It devises political space among clearly outlined and exclusive political units, the 'ins' and the 'outs', and does so without really allowing multiplicity and overlapping configurations to unfold. It is criticized, more generally, for its rather modern aim of exporting some of the already existing qualities further on towards the fringes of the European configuration.

This view appears to be justified in the sense that the EU assumed, at least initially, a rather passive stand *vis-à-vis* the Baltic and Barents configurations, although this position has changed over time. A somewhat similar pattern is to be traced in the case of the ND. The European Commission appeared, at first, hesitant about the initiative but then made it part of a policy that aims at the construction of common European space by a flow of pre-given policies that merely flow in one direction, from the core outwards.[14] Regionality has been used to bolster the EU

---

[13]   Christopher Browning, *The Construction of Europe in the Northern Dimension* (Copenhagen: Copenhagen Peace Research Institute, 2001), pp. 30 and 48.

[14]   *Ibid.*, pp. 17–29.

as it is, not in terms of alterting the Union.

The 'school' may also find an excuse in that the Northern Dimension is too much top-down and pre-given in character. It can be seen as quite establishment-driven, openly political, centralized and distinctly lacking in the bottom-up type of engagement and activism that has been to some extent present both in the cases of the Baltic and Barents Cupertino. A further reason could be that there is uncertainty about whether such an umbrella concept is really needed. Why to advocate a departure that is not explicitly tied to regionalism or linked to a specific regionalist project? It may also be that representations such as the North are taken to be too imbued in peripherality and emptiness in order to have a chance. It is too cumbersome to problematize their ontological status that tends to be perceived as given and natural. Being located at the deepest level of discursive structures, it is not conducive to change. The 'school' has a reputation of focusing on "change within continuity", and the "branches" of the discursive "tree" more than the "trunk", the latter being regarded as too sedimented and frozen.[15] There might, due to such a tendency, exist a rather profound epistemological kind of disbelief in northernness in the first place.

It has to be added, however, that the discursive rules underlying the postmodern approaches in no way prohibit a probing into the larger issues underlying the formation of political space that loom in the background. The theories of the 'school' are, in principle, able to cope with change and deal with cardinal markers of political space – as exemplified by Iver Neumann's work on the relationship between Russia and the eastern marker.[16] A tackling of the EU-related initiative by elevating northernness into a core issue is by no means off limits, although the extension of the 'liberating moment' of the postmodern turn is still to be extended to include also the neo-North.[17]

Mainstream Voices

Be it as it may, the scholars that have taken an active interest in ND tend to be part of the academic mainstream. They have, for the most part, a background in various

---

[15]   See Thomas Diez, 'Europe as a Discursive Battleground: Discourse Analysis and European Integration Studies,' *Cooperation and Conflict*, Vol. 36, No. 1 (March 2001), pp. 5–38.

[16]   Iver B. Neumann, *Uses of the Other: The 'East' in European Identity Formation* (Minneapolis: University of Minnesota Press, 1999).

[17]   See, for example, Browning, *The Construction of Europe*, p. 3.

international and foreign affairs institutes.[18] This state of affairs has contributed to the ND being largely trapped within a modernist discourse. Regionality has not figured as an issue and constitutive principle worth serious attention. Those parts of mainstream research that have engaged themselves actively in the theme have even shown some understanding *vis-à-vis* the initiative, though within bounds. Their aim has not been one of contributing to a disempowerment of the initiative from the very start and the verdict is still out as to whether the record is one of success or failure. The standard position appears to be that it is too early to tell.[19] Some progress has been detected, although the achievements are taken to be rather modest. There has been little to report once the searchlight has been focused on various ND-related processes and the measurable outcomes. For example, with money being used as a yardstick of success, the Northern Dimension has basically been depicted as 'cheap talk'.

These 'yes-but' answers seem to originate with researchers joining other types of commentators in focusing on various EU- and state-related processes pertaining to the initiative. Scholars, then, largely share the view that judgement has to be related to interests, actorness, processes and measurable outcomes. The ND undoubtedly has some potential, albeit it has to mature – so the argumentation goes – into a budgetline of its own, provide some concrete structural outcome or amount to tangible projects. In other words, the regulative rules of the discourse tend to downgrade the more long-term, visionary and framework-related aspects of the ND and instead direct attention towards the more concrete, short-term and project-like aspects. A typical claim consists of arguing that the initiative remains 'stillborn', 'a bag of hot air' and stands out as 'lip service' if it does not yield concrete and empirically measurable results. The politics of naming and the 'talking' of the ND into existence, *i.e.* that of exploiting the symbolic potential of the North in the EU's establishment of frames of representation, is not seen as constituting a policy in any true sense of the word.

What lurks behind an elevation of the initiative's various institutional and material aspects and, conversely, a downplaying of its symbolic and ideational aspects pertaining to policy perspectives, seems to be an anchorage in the usage of a rationalist and rather individualized frame of analysis. The North is seen as given

---

[18] For a couple of contributions from the critical geopolitical school that go against the tide, see Anssi Paasi, *Territories, Boundaries and Consciousness: The Changing Geographies of the Finnish-Russian Border* (Chichester: John Wiley & Sons, 1996) and Sami Moisio, 'Pohjoisen ulottuvuuden geopolitiikka: pohjoinen periferia ja uuden Euroopan alueellinen rakentaminen,' *Terra*, Vol. 112, No. 3 (2000), pp. 117–28.

[19] See, for example, Esko Antola, 'The Presence of the European Union in the North,' in *Dynamic Aspects of the Northern Dimension*, pp. 115–32; Hiski Haukkala, 'Introduction,' in *ibid.*, pp. 9–20; Arter, 'Small State Influence within the EU'.

to experience in terms of objective reality, and perceived as being open to investigation as such. The stress on objectivity is understandable in the sense that it guards against accusations of siding with some particular political agenda and engaging in 'speculative' type of endeavours. The assumption tends to be, it seems, that there has to be a pre-set package of interests pertaining to some specific actors, and this, then – once detected and analyzed – yields the ground for unbiased evaluation. And in order to be on the safe side in the arriving at conclusions, there has to be a final desired outcome that is in line with what is to be predicted on the basis of such a package of interests. Void of any final outcome, analysis focuses on the various processes and the way the interests involved materialize in the context of these processes.

Such an objectivist fixation implies, in one of its aspects, that there has been little interest in looking beyond the formal diplomatic process and the states as rather contained and pre-defined actors. The cardinal markers of political space and constitutive principles are considered as settled and unchanging. They are seen as *fait accompli*. No efforts of comparing the workings of northernness in an EU-context with those discernible in some other fields, for example tourism, advertising, photography or literature (recent experiences indicating the North marks a growth industry) have so far seen the light of the day. There has, in a similar vein, been a lack of interdisciplinary approaches using cultural geography or anthropology as inroads into enquiry – to name two disciplines that have been more sensitive to borders, bordering and the employment of non-statist representational frames. Should the traditionally derogatory image of the North be preserved in order to secure the dominance of the Western marker, or should the "dead lands" of the North be allowed to be part of "the new Europe"?[20]

There are, along the more objectivist and interest-oriented lines, studies for example on the Danish, Estonian, Latvian and Swedish attitudes *vis-à-vis* the Northern Dimension.[21] However, northernness has, as a representation, been of

---

[20] See W. Brian Newsome, '"Dead Lands" or "New Europe"? Reconstructing Europe, Reconfiguring Eastern Europe: "Westerners" and the Aftermath of the World War,' *East European Quaterly*, Vol. XXXVI, No. 1 (March 2002), pp. 39–62.

[21] On various country-specific contributions, see, for example, Bertel Heurlin, *Denmark and the Northern Dimension* (Copenhagen: Dansk Udenrigspolitisk Institutt, 1999); Nils Muižnieks, 'A European Northern Dimension – A Latvian and a Swedish Perspective,' Report from a seminar organized by the Olof Palme International Center and the Latvian Institute of International Affairs, Riga, Latvia, 10–11 December 1999; Kristi Raik, 'Estonian Perspectives on the Northern Dimension', in *Dynamic Aspects of the Northern Dimension*, pp. 151–66; Jennifer Novack, 'The Northern Dimension in Sweden's EU Policies: From Baltic Supremacy to European Unity?' in *The Northern Dimension: Fuel for the EU*, pp. 78–106. For various national views, see also *The*

little interest within such a setting. It has been objectified by being depicted as a wrapping or by viewing the marker as an element needed in the marketing and selling of the ND-package. The factual approach implies that the constitutive aspects of the Northern Dimension have remained largely unnoticed.

It seems that the frame used has, in general, been constructed in a manner that does not allow for the empowering of the North as a lens or a prism, *i.e.* a representation with an impact that reaches beyond the instantly factual, institutional and process-oriented. The linkages between the level of symbols, facts and interests have remained poorly developed. Representations such as the North have not been singled out and placed in perspective as something that also incorporates and influences the way interests are formulated in the first place. This lack of comprehension appears to explain why the northern element, though an integral part of the initiative, has for the most part been side-tracked. It has stayed void of attention and does not gain the eminence it deserves as a departure allowing, at least in principle, speaking 'Europe' also from a peripheral perspective. The usage of the Northern marker in an EU-related context appears to be largely buried in silence. It is seen in factual terms instead of being tackled head-on by enquiring into its role and meaning, including the denaturalization and re-politicization underway that paves the way for different ways of conceptualizing boundaries and political space, and thereby also a variety of 'Europes'.

Such a down-grading of the initiative's symbolic aspects and endeavours of policy framing also clarifies why there has been scant interest in the numerous clashes that emerge once northernness gets elevated and is invited to compete for space and attention at the expense of some more established representations. What gains the upper hand, the Nordic/Scandinavian or the northern departure, what is the North's relationship to representations such as the Baltic, Barents or the Atlantic and how do the various relevant actors devise their own favourite version of the North once it has been established that the North has turned broader in scope than previously and actually constitutes one of the key sites in the debates on the unfolding of the post-Cold War Europe? If not a binary code (and the other of the South), how does the North relate and resonate with the other cardinal markers of the East, West and South? Finland – and perhaps also Norway and the US – favours representations that open up the North towards the East and Russia.[22]

Northern EU: National Views.

[22] See Peter van Ham, 'Testing Cooperative Security in Europe's North: American Perspectives and Policies,' in Dmitri Trenin and Peter van Ham, *Russia and the United States in Northern European Security* (Helsinki/Bonn: The Finnish Institute of International Affairs/Institut für Europäische Politik, 2000), pp. 57–88; Edward Rhodes, 'The United States and the Northern Dimension: America's Northern European Initiative,' in *International Perspectives on the Future of the Barents Euro-Arctic Region*

Sweden, Germany and the Baltic countries tend, for their part, to prefer a Baltic North instead of the northern North, and Denmark has also opened up for an Atlantic (western) vision of the North (although Denmark primarily operates with a Baltic North) as indicated by the ministerial conference to be organized on Greenland during the Danish EU-presidency towards the end of 2002 on the Northern Dimension under the rubric 'the Arctic Window'. Denmark also intends to ask the Commission to prepare a report of the Arctic Region and revise the Action Plan. This contest and the consequent diversification of the North has so far largely escaped analysis as northernness as a representation or a frame has ranked relatively low to start with.

A High Level of Ambitions

For sure, also the empiricist, individualist and rationalist approaches carry some weight. The initiative has, after all, been launched by Finland and there are, indeed, national interests in the background. The statist actors involved have for a large part seen – as testified by Nicola Catellani – the initiative as something short-term, interest-oriented and a policy tool to be utilized particularly in the context of issues pertaining to the financing of various projects relevant for northern Europe.[23] The EU deals with the initiative in intergovernmental terms, it is part of the Union's external affairs, and the ND has, in some of its aspects, a project-like nature either stressing the need to fix various economic, social and environmental problems residing in northernmost Europe or focusing attention on the region's exceptionally rich natural resources. There is much in the logic and the unfolding of the initiative that lends itself easily to a rationalist, empiricist and interest-oriented scrutiny.

Hanna Ojanen, for one, follows such a line of enquiry.[24] However, it is worth noting that she also extends her analysis in some ways beyond these more ordinary departures by stating that the gist of the initiative consists of a Finnish effort to "customize" the EU, *i.e.* to mould it to one's own liking. The effort is to provide "the Union with a more Finnish face". Notably, she does not downplay the role of Finland, a peripheral actor and a newcomer to the EU, and narrow the move

---

and the Northern Dimension, Report from a think-tank seminar, Björkliden, Sweden, June 2001, pp. 41–56.

[23] See Nicola Catellani, 'The Multilevel Implementation of the Northern Dimension,' in *The Northern Dimension: Fuel for the EU,* pp. 54–77.

[24] Hanna Ojanen, 'How to Customize Your Union: Finland and the Northern Dimension of the EU,' in *Yearbook of the Finnish Institute of International Affairs* (Helsinki: Finnish Institute of International Affairs, 1999), pp. 13–26.

down to interests, specific projects, financial schemes and short-term policies. Instead, room is provided for issues pertaining to policy framing. The initiative is upgraded and approached as something rather ambitious with Finland endeavouring at influencing the very figure of EU. Ojanen furnishes the initiator with exceptional subjectivity as to the possible outcomes – as does also David Arter in a somewhat similar study on Finland and the Northern Dimension. Finland's aim in launching the initiative has, in his view, been one of "building the political capital with which to relocate Finland from being a new and geographically peripheral to become a core EU member."[25] Finland is seen as having turned innovative and influential "on a wider agenda". More broadly, Arter explores a theory of change in arguing that there is a comparative advantage in being "small but smart" (in relation to large countries that tend to ride on continuity and strength in terms of resource-power). This combination of small and smart might account for success encountered in the process of agenda-setting, although Arter deems that Finland has been less effective when it comes to implementation of the ND in more concrete manner.

Both Ojanen and Arter provide the impression, although without exploring the theme in detail, that an unusual twist is visible in the power-relations between the core and the periphery. Finland has been empowered to think and operate in a long-term fashion and to pursue policies on the level of symbols and broader policy frames that pertain to the overall figure of the Union. It has become conceivable that a peripheral actor located at the Union's outer border may exercise considerable influence. The initiator has been able to gain leverage due to the fact that the EU's new border overlaps with Finland's own external border. A small and peripheral (*i.e.* not central in terms of westernness) member – and therefore *a priori* powerless according to the standard conceptualizations and theories – is not merely small but may potentially (if also smart) turn so influential that it becomes somewhat difficult to account for such a leverage. Yet Ojanen claims, through the terminology she uses, that Finland is not only adapting to the EU but also influences – by combining location with a load of ideational resources – what the EU is about. Finland is, in other words, interfering with the core's sovereign rights of ordering, and has been able to do this without being immediately rejected and penalized (which would have been an outcome much more easy to comprehend and explain after Finland having committed the 'sin' of breaking with the discursive rules that largely underlie the construction of 'Europe' as political space). It has been allowed to disturb the core and meddle with the key markers and constitutive issues pertaining to policy framing, and thereby drag its 'provincial' concerns far beyond the usual limits.

---

[25]	Arter, 'Small State Influence,' p. 695.

The term 'customizing' appears to indicate that the ambitions of the initiator are quite far-reaching and that Finland has a vision and an image of what to aspire for. The policies pursued do not just pertain to something narrowly Finnish or entail Europe's northern North. Both these aspects are included, but the initiative is more ambitious. The reading could be – in reflecting upon the 'mystery' articulated by both Ojanen and Arter – that the legitimacy gained by the initiative is tantamount to the core having lost – or abdicated – some of its constitutive power. The EU has less of a monopoly of classifying, exercising agency in the defining of order and deciding upon the choice and application of basic departures in terms of the cardinal signifiers of geography, this bolstering the position of the margins and opening up for a more multiperspectival Europe.

Ojanen's key concept points to that by operating in the margins and leaning on regionality – by changing one's own as well as the EU's approach to bordering and thereby influencing what the margins are about – the horizon of what may be politically achievable alters. A different EU – one being more democratic and de-centred – becomes imaginable, and actors located at the margins are therewith provided with the option of taking the lead. Moreover, the EU unfolds differently with the installation of a new prism, and within such a setting – concerning the future figure of Europe – Finland gains in subjectivity. The neo-North removes, as to Europe's symbolic order, Finland from being located at a fixed edge and places it, instead, at the uncertain and changing margins. With these margins not yet categorically defined, actors such as Finland are equipped with the legitimate right of influencing what their borders are about and how they function. However, it is above all the Union's tolerance towards rapid regionalization as well as its enlargement and reaching out to the applicants and some of the non-members that unsettles these borders in the first place. In impacting on the openness that is there, and utilizing their own location at the margins, even small actors may gain an opportunity of influencing what the broader European constellations are about. They can turn into 'Europemakers' by fusing a small amount of themselves into the Union, thereby moulding a Union more to their liking, that is 'customizing' the Union in Ojanen's terminology.

This is to claim that marginality does not amount to powerlessness. It may on good grounds be theorized, as has been done by Noel Parker, as a site of power.[26] Such a perspective offers insight into the change underway and some explanation concerning the question why it has become possible to talk about Finland 'customizing' the EU in the first place. The empowering rather than disempowering of the initiator is in tune with such a theory. However, the options

---

[26] Noel Parker, 'Integrated Europe and its "Margins": Action and Reaction,' in Noel Parker and Bill Armstrong, eds., *Margins in European Integration* (Wiltshire: Macmillan Press, 2000), pp. 3–27.

opening up are left without further exploration as Hanna Ojanen and David Arter both refrain from elaborating and probing into the workings of marginality, the impact and meaning of regionality or analyzing the visions that Finland might be utilizing in its 'customizing' of the Union. What kind of Europe and EU does Finland actually aspire for? To what extent should the EU be premised on regionality? What are the alternative models and how do they play out in the context of the ND? The claim is there that Europe is changing and that even peripheral actors such as Finland may, at best, succeed in imposing their mark on the way the European configuration is unfolding (or is it actually the EU taking over a previously oppositional and uncontrolled North?). Stepping beyond some of the ordinary discursive rules of EU-research allows Ojanen and Arter to grasp these questions, although the more precise workings of such a 'miracle' is left largely unexplored and there is no further clarification as to the potential outcomes of the initiative.

In one of its aspects the background to such a shortcoming pertains to that also Ojanen focuses largely on outcomes and processes (she has also published a different, broader and EU-centred article on the ND).[27] The idea of northernness working primarily as a lens or a prism providing for different visions is exempted from her analysis. The claim that Finland gains influence above all by playing with representations that provide new visions, and uses marginality as a platform and site from which to inject these insights, is not sufficiently accounted for.

Yet 'customizing' hints properly at a turning of tables. It pinpoints, as a concept, a reversion in the core-periphery relations, albeit the general departure of Ojanen's analysis does not back up the usage of this rather intriguing term that is elevating Finland more or less onto a level *or par* with the Commission as to influencing the figure of the EU. Her factual, interest-oriented, statist and somewhat static approach is not conducive to any further exploration of the break-through and its background. The analysis does not focus on the play with symbols, the re-installation of northernness as a marker of Europeanness or the role of region-building, thereby omitting an essential aspect of the initiative. In any case, it seems clear that the country is not trading a previous *Ostpolitik* for a *Westpolitik* and riding on the conclusion that it has to turn western after having been liberated by the end of the Cold War from its previous associations with easternness. The point of departure is not that of Finland's national lion waving a (western) sword while standing on an (eastern) semitar. The representational frames are not seen as fixed, frozen or given beforehand. They are not cast in a binary fashion to start with, and hence the option of inventing a third way, one based on northernness, emerges. The move is strategic in essence as it entails an effort of influencing the

---

[27] Hanna Ojanen, 'Conclusions: Northern Dimension – Fuel for the EU's External Relations?' in *The Northern Dimension: Fuel for the EU*, pp. 217–37.

constitutive rules themselves by intruding into the very order-making, and thereby also stake out a posture that is more to one's own liking. More specifically, the northern representation includes the Finnish-Russian border that has for long been an essential aspect of Finnish identity politics as outlined by Anssi Paasi.[28]

The Ojanen-type of analysis overlooks, it seems, that the initiative yields a different Finland, one with new interests, a new location and new relationship to the core constitutive principles of political space. In other words, 'Finland' is not a pre-set entity with a fixed location but one aspiring to do away with its outer border as an edge and to strengthen the features of being located at the margins. The upgrading of the North into a representation that pertains to the European Union has been preceded, it appears, by an acceptance of Finland's own northern credentials. Such an enrolment in northernness then furnishes Finland with the power to catapult the same marker (and a piece of itself) into the discourse on the EU's approach *vis-à-vis* its margins. This is to say that an obscure representation, one that Finland was previously not at ease with, becomes recognized and consequently instigated – in the form of advocating a broader northernness – into the contest between different departures used in positioning the EU and Europe more broadly.

The move entails a breach in the symbolic order and the representational framework underlying Europe: two representations that have stood apart for long are invited to meet, clash and intermingle. There is a reversal of trends in the sense that Europe's traditional going North is called to a halt and the North is instead pushed towards Europe. It thus turns less insulated than it used to be. The North gains in leverage in being spoken of as constitutive departure (linked to regionality) and core signifier of political space. The effort challenges the long-standing divorce between the narrow, northern 'we' and a European one, thereby devising and opening up for a different, and joint 'we'. This broader 'we' may then also include actors that have for long been outside the borders of the more narrow and well-bordered 'we' (Russia being potentially a case in point). The move entails a new type of Europe, EU and Finland, all with flexible – and not fixed – external borders. And consequently, images of in-between spaces that add to the "fuzziness" of current Europe.

The choice of 'speaking' North is potentially quite powerful, although not immediately measurable in terms of short-term gains, projects, financial schemes, institutions etc. A Europe co-figured by the North is different from one without such a dimension in terms of basic policy frames and symbolic departures. Finland has perhaps not encountered immediate success in the efforts of installing a new representational frame and in moulding the Union's northern borders to its liking along the lines that the frequent talk in the context of the ND about "a Europe

---

[28] See Paasi, *Territories, Boundaries and Consciousness.*

without divisive borders" calls for.[29] However, already the achievements so far leave much to account for. Finland has managed to provide additional space and legitimacy for a vision of a Europe to its liking and to place the issue of representational politics on the EU's agenda. The endeavour to diversify the EU and to provide it with an additional constitutive marker has not been rejected out of hand. Quite to the contrary, the North has turned into an established and broadly agreed framework qualifying what the EU is about and where it is. The northern 'lens' is in place (although on terms very much dictated by the EU). The visions it provides are at work, although they are not immediately visible and measurable with the tools used by mainstream research.

Clearly, 'customizing' has much that speaks for it, but it also has connotations of the Union being somehow tricked into something that is more or less against the interests and the 'true' nature of the EU. It is to be observed, however, that the Union has - after some deliberations - accepted and incorporated the term. To describe the process as 'customizing' projects a rather passive EU. The approach downplays the question whether there could also be some profound reasons for the Union to accept the introduction of an additional ordering device and a change in its representational framework. After all, the EU is rather postmodern in essence and region-building constitutes an essential part of the spatial policies pursued. Yet it is purported as an entity not really in control of its representational politics and one that an equally permanent Finland then 'milks' by her assumedly skilful policies. The question is never put whether Finland has actually succeeded in proposing a change in the Union's representational framework that corresponds most profoundly to the Commission's own visions, needs and interests at this current juncture in international relations.[30] Does Finland aspire for an EU that is different from the current one in being increasingly premised on regionality, or is the initiative merely about giving a name to policies pursued by a union that is already there?

Notably, there is much that begs for the question of what has allowed the northern prism to gain in space in the first place. What explains the reversal underway with Europe no longer narrowing down and de-politicizing the North but the North taking an opposite turn by going Europe? Why do the discursive rules underlying 'Europe' suddenly grant space for deliberations based on North-speak? A considerable change seems to have taken place as to the relative weight of the different cardinal markers of European political space, and this is hardly anything

---

[29]   See Chris Patten, 'Statement on Transatlantic Relations. Speech to a Plenary Session of the European Parliament,' Strasbourg, 16 May 2001, at < http://www.Europa.eu.int/comm/externalrelations/patten/speech01204.htm >.

[30]   This view has been put forward by Catellani, 'Multilevel Implementation,' p. 71.

isolated but part of larger discursive whole that has to be accounted for in providing the ND with a more extensive and balanced background.

A State of Passage

The initial success of the Northern Dimension seems to pertain to some broader alterations on the European scene. The initiative takes advantage, one might argue, of what Jean-Francois Lyotard calls "the end of the grand narratives",[31] Edgar Morin characterizes as "the low ideological tide"[32] and some poststructuralists label as the formative moment. Europe's division, based on the idea of two mutually exclusive spheres, has taken a considerable blow. A rather totalizing symbolic order has lost, it seems, in credibility with the end of the Cold War.

But it is not just the binary divide between the East and the West that has encountered difficulties; also the Eastern and Western markers themselves appear to be under pressure. The Eastern one is ominously absent from the current debates on Europe's future. It is there but as something pertaining to the past or a marker that has moved so far to the East that it no longer qualifies in the positioning of a number of countries hence to be regarded as being part of Central Europe. Easternness has been narrowed down outlining the position of actors such as Russia, Ukraine and Byelorussia. It increasingly signals marginality and is also inflated as it plays, far less than it used to do, the role of West's constitutive Other. The West emerged, for its part, as a 'winner' of the East/West contest, but it has nonetheless been increasingly absent – as noted by James Kurth – from the more recent discourses in international affairs.[33] It does not frame, motivate and legitimate political action to the same extent as before, and 'Europe' appears to gain features somewhat distinct from 'West'. The Western marker is certainly on the scene and much of its constitutive power is intact, but the marker may no longer be constructed to the same extent as previously through a binary move by playing it against some assumed East. These changes seem to imply, on a more general note, that space has opened up alongside the two representations that previously outlined and framed rather categorically what the European political order was basically about.

Europe's new situation actually resembles, one may claim, the one that prevailed in Romania in 1989. The rebels then waved a national flag with a red star, the Communist symbol, but one cut out from its centre. This cutting left a

[31]   Jean-Francois Lyotard, *La condition post-moderne* (Paris: Minut, 1979).

[32]   Edgar Morin, *Penser l'Europe* (Paris: Gallimard, 1990).

[33]   James Kurth, 'America and the West: Global Triumph or Western Twilight?' *Orbis*, Vol. 45, No. 3 (Summer 2001), pp. 333–41.

wide hole in the very middle of the flag, the empty space epitomizing a brief period of openness and a wavering order.[34] The current state of passage and the uncertain standing of the long-dominant symbolic order appear to strengthen the hand of the margins. Essential borderlines have become unsettled between the core and what used to be the edge, as well as between the European configuration as a whole and its nearby areas. The question emerges: where is the core, where does Europe end and what are the constitutive rules applied to the fringes? This again appears to call for approaches that are in some case off-centre. There is suddenly value in viewing Europe from the margins instead of dwelling exclusively on the dynamics at the core. Focusing on the margins and the demise of essential constitutive borderlines such as the one between the concepts of Europe and the North, reveals that the 'hole' in the middle is accompanied by a considerable disorder and openness at the outer spheres. Exploring what happens once a previously rather detained North, one engulfed by its particularity, is suddenly set free from a number of constraints and allowed to go Europe, acquires actuality.

A focusing on the workings of marginality complements other studies as to the configurations emerging with the breakdown of previously well-guarded mental borders. The implosion of the old order certainly impacts on the established cardinal markers of political space, including the North. The search for meaning and durable ground that has followed the demise has mandated the incorporation of a marker that has, over a long period of time, stood out as being largely void of political meaning and constituted something of a blank space. There is thus room for the northern marker to grow in pre-eminence, and to do so precisely because of being light in content (in contrast to the South which is heavily loaded with pre-given meaning and burdened by history). It may increasingly outline a sphere of its own without just remaining a reflection of the East/West dominance and is provided with the option, it seems, of re-entering the European political scene by competing for the position of a 'quilting point' or '*point du capiton*'. Questions emerge whether this happens through a process of normalization (that is some parts of previous history re-emerging) or if the increasing pre-eminence of the North flows out from the particularity of the northern marker, and whether that particularity prevails. More generally, if the North is viewed as a mirror, what does it reveal and what are the messages written on its surface?

The North certainly remains engulfed with a considerable dose of scepticism as to its Europeanness, and it still lacks a naturalized position along the lines of the more permanent 'European' markers. However, these properties also render it quite conducive to a constructivist approach. The very processes of bolstering the

---

[34] See Slavoj Žižek, *Tarrying with the Negative: Kant, Hegel and the Critique of Ideology* (Durham: Duke University Press, 1993).

position of the North in regard to Europe provides– due to their openness and the uncertainty of the outcome – insight not only into the political employment of the key directional arrows of the compass but also into the more basic constitutive rules at work in the context of the unfolding of the new post-Cold War Europe.

The effort of reshaping the order amongst the dominant markers of political space obviously raises a host of questions. In the first place, is an obscure marker such as the North really able to expand and provide anchorage by knotting and subsuming the other, more established signifiers to such a degree that the North qualifies for a prominent position in the construction of a post-Cold War Europe? Is it up to the challenge as an increasingly significant representational frame?

The task is thus, in order to provide some answers, to furnish the northern marker with a background. It has to be contextualized in view of its current de-bordering and expansion. Answers are required as to the qualities and symbolic resources that may – or may not – allow northernness to link in, stretch out and regain some of its standing as a signifier that is in some sense constitutive to a political order no longer laid down and exclusively defined by the dominance of the binary East/West narrative.

A rather profound question consists of settling the basic nature of the North as a political marker. The story could basically be about the North contributing for its part, and alongside some even more central markers, to filling the 'hole' caused by the crumbling of a previous symbolic order. It would hence stand out, by providing anchorage, conducive to a restoring of the damage caused by the implosion. It could be about the return of normalcy, although the story is perhaps more about the increasing weakness of the previously dominant markers than the strength of the neo-North. In any case, the added weight of the northern marker appears to resonate with the demise of the old order. The North has been able to re-emerge, it seems, from the fringes and take advantage of being liberated from the dominance of the Eastern and the Western representational frames. The North has gained in subjectivity to the extent that it potentially constitutes an increasingly crucial element in remedying the European configuration.

The Heritage of the North

In fact, two very different explanations seem to be on offer in accounting for the recent inroads of the North. For one, the story could be one of recrudescence with the North being able to take advantage of its historical record. It serves, one may claim, as a reservoir of meaning to be employed in order to arrest some of the fluidity caused by the end of the Cold War.

And, as to this line of argumentation, there might indeed be good reasons for comprehending the North as an untapped storage of meaning. It is, in principle, well furnished with the option of appearing as a contender for the position of a cardinal signifier, *i.e.* a standing that it already had up to the Napoleonic wars. The historical North then outlined a far broader sphere than just the northernmost North – and some of the symbols involved have survived up to this day as indicated for example by the iconography related to the Russian bear or the Russian winter.

The North reaches, as to its temporal aspects, far back in defining otherness already in the ancient Greece and Rome, and centuries after that. Northernness was created to complement the South and had the function of delineating true cultural and economic backwaters. It ordered political space in constituting the South's Other, and stood out as an ambiguous and hostile sphere inhabited by uncivilized and rough barbarians. The North was comprehended as the land of the dark and unholy forces. Over time, the peripherality of the North turned milder and more positive images surfaced. Northernness became usable as a resource in the identity-building processes of the northern realms and nations located in the area. The North got contours as something Europeanized and was later also nationalized.

It was increasingly depicted – since the mid-17th century – as a political marker and seen as an organizing principle that outlined a playground of power politics. The prime actors of the game consisted of the then European major powers: Sweden, Denmark, Prussia, Poland and Russia. The (statist) usage of the North in configuring political space stood out as an integral part of the dominant discourse on international politics, one premised on a balance of power. The naming of 'the Great Northern War' demonstrates such tendencies quite clearly. International relations were comprehended as kind of Newtonian system that seeks rather mechanically its own equilibrium. This comprehension was applied to Europe as a whole but included also a number of subsystems, among them the northern one.

The North had, in its broadest form, a considerable coverage. It spanned half of Europe and it outlined – as indicated for example by August Ludwig Schlözer's *Allgemeine Nordische Geschichte* (1771) – the position of Germany, Poland, the Low Countries, Russia as well as the Nordic and the Baltic countries.[35] Such a reach had ancient roots within the Greek as well as the Roman worlds. Both these configurations were divided along a North-South rather than an East-West dichotomy.[36] The northern qualities were used to bolster the position of kings as well as tsars in European politics. Sweden's Karl XII carried the name of "a northern hero", Russia's Catrine the Second was seen as "semiramis des

---

[35]   See August Ludwig Schlözer, *Allgemeine Nordische Geschichte. Aus den neuesten und besten Nordischen Schriftstellern und nach eigenen Untersuchungen beschrieben und als eine Geographische und Historische Einleitung zur richtigern Kenntniss aller Skandinavistischen, Finnischen, Slavischen, Lettischen und Sibirischen Völker, besonders in alten und mittleren Zeiten* (Halle, 1771).

[36]   See Lehti, 'Competing or Complementary Images'.

Nordens", Nicolai I was interpreted as standing for "the Northern Star" and a Polish legion fighting the Napoleonic forces was known as the "Northern koloss".[37]

Northernness thus had a crucial role in outlining some aspects of Europe. This kind of usage of northernness as one of the master-signifiers of European political space dominated the political language of the entire 18th century, to fade out during the mid-19th century. For the contemporaries, St. Petersburg, Berlin, Copenhagen and Stockholm formed more or less one political scene. For example, when Alexander I appeared to help Europe to subdue Napoleon, he was seen as arriving from the North and not from the East.[38]

A contender emerged on the scene once the position of easternness was bolstered during Enlightenment. The eastern marker was extended to blur interpretations concerning the North. Yet, the transition from the dominance of the North to the broadening of the East spanned decades. A considerable amount of people, having the choice between the North and the East, saw themselves as belonging to the North still at the beginning of 19th century. The North dominated also the writing of history that remained quite state-centric up to the early 19th century. It seems that easternness, as an attribute of Europe, entered the discourse prior to the East being perceived as constituting a prime political – and not just a cultural – scene. The East and the North were initially not exclusive poles and their boundary remained vague for some time. Larry Wolff points out that these two cardinal markers overlapped, but he claims that there was a rather quick retreat of the North towards the position of a northern North.[39] Scholars such as Robin Okey support the claim that there was a considerable period of transition.[40] The Crimean War spurred, it seems, the emplotment of Russia as an eastern and not a northern actor, although the process remained incomplete at least up to the First World War with Soviet Russia being excluded – and excluding itself – from the rest of Europe. The Second World War to some extent blurred the picture but the Cold War re-confirmed the easternness of the Soviet Union. The resulting primacy of the East-West divide contributed, in one of its aspects, to a further northern-

---

[37] Hans Lemberg, 'Zur Entstehung des Osteuropabegriffs im 19. Jahrhundert. Vom "Norden" zum "Osten" Europas,' *Jahrbuch für Geschichte Osteuropas*, Vol. 33, No. 1 (1985), pp. 48–91.

[38] On the Eastern marker, see Robin Okey, 'Central Europe/Eastern Europe: Behind the Definitions,' *Past & Present*, No. 137 (1992), pp. 102–33; Mikko Kivikoski, 'Onko Itä-Eurooppaa enää olemassa?' in Pauli Kettunen, Auli Kultanen and Timo Soikkanen, eds., *Jäljillä. Kirjoituksia historian ongelmista. Osa 2* (Turku: Kirja-Aurora, 2000), pp. 165–6.

[39] See Larry Wolff, *Inventing Eastern Europe* (Stanford: Stanford University Press, 1994).

[40] See Okey, *Central Europe/Eastern Europe*.

ization of the North in a European context while the previous North-South axis was removed from European affairs – to be later projected on global politics as a whole.

The historical record could, one may argue, lay the ground for the argument that the North is on its way back. The claim could potentially be made that the North is returning in order to re-conquer the position of a core constituent marker of Europeanness, one loaded with relevant political meaning. However, such arguments are basically missing from the discourse. The neo-North does not seem to ride on claims based on historical grandiosity and also research has by and large pushed such arguments to the sidelines. Images of the former Hanseatic League played a crucial role in the debate about Baltic Sea Cupertino and stories about former Pomor trade surfaced in the context of initiating the Barents one, but the discourses on the neo-North have at least so far taken a different route as there appears no major need for transcending some previous negative emplotment of northernness.

A Constitutive Outside

Actually, the historical record could speak for a rather different interpretation. During the 19th century the North was gradually pushed to the sidelines by the increased prominence of both the East and the West. It declined in rank, ceased to be categorized as negative difference on a grand scale and turned, over time, into a marker of outmost peripherality. It was severed from being part of Europeanness in time (representing something left behind) and place (being a liminar to Europeanness). Whilst 'Europe' gained – with the maturing of the modern period – connotations of centrality and progress, northernness was restricted in meaning outlining a barren and hostile no-man's land. It was associated with remoteness as well as primitivity and furnished with perceptions of lagging behind. The North was seen as a void located at the fringes of what truly matters, *i.e.* a marker distinct because of its perceived emptiness and lack of meaning rather than being a reservoir of some particular meaning. The constitutive impact of the North declined and changed in essence as the North was primarily comprehended as a stranger to be kept within bonds, mastered and subordinated to the power of the constitutive markers of East and West. It was comprehended as an object to be consummated by the main cartographic attributes, and in general to be sorted out by conquering it to modernity. The dominance of the East and the West gave ground, for example, to the North being comprehended in Soviet eyes as "the Red Arctic". It was depicted as a sphere not to be let to its own devices but mastered, engineered and remade so as to be heroically readjusted to human (socialist and modernist) needs.[41]

The prevailing image has thus been, until recently, that Europe's North harbours – to the extent that these two mental constructs could be linked to each other in the first place – little subjectivity. It is not seen as being comparable with the core constituent markers of European political space. At its low the North was comprehended as something of a *terra incognito* that came to life only by being discovered, mapped, named and spoken into existence by explorers arriving from the civilized and developed areas, those that really count and are central in the process of providing meaning. It was allowed to stay on the scene but merely in the form of a supplement and by remaining external to the dominant markers of political space.

The North therewith performed, during the peak of the modern period, the role of forming what could be seen as the constitutive outside of the dominant constellation. It had the posture of a liminar marker reflecting merely local dynamics and, so to say, a rhythm distinct from the modern beat. Yet, although weak and subjugated, it could not be totally expelled. The North was, in fact, defined by a double move: It was excluded from centrality and prevented from acquiring a position *on par* with the dominant markers, but at the same time it had to be included and allotted some space of its own. Basically the North was there in order to define Europe by contrast; it was the 'moon' reflecting the light of the European 'sun'. There was a process of negation at work: the North was what the South was not. The North had to be included and accounted for – despite its nothingness, undifferentiated nature and potentially subversive character – as the presence of such an obscure marker testified to the completeness of the overall configuration formed jointly by the core directions on the compass. The dichotomic relationship was one of dependency as the confirming of the notion of completeness as well as the impression of firm foundations was rather essential to the standing of the more dominant markers themselves. The North had to be there for the others to have a relative advantage. The overall constellation hence enabled the northern marker to gain some standing of its own, but at the same time anything northern was pushed into an inessential and fringe position, remote both as to time and place.

The recent changes imply, if read against this background, that the North no longer remains as northernized (*i.e.* naturalized and just limited to the northernmost North) and subjugated as previously. Instead, it serves as a reservoir to be employed in order to arrest some of the fluidity and turmoil caused by the end of the Cold War. Northernness may become less of a supplement; it can expand in

---

[41] The story is given by John McCannon, *Red Arctic: Polar Exploration and the Myth of the North in the Soviet Union 1932–1939* (Oxford: Oxford University Press, 1998).

scope and be attached to themes that do not just pertain to Europe's past but also its future. Or to be more precise, it may be cast in a futuristic light as its anchorage in history – due to a profound marginalization – has turned quite thin. Despite its 'nothingness', or precisely because of it, the expansion of the North has profound consequences. Signifiers such as the North do not merely describe and position subjects that are already there. They provide – as already argued – a frame of interpretation and a departure to be used in the process of positioning oneself. This is far from a technical act as it is part of a process that produces a series of symbolic deeds around which identities as well as political and social reality can be constituted. To be defined by northernness – instead of some other cardinal marker – is a process that might pertain quite profoundly to the subjectivity of the different actors and institutions on the scene. For example, the position and the pre-eminence that the 'Nordic' concept enjoyed during the peak of the modern era as a kind of developed and rational non-North might be at stake (although in every-day language the difference between the Nordic and the northern has sometimes remained diffuse). The challenge appears to be there to a certain extent enforcing for example Nordic cooperation to subjugate itself to the EU's Northern Dimension – as advocated in a recent Nordic Council report on the Northern Dimension.[42] The challenge seems to be there despite of that the neo-North basically builds on the postmodern logic of multiplicity. There might also be some other identities, cultures and delineations that, dependent on a strict and well-bordered northern North, suffer from the Northern marker being set free and extended beyond its previously rather unpolitical features and limits of 'a last Frontier'.

Paradoxically, also the northernmost North might come out as one of those resisting the europeanization of northernness. Northernness has for long been furnished by connotations of being 'strong and free', thereby providing departures helpful in defying incursion and resisting conquest by those seen as harbouring exploitative and 'foreign' ambitions. The Union's Northern Dimension might be interpreted as yet another effort of undermining the North's capacity to stay aloof and counteract disciplining. The initiative is thus not seen as emancipatory in essence but rather interpreted as a move endeavouring at further undermining the very marker that has provided ground for the northern North's distinct and separate existence. In other words, it is not the northern North marketing its own qualities in terms of time and space to a broader audience. There are, in general, good reasons for examining closer who are the founding fathers of the neo-North, what functions the marker occupies in various contexts, and how the discursive power embedded in the marker works in relation to established forms of spatialization.

---

[42] See Nordic Council, *Öppet för världens vindar. Norden* (Copenhagen: Nordic Council, 2000).

The struggles that the neo-North introduces also pertain – even if tuned down in the context of pushing forward the Northern Dimension initiative – to the figure of 'Europe'. The initiative potentially invites for a Europe that revolves along new and different lines. It may thus be argued that it is not just the 'hole' in the middle of the post-Cold War Europe that is being remedied by providing northernness with an enhanced position. The northern contender is not to be compared with the effort immediately after the Cold War to re-launch the idea of a separate *Mittel-Europa*, an idea far more offensive and openly political by design. The latter coinage shattered, both intellectually and politically, the idea of Eastern Europe as well as raising troubling questions about centrality. It intruded into a variety of sensitivities related to who defines, and thereby dominates the centre. In addressing core constitutive questions in a far too direct manner, the idea of a narrow *Mittel-Europa* failed to gain ground, whereas the EU's Northern Dimension constitutes a far more subtle operation. It does not raise questions of centrality but pertains clearly to marginality, *i.e.* a less guarded and fixed discourse, although also the defining of marginality holds the option of having an impact on the overall European configuration.

A Possible Trajectory

The neo-North embedded in the ND-initiative seems to operate in a distinctly low-key fashion, and some of its working may hence escape attention. It aspires to stay at the margins and off-centre without raising questions about the core. There are no outright revanchist themes attached to it that would spark off any immediate alarm.

Moreover, the discourse pertaining to the neo-North does not have the form of either-or. It does not claim authenticity and aspire to lay the ground for some organic and exclusive community. Consequently, the problem of explaining the Northern Dimension tends to pertain to questions like 'where is the beef' and is there really anything to it, not suppositions about some heavy and disturbing content. It is merely purported as endeavouring at adding to plurality by introducing yet another trajectory to be explored in knitting together the Baltic, Barents and Arctic regional formations. The northern marker appears to re-enter the European scene as one option out of many without doing so in any categorical, explicitly political, revanchist or conflictual manner. It urges, no doubt, for subjectivity but does not aspire for such a standing by ousting some other markers. The aim is not that of swapping peripherality for centrality but rather one of proliferating peripherality through what has been aptly labelled as the "policies of emptiness".[43] Northernness flies, one may argue, in resonating with the multiplicities, fragmentations, overlappings and contingencies of so many other contemporary claims to political subjectivity. It feeds on the image of growing in the cracks of the previous East/West order and advances gradually as a domain that remedies these cracks without causing any broader disturbances and outright instability. It operates by being exotic, idealistic and harmless rather than purported as offensive, heavy in substance and thereby challenging.

There are thus reasons to argue that the overall process is not just about reshuffling the various constitutive elements within a fixed and pre-given symbolic order. The neo-North appears to be about deconstruction more than construction, and it operates in the context of globalization, regionalization, networking and localization rather than any domain defined by traditional statist departures. The insertion of northernness is not, if viewed in the perspective of change conceptualized as 'glocalization', only about slipping out of the East/West bind and assuming the role of a third. It basically acquires a looser relationship to the very frame of space and time expressed primarily through a positioning along the lines dictated by main cartographic markers. Or, stated in relation to the flag/hole metaphor: the problem does not just pertain to a 'hole' in the middle of the 'flag' that is on the way to being remedied. What could be at stake are the core constitutive principles and departures of the whole symbolic order that is the 'flag' as a cultural construct. The nature of the overall figure is less self-evident than it used to be. And, in such an indirect manner, also the neo-North pertains and feeds on broader questions of time and space that have popped up in the post-Cold War period.

Exit from the Bipolar Divide

It appears that northernness is quite slippery in content. The marker has – as observed by Sergei Medvedev – considerable features of obscurity and anonymity: "Whereas the East, West and South have more or less fixed meanings, and are interpreted as relatively populated and explored, the North appears as a mythological domain, a semiotic project, a constructed identity."[44] He claims that it "is more often communicated than experienced, imagined rather than embodied." The North constitutes, he argues, the furthest of all corners in the world: "It is the most elusive and the least circumscribed, an ill-defined space rather than a delineated place." It lacks, he asserts, in locality, territoriality, borders and other

---

[43]   Sergei Medvedev, 'The Blank Space: Glenn Gould, Russia, Finland and the North,' *International Politics*, Vol. 38, No. 1 (March 2001), p. 91.

[44]   *Ibid.*, p. 92.

accoutrements characteristic of what is labelled as "our rational geometrical civilization", and does this far more than the other master-signifiers of political space.

There are also other authors offering similar observations. For example Kenneth Coates states, in an article on the northern cultures, that the North "is as much a creation of the imagination as it is a physical or human reality."[45] This is, of course, not to deny the mythological aspects – and the constructed nature – for example of the West or the East. Both are rich in mythological content and have been coined mentally and intellectually before they have taken root as geographic and political departures. What singles out the North, however, appears to be the relationship to the modern project. According to Medvedev:

> Both the West and the East have been explored and assimilated by modern culture. In general, the East-West binary opposition is essentially a modern project, originating in the era of geographic discoveries, legitimized in Eurocentrism and colonialism (cf. Kipling's "never the twain shall meet"), and consummated in the Cold War division. The oriental East and the political East (the Byzantine, Russian and later the Soviet Empires) have all intertwined in the western mind, playing the role of Europe's Other. East and West have been filled with political practices and therefore demystified, attributed to nations, ideologies and institutions. The intensity of the East-West dichotomy has obscured the signifiers of North and South. Whereas in the premodern times North had been marginalized by centrality of the South, in the modern era North was marginalized by centrality of the East-West narrative.[46]

One could claim, along similar lines, that out of the cardinal signifiers of political space, it is only the North that has been clearly allotted with the position of a constitutive outside. It stands for a sphere where the social domination over nature is less obvious; North is thus easily depicted as not only a product of history but also outside history and, therefore, also a rather spiritual sphere. Europe's South has been able to utilize memories of a glorious (Hellenic) past. It is thick in political and cultural meaning whereas the North, with its connotations of raw nature, has been unable to draw upon anything similar. The South has been so intimately linked to its former civilizational position – and still remains territorially attached to a self-understanding that builds on a formidable past – that it has been impossible to marginalize it in the same manner as the North. The South remains, as is often stated, 'the cradle of civilizations' and is thereby less able to control – if not generate – meaning in the context of central constitutive processes.

---

[45] See Kenneth Coates, 'The Discovery of the North: Towards a Conceptual Framework for the Study of Northern/Remote Regions,' *The Northern Review*, No. 12/13 (1993/4), p. 15.

[46] Medvedev, 'The Blank Space,' p. 91.

Meaning is given by rather fixed interpretations of history and hence less permeable and maniputable. The South is, despite its rather weak social and economic position, culturally 'the source of light', an icon instead of an anti-icon.[47]

The pushing of the North into the position of what Medvedev characterizes as "a generic outback, mother of all peripheries" has succeeded as the North is far more labelled by periods of marginality in being void of the qualities that provide meaning other than in terms of pure nature and outsidedness. It harboured such a role of the land of nothingness already in relation to the ancient South. As the land of darkness and one inhabited by barbarians, the North figured as the South's constitutive outside. Undoubtedly also the South has been marginalized during the modern era. It has lost in standing compared to its 'golden days', albeit the Southern marker has been able to preserve some of its character as a source of meaning. The South thus has a position that is to some extent different from the one held by the North with the latter being imbued with connotations of an object void of generating any relevant meaning (*i.e.* 'a source of light') of its own.

The character of a liminar in the context of the rather rationalist modern project has made northernness rich, as argued, in myths and hidden meanings. This is one reason why it appears to fit quite well with the current conditions that favour diversity and inventiveness. More recently it seems to be on its way of transforming into a brand name: a kind of North®. It is light and less burdened by images of permanency than the more established markers of political space. The neo-North resonates with what Zaki Laïdi has characterized as "a pure play of signs".[48] The borderlines between meaning and non-meaning have turned thin, as have those between the sphere of politics and that of nature.

A Play of Signs

Quite similar questions seem to pertain to the neo-North in the context of the EU's Northern Dimension. The coinage is important on its own terms and does not necessarily have to translate – in tolerating a considerable gap between experience and expectations – into something societally tangible in order to count.

This begs the question whether it is the symbols used or the more concrete project-oriented results that the ND may yield that counts in passing judgement

---

[47] For a comparison between the North and the South, see Ulla Holm and Pertti Joenniemi, *North, South and the Figure of Europe: Changing Relationships* (Copenhagen: Copenhagen Peace Research Institute, 2001).

[48] For a general interpretation along these lines, see Zaki Laïdi, *A World Without Meaning: The Crisis of Meaning in International Politics* (London and New York: Routledge, 1998).

concerning whether the initiative is for real and if it can be viewed as having encountered success. Will it be able to generate meaning and serve as a frame in depicting political space in the new, post-Cold War Europe already as a set of symbols, thereby to be viewed as having resulted in a positive outcome, or does it have to amount to something 'real' and tangible in the context of the EU? Can the play with the 'nothingness' of the North serve as an ordering device and representational frame in the first place and is it at all possible to talk about success in the context of an implosion and the demise of a set of previous structuring oppositions? Is the ND not doomed to remain an exercise in naming, a PR-intensive, media-effective and low-cost move but one which in the end yields very little if anything in terms of concrete exchange and communication due to its very emptiness?

Mostly the view has been that there has to be correspondence between the sign and the social sphere, and this requirement of correspondence has also underpinned most of the ND-related research. The judgement passed has in general followed a distinctly modern line of interpretation. This seems to account for at least some of the pessimism, explain the focusing on various diplomatic processes as well as the down-playing of the symbolic aspects of the move of advocating northernness. Poststructuralist approaches – along the lines opened up by Zaki Laïdi – allow for a different interpretation. The EU's Northern Dimension does not have to be underpinned by objectivist and functionalist arguments in order to be alive and it does not have to amount to an explicit project and short-term results in order to be credible. The rhetorical qualities and representational aspects – appearing for example in the weather forecasts – of the move are important as such. The initiative is not compelled to succeed in some finalistic manner or turn into a 'grand design' for it to be interesting and worth analysis. It counts, according to the latter frame of interpretation, already as signifier, a play of signs and a frame used in various contexts.

The Northern Dimension is there above all as a joint field of communication, one furnishing a number of previously suppressed actors and perspectives with a constitutive voice and standing of their own. It figures as imagery and may serve as a frame that privileges one set of subjectivities and perspectives while downgrading other political visualizations of political space. It has value above all in offering a platform and a meeting-place that is neither premised on easternness nor westernness. In bringing about such an option that goes beyond the dominant binary divide, northernness offers the ground for a genuine dialogue. It is, no doubt, related to the region-talk that has, during the post-Cold War period, proliferated in the form of the Baltic Sea, Barents and the Arctic regions, or – for that matter – in terms of a great number of transborder and cross-border initiatives.[49] However, it does not boil down to any distinct regional arrangement. What it does, instead, is to influence the constitutive principles and departures at large by legitimating multiplicity in terms of construction of in-between spaces. The Northern Dimension provides the more specific regionalist initiatives with a common frame, thereby adding to their standing. Instead of appearing as oddities and deviations, they gain the position of being legitimate constructions among others. Moreover, the ND links them further to the EU and the policies of the Union, and more broadly to the contest on the Europe to come.

The initiative allows the Union – and the various other actors within its purview – to perceive themselves from a new and different perspective. It liberates them from a number of constraints introduced by the dominance of the East-West divide and, more generally, the modern way of outlining political space. Northernness thus invokes a new point of anchorage, albeit in a rather flexible and light manner. It does not aim at creating disunion by offering a new grand way of dividing Europe along a North/South axis. It is not about the return of history or the introduction of a new binary divide. One modern configuration and way of devising political space is not traded for another. Instead the move lays down a vision that contributes to an ordering of the rather uncertain and volatile post-Cold War conditions in a rather subtle and emancipatory manner by broadening the range of choices. It thereby represents a step forward rather than a return of history, and this disregarding of the final, instrumental and structural outcome. As argued by Sergei Medvedev, "narratives of 'great Europe' and 'great Russia' are estranged in the Northern fringes, and their opposition is made relative." More particularly, Medvedev claims:

> The North is less influenced by the "vertical" discourses and structures of subordination. It has never been strongly subjected to its disciplining projects of Catholicism, Russian imperialism, Soviet Communism, Atlanticism or Europeanism, while its Lutheran legacy has never amounted to creating a dominant type of supranationalism. Therefore, over the centuries, North has developed a sort of cultural and political permissiveness, an allergy to grand narratives and various forms of collectivism, and a healthy pragmatism on Lutheran individualism and Hanse-type of liberalism.[50]

The Northern Dimension represents, in this perspective, an effort to pave room for a Europe on northern terms. It indicates that the northern marker no longer subordinates itself to the new Europeanness – as it did in the context of the East/West Europe. It aspires, instead of accommodation and occupying an obscure

---

[49]   See Wæver, 'The Baltic Sea,' p. 293.

[50]   Medvedev, 'The Blank Space,' p. 99.

position at the edges or outright opposition for that matter, for subjectivity on terms of its own by assuming the position of a co-determinant and a departure that brings together aspects that have remained apart from each other for so long. The aim is one of bringing about a Europe, if seen from the northern latitudes, which is not just 'there' (Brussels or Moscow for that matter) but also 'here'. For this to come about northernness has to liberate itself from the detainment caused by the deeply entrenched and intellectually rather well rehearsed barriers that over a long period of time allotted the North with a position of a kind of outcast.

The Northern Dimension goes some way in doing this, although the outcome is dependent on whether the artifice coined sinks in and is found largely acceptable. Public discourses may be as decisive as the more formal, diplomatic processes pertaining to the fate of the initiative. In any case, the launching of the initiative – and the underlying discourse – challenges the way northernness has been configured over two centuries. It is a move away from just passively inheriting and swallowing whatever is coined at the traditional core. Moreover, it seems to provide clues of self-liberation and testifies to an ability to translate one particular aspect of the postmodern play of signs, comprehended as politically usable, to be employed in the course of Europe-making.

Concluding Remarks: What is the Core Constitutive Principle?

Clearly, constitutive questions about politics have escaped the previously well 'guarded' confines of Europe-talk. Such a breach requires, it seems, a response from the EU as a whole, particularly the Commission being charged with the task of framing the Union. However, the truly crucial issues do not just pertain to governance or the erection of regimes, *i.e.* responses that the EU has been quite familiar with. They also invite for choosing between different representational frames, each with their own delineation and identities. The Union is called – or challenged – to go beyond its ordinary reactions and it is confronted with such constitutive issues particularly in the northern part of Europe.

That the challenges arise from the northern margins comes, in a sense, as a surprise. This is so as the northernmost Europe has usually been equated with the predominance of geopolitical, statist and sovereignty-oriented understandings of political space. Sovereignty has been the unchallenged principle applied in the construction of political space. The northernmost North has been quite militarized, with a high level of tension and strict lines of territorial demarcation. Much of the old has been more alive in the 'High North' than elsewhere in Europe. The construction of order and the representational frames used have led to tight measures of inclusion and exclusion and firm classifications into 'we' and 'they'. There has been little reason to ponder change as the constitutive principles seem to have enjoyed a certain permanency. Considerable homogeneity has reigned in this regard, and overlapping configurations have been depicted as a source of confusion and strife.

Yet, during the recent decade, northern Europe has also been quite conducive to region-building. The northern part of Europe has turned, within a rather short period, rich in various regional arrangements such as the (reformed) Nordic Council, Council of the Baltic Sea States, Barents Euro-Arctic Council and a broad variety of various euroregions, transborder and cross-border arrangements. Moreover, the challenge is not just that of some particular regional configurations adding plurality to the political landscape. The Union is, above all, compelled to sort out questions pertaining to the legitimacy of different constitutive principles as regionality has been applied more than in most other parts of the continent. The interpretative horizons based on impermeable boundaries have quickly declined in importance and, as a consequence, the northern part of Europe has defied and circumvented previous lines of demarcation in capitalizing on the demise of the bipolar order. There has been a pooling of resources in order to improve the region's relative weight in the contest between centrality and marginality in the new integrative Europe.

Various regional entities, ranging from small and local ones to very large cross-border arrangements, have emerged leaving a considerable imprint on the political and economic landscape. Projects have been launched despite that they are not guaranteed by any secure origins or, for that matter, any known outcome. Trajectories have been explored with advantages, costs and uncertainties. There has been openness and good chances of 'speaking' various representations such as the Baltic and the Barents ones into existence in order for them to transcend the previous sovereignty-based delineations of political space and to reach beyond the Union's ordinary policies based on sectoral and functional approaches. New formations have been imagined, visions launched into the public sphere, and they turned rather quickly to political realities and concrete regionalist projects.

The appearance of the Northern Dimension seems to indicate that such policies have now been brought one step further. Thinking has been liberated – one may argue – from being tied to distinct regional configurations, and the policies pursued have turned more spatial in terms of being less bound to specific territorial constraints or bordering along strict lines. As a consequence, questions pertaining to the more general constitutive principles involved in Europe-making are rendered open.

The broader talk of the recent years does not only aim at bringing about permeable borders and install various border-crossing arrangements; it focuses above all on regionality as such with border regions and spaces in-between gaining in authority and legitimacy. In fact, there appears to be two rather different constitutive logics present in the current Europe-talk, each with an impact of its own. There are the more traditional sovereignty-related departures present, but also representational frames are used, formations advocated and performative discourses waged that do not resonate with the assumed predominance and exclusivity of sovereignty. And yet, there are few explicit signs of a confrontation to be discerned. There is no clear-cut clash to be evidenced despite the fact that sovereignty and regionality stand out as two rather diverse logics and constitutive departures.[51]

There are tensions, though, in the sense that with regionality as a new foundational principle, one challenges the other. As regionality does not present itself as being based on sovereignty and no longer figures as something derived, there must be a clash and confrontation evolving. Focusing on networks, deregulation and flows, going away from security, statist concerns, divisive borders and national economies (as region-talk tends to do) is quite different from focusing on the re-constructing of nation-states or endeavouring at establishing strict and uniform criteria for citizenship.

However, this clash seems to be comparatively mild as the region-builders to the North clearly attempt to tune down any images of a dichotomic relationship. Images of a confrontation are evoked by the claim that the installing of representational frames such as the Northern Dimension is about something else. It may be argued – if the issues are pinpointed in the first place – that these representational departures operate at a different wavelength and endeavour at assisting the (sovereignty-based) states or doing things that states do not engage themselves in. Besides, there is the message that they will remain light as regards institutions and thereby insignificant as to their regulatory capacities. They *are* more than they *do*. The argument seems – in terms of endeavouring at finding a way around the discursive limits set by the predominance of sovereignty as a core constitutive principle – to be that the unfolding of regionality follows a 'post-sovereign' path. The formations that follow tend to multiply authorities and identities in modes that overflow sovereignty, and thereby contribute to multiplicity in a manner where the order of sovereignty becomes one reality among many. As noted by Ole Wæver, this is of course against the logic of sovereignty, which claims to be the one and only.[52] But yet it can go on, he adds, as regional formations are marginal, off-centre and light in authority.

---

[51] See Wæver, 'The Baltic Sea'; Pertti Joenniemi, 'Regionality? A Sovereign Principle in International Relations?' in Heikki Patomäki, ed., *Peaceful Changes in World Politics* (Tampere: Tampere Peace Research Institute, 1995), pp. 337–79; Mario Telò, 'Introduction: Globalization, New Regionalism and the Role of the European Union,' in Mario Telò, ed., *European Union and New Regionalism: Regional Actors and Global Governance in a Post-Hegemonic Era* (Aldershot, Burlington, Singapore: Ashgate, 2001), pp. 1–21.

[52] See Wæver, 'The Baltic Sea,' pp. 301–303.

What allows markers such as the North to grow in eminence is that they seem to be related to non-sovereignty shaped processes. They entail experimenting with new principles and departures that do not directly challenge sovereignty and the related East/West logic. They are in tune with globalization as unfolding in the forms of the cyber-economy, in the virtual reality of (war) games, in the non-territorial politics of technological innovations and such like. The northern part of Europe appears not only to be quite receptive to these trends and such experimentation; there is sufficient actorness to translate the new developments and clear evidence of change into a constitutive debate as to the basic departures of Europe-making. There has been an upsurge of regionality-based configurations, the reason perhaps being that sovereignty is quite firmly anchored and self-evident in the region. The modern period – with sovereignty in the front seat – has unfolded in a relatively unproblematic fashion. There is, in consequence, also courage to provide space for some other principles as long as they do not directly challenge the dominance of sovereignty. Or to phrase it differently: the sovereignty-based discourse is taken to be so strong that minor deviations for example in the form of North-speak do not seem to matter.

At best sovereignty might even search for an alliance with regionality instead of just wanting to be contained and circumscribed by a competing approach. The ND could indeed be comprehended as exemplifying such an alliance and thereby leaning on a certain duality as to the underlying constitutive logics. It is not just acceptable from the point of sovereignty but also something interesting to experiment with. Applying a northern frame is viewed as a positive trajectory as long as it stays in a somewhat obscure and harmless form and remains sufficiently under the control of the respective states and ties in with their modern, sovereignty-based discourse.

The situation also looks interesting from the perspective of the European Union. The EU has often been the driving force in devising an increasingly complex landscape by its various strategies, embedded above all in the structural funds. The Union harbours a plurality in the sense that there are trends pointing to a uniform, state-like Union with relatively hard and impermeable borders akin to the modern state borders associated with a Westphalian sovereignty-based system. However, there is also a Union that draws on debates concerning the development of an Empire-like Europe of concentric circles centred on Brussels, *i.e.* a configuration in which power and influence decline the further one is from the centre. This configuration has a distinct core but rather obscure borders towards the edges. A third model builds upon a variegated conception of Europe and the EU in which there is not one but several centres, power is dispersed throughout interlocking and overlapping regionalist formations with rather fluid external borders. Each of these three potential configurations/metaphors is driven by a logic of its own, *i.e.* a modern (a concentric EU), less modern or perhaps even a premodern (an EU of concentric circles) and a postmodern (a clearly decentred EU of the Olympic rings) one.

Particularly in this latter case – within such a trilogy of heuristic models outlining potential trends within Europe – the EU would be unfolding as a supreme example of postmodern politics.[53] It is interesting to note, against this background, that it is particularly around the Baltic rim and in northern Europe that the Union is called to tune in to something that has been infused with features that deviate from modern clarity and endeavours of unambiguous bordering already prior to the Union establishing itself in the region. Such a situation allows the Union to build further on what is already there and offers the perspective of bringing the non-modern aspirations further than in most other regions. The Northern Dimension could be seen as the latest step along such a road, one that also contains the idea of strengthening the position of regionality in the context of the EU and Europe-making at large.

The North, as a major marker, undoubtedly frames something. Hence the question arises what this something is and how does it tie in with the dominant discourses pertaining to the construction of political space. In the case of the EU's Northern Dimension the subsets consist, above all, of the various regionalist entities that are already there. In informing policy responses at the margins, the ND does not seem to pertain to a 'Europe of concentric circles' and a sovereignty-based logic conducive to rather hierarchic and unicentred structures. Instead, it forms co-space rather than sub-space within the broader European configuration. The initiative – although contested and open to various efforts of deflating its meaning – appears to aim at legitimizing in-between type of spaces. It does not represent the voice of the core but stands out as a proactive move with the margins speaking and contributing to a visualization of the Union that strengthens images of a 'Europe of the Olympic rings', a more variegated Europe with an increasing amount of horizontal aspects, cross-pillar features and policies that transcend the borders of the different directories within the EU. The constitutive logic advocated consists above all of regionality and de-bordering.

The different configurations outlined above hinge on the question whether Europe will constitute itself as a single centre, will it develop around multiple

---

[53] For arguments along these lines, see Ole Wæver, 'Territory, Authority and Identity. The Late 20th Century Emergence of Neo-Medieval Political Structures in Europe,' paper presented at the first EUPRA conference in Florence, November 1991; John Gerald Ruggie, 'Territoriality and Beyond: Problematizing Modernity in International Relations,' *International Organization*, Vol. 47, No. 1 (Winter 1993), pp. 139–74; Markus Jachtenfuchs and Beate Kohler-Koch, *The Transformation of Governance in the European Union* (Mannheim: Mannheimer Zentrum für Europäische Sozialforschung, 1995).

centres as a 'Europe of regionalities', in what way will it be bordered and what representational frames are going to be utilized in outlining the overall European configuration. All the recent federalism-talk points to rather a modern ambition of aspiring for an orderly, uniform, basically vertical and rather controlled Europe. The effort is one of devising borders that provide a clear-cut division between an inside and an outside, and thereby also offer the ground for a coherent EU-identity that could prevail everywhere within the Euro-polity. Such an EU, one with a rather uniform representational frame, would provide little if any room for questioning the dominance of the core. There is, if this is the way the EU goes, little space for the unfolding of a polity premised on northernness as one of its constitutive markers. Nor would the configuration allow for debordering or engagement in a dialogue premised on equality with non-applicants such as Russia, Norway or Iceland. There would be no genuine meeting-ground, that is space furnishing those to be met with some subjectivity and voice of their own. The Northern Dimension would shrink to very little within such a context. It would basically turn into a vehicle of spreading a pre-given homogeneity both within and in the 'near abroads' of the Union, as indicated by Christopher Browning in his analysis on the way the ND has unfolded in the context of the EU.[54] The question is thus not whether the Northern Dimension is something and whether it stands out as a success or a failure, but what are the interpretations imposed and the context in which the marker is utilized.

It might be added, however, that the dynamics of the Northern Dimension also allow for a different interpretation. The initiative may also be seen as speaking for a rather polycentric EU and seen as a move contributing to a configuration bordered in an increasingly fluid and "fuzzy" manner. It does so above all by raising questions about representational departures in its installing of an additional frame, one premised on northernness into the discourses pertaining to the essence of the EU. The North, as marker and a principle of legitimation, is explicitly seen as strengthening the unfolding of region-building in a distinct part of Europe. It injects a departure that invites the margins to carve out space of their own. Moreover, the northern marker is not just brought in as a boundary marker; it could also be seen as contributing to the outlining of a meeting-place that mediates between the internal and external aspects of the Union. It allows, the argument often goes, the boundaries to fade in significance thereby turning them into administrative rather than tightly statist borders. As to the Union as a whole, the ND strengthens – if allowed to mature and unfold in a radical manner – a configuration that has not just one but several focal points. There is so far no major development towards such a direction to be traced, but the potential of drastic restructuring is in principle there.

---

[54]   Browning, *The Construction of Europe*.

The common bind as to the three heuristic models outlined above consists of the EU itself, whether basically unicentric or polycentric. The Union provides the framework within which various constellations relate to each other. With respect to the territorial aspects of politics, implementing the ND-initiative would mean more emphasis on regionality-steered configurations. In such a 'Europe of regions' intermediate structures and spaces in-between would provide the crucial building blocks in terms of constitutive politics. The multiplicity of governance structures and the multiple identities of the actors – *i.e.* features pertaining to regionality – would become the norm in considerable parts of Europe. Overlapping membership of actors in various policy-networks would link the various parts of Europe, but neither the state nor the EU would emerge as the dominant level of governance.

Yet, region-building may have its place within all three scenarios. Regional cooperation is quite possible without regionalism, *i.e.* it may take place without undermining modernist notions of state sovereignty. Along these lines, regional cooperation can also take place in a unicentred EU with hard and impermeable borders, although by remaining something derived and administrative. The two other scenarios both move away from the more simplistic conception of regions as mere subsets of statist and sovereignty-governed spheres of political space, and allow for and invite a less restricted unfolding of regionalist formations.

The evolution on the continent seems to indicate that the development underway is neither based on the traditional image of modern state systems (and regions as administrative and derived entities therein), nor that the European Union is emerging into a state-like super-structure with clearly delineated borders and an internal hierarchy. And what makes northern Europe interesting to study is that it seems to allow for regions to gain considerable space alongside the more traditional configurations, and the more recent development points to even broader shifts in elevating regionality – in a broader and more principled sense – to stake out a position as an agreed constitutive principle, one with its own representational frames and markers. Regional entities do hence not just stand out as islands within political space governed by sovereignty as a core departure. They appear to be supported by regionality, *i.e.* a departure not fully premised on sovereignty. The Northern Dimension could be seen in this perspective as representing and riding on the bolstered position of regionality. However, success is by no means guaranteed, and it may well be that the marker is embedded with qualities implying that progress does not stall from the very start.

However, one could think that the increased eminence of regionality contributes, in the short run, to a Union of concentric circles, an Empire-like configuration with regionality providing, in particular, shape to the edges. It would do so primarily in the form of specific regionalist configurations, formations that would be there without the backing of a strong and broadly agreed policy frame premised on regionality as a constitutive principle. However, in the long run the figure could gain features of a Europe of 'the Olympic rings', a construction more easily defined in terms of flow rather than some specific place. It would then evidence the potential of the regionalist forms of differentiation in regard to other contending trends, and do so against the background of an increasingly global constellation.

The undisputed plurality of the current-day Europe has already led to an upsurge in border studies and the emergence of new conceptualizations as well as theories on political space. Many of the previous silences have vanished. It seems, however, that the question of representational frames as to the usage of major cardinal markers has thus far remained at the fringes of scholarly interest, including the one focusing on the ND. Quite clearly the task for research is to take on the challenge, to make these departures far more visible and perceive them not as given but as changing and constructed entities containing relationships of power. They too serve as sites and agents of order and disorder in an increasingly dynamic global landscape, and hence the newly-important North with its different faces and expressions may serve as an inroad to a broader and so far insufficiently explored problemacy. Some initial efforts notwithstanding, the state of the art in the field of regionality and the various Europes that representations such as the North turn visible, remains at the level of pre-theorizing.

Bibliography

Adam, Barbara, *Timescapes of Modernity. The Environment and Invisible Hazards* (London: Routledge, 1998).

Agnew, John, *Geopolitics: Re-visioning World Politics* (London and New York: Routledge, 1998).

Albrecht, Ulrich, *Internationale Politik. Einführung in das System internationaler Herrschaft* (Munich/Vienna: Oldenbourg, 1986).

Allardt, Erik, 'Representative Government in a Bureaucratic Age,' in Stephen R. Graubard, ed., *Norden - The Passion for Equality* (Oslo: Norwegian University Press, 1986), pp. 200–225.

Alvarez, A., 'Ice Capades,' *The New York Review of Books*, Vol. XLVIII, No. 13 (9 August 2001), pp. 14–17.

Ambjörnsson, Ronny, *Öst och Väst. Tankar om ett Europa mellan Asien och Amerika* (Stockholm: Natur och kultur, 1994).

Amundsen, Roald, *Luoteisväylä. Kertomus Gjöan matkasta 1903–1907* (Porvoo: WSOY, 1908).

Andersen, Bent Rold, 'Rationality and Irrationality of the Nordic Welfare State,' in Stephen R. Graubard, ed., *Norden - The Passion for Equality* (Oslo: Norwegian University Press, 1986), pp. 112–42.

Andersen, Svein S., *Norway: Insider AND Outsider* (Oslo: Advanced Research on the Europeanization of the Nation-State, 2000), at < http://www.arena.uio.no/publications/wp00_4.htm >.

Anderson, Benedict, *Imagined Communities: Reflections on the Origin and Spread of Nationalism* (London: Verso, 1993).

Antola, Esko, 'The Presence of the European Union in the North,' in Hiski Haukkala, ed., *Dynamic Aspects of the Northern Dimension* (Turku: Turku University, Jean Monnet Unit, 2000), pp. 115–32.

Anttila, Lauri, 'Tents and Fireplaces in Labrador,' *N66. Culture in the Barents Region*, No. 8 (1999), pp. 6–9.

Archer, Clive, 'Arctic Cooperation: A Nordic Model,' *Bulletin of Peace Proposal*, Vol. 21, No. 2 (June 1990), pp. 165–74.

Archer, Clive and David Scrivener, eds., *Northern Waters: Security and Resource Issues* (London: Croom Helm, 1986).

Arctic Monitoring and Assessment Program (AMAP), *Arctic Pollution Issues: A State of the Arctic Environment Report* (Oslo: AMAP, 1997).

Arter, David, 'Small State Influence within the EU: The Case of Finland's "Northern Dimension Initiative",' *Journal of Common Market Studies*, Vol. 38, No. 5 (December 2000), pp. 677–97.

Ashcroft, Bill, Gareth Griffiths and Helen Tiffin, *The Empire Writes Back. Theory and Practice in Post-Colonial Literatures* (London: Routledge, 1989).

261

Baerenholdt, Jørgen Ole, *Bygdeliv* (Roskilde: Roskilde University Center, Department of Geography, Social Analysis and Computer Science, 1991).

Baerenholdt, Jørgen Ole, *Innovation and Adaptability of the Murmansk Region Fishery Kolkhozes* (Roskilde: Roskilde University Center, Department of Geography and International Development Studies, 1995).

Balasinski, Jystine, 'Culture and Politics in Transition Régimes: The Case of Theatre in Poland in the 1980s and 1990s,' Paper presented at the Politics and the Arts Group Symposium 'Politics and the Arts: Making Connections in Theory and Praxis,' Berlin, 23–25 May 2002.

Barnes, Trevor and Derek Gregory, 'Place and Landscape,' in Trevor Barnes and Derek Gregory, eds., *Reading Human Geography: The Poetics and Politics of Inquiry* (London, New York, Sidney, Auckland: Arnold, 1997), pp. 292–8.

Barnett, Clive, 'The Cultural Turn: Fashion or Progress in Human Geography?' *Antipode*, Vol. 30, No. 4 (1988), pp. 379–94.

Barr, Susan, *Fram mot Nordpolen. En hundreårsbragd Fram-ferden 1893–1896* (Oslo: Schibsted, 1996).

Barr, Susan, 'Norge inntar arenaen,' in Sigmund Nesset and Helge Salvesen, eds., *Ultima Thule* (Tromsø: Universitetsbiblioteket i Tromsø, 1996), pp. 202–12.

Barthes, Roland, *Camera Lucida: Reflections on Photography* (New York: Noonday Press, 1990).

Benjamin, Walter, 'Das Kunstwerk im Zeitalter seiner technischen Reproduzierbarkeit,' in Walter Benjamin, *Das Kunstwerk im Zeitalter seiner technischen Reproduzierbarkeit* (Frankfurt: Suhrkamp, 1963), pp. 7–44.

Benjamin, Walter, 'Kleine Geschichte der Photographie,' in Walter Benjamin, *Das Kunstwerk im Zeitalter seiner technischen Reproduzierbarkeit* (Frankfurt: Suhrkamp, 1963), pp. 45–64.

Benkow, Jo, *Folkevalgt* (Oslo: Gyldendal, 1988).

Berdoulay, Vincent, 'Place in French Language Geography,' in John A. Agnew and James S. Duncan, eds., *Power of Place* (Boston: Unwin Hyman, 1989), pp. 124–39.

Berger, John, 'It Can Happen,' *N66. Culture in the Barents Region*, No. 1 (1997), p. 28.

Berkes, Fikret, 'Traditional Ecological Knowledge in Perspective,' in Julian T. Inglis, ed., *Traditional Ecological Knowledge* (Ottawa: International Development Center, 1993), pp. 1–9.

Bernes, Claes, *The Nordic Arctic Environment – Unspoilt, Exploited, Polluted?* (Copenhagen: The Nordic Council of Ministers, 1996).

Berton, Pierre, *The Arctic Grail. The Quest for the North West Passage and the North Pole, 1818–1909* (New York: Viking, 1988).

Betänkande. Svensk-Norsk Renbeteskommissionen av år 1997. Ruota-Norgga Boazoguohtunkommišuvdna 1997 (Alta: Björkmanns, 2001).

Bialostocki, Jan, '"Studien zur Ikonologie" nach vierzig Jahren,' in Erwin Panofsky, *Studien zur Ikonologie der Renaissance. Zweite Auflage* (Cologne: Dumont, 1997), pp. 7–16.

Birkeland, Inger J., 'Nature and the "Cultural Turn" in Human Geography,' *Norsk Geografisk Tidsskrift*, Vol. 52, No. 4 (1998), pp. 229–40.

Birkeland, Inger J., 'The Mytho-Poetic in Northern Travel,' in David Crouch, ed., *Leisure/ tourism Geographies: Practices and Geographical Knowledge* (London: Routledge, 1999), pp. 17–33.

Bjerkli, Bjørn, 'Landskapets makt: Sted, ressursutnyttelse og tilhørighet i en sjøsamisk bygd,' *Norsk Antropologisk Tidsskrift*, Vol. 8, No. 2 (1997), pp. 132–56.

Bjørklund, Tor, 'The Three Nordic 1994 Referenda Concerning Membership in the EU,' *Cooperation and Conflict*, Vol. 31, No. 1 (March 1996), pp. 11–36.

Bjørklund, Tor, *Om folkeavstemninger: Norge og Norden 1905–1994* (Oslo: Universitetsforlaget, 1997).

Bladh, Gabriel, *Finnskogens landskap och människor under fyra sekler* (Karlstad: University of Karlstad, 1995).

Bloom, Lisa, *Gender on Ice* (Minneapolis: University of Minnesota Press, 1993).

Bomsdorf, Falk, *Sicherheit im Norden Europas. Die Sicherheitspolitik der fünf nordischen Staaten und die Nordeuropapolitik der Sowjetunion* (Baden-Baden: Nomos, 1989).

Bonvicini, Gianni, Tapani Vaahtoranta and Wolfgang Wessels, eds., *The Northern EU. National Views on the Emerging Security Dimension* (Helsinki/Bonn: The Finnish Institute of International Affairs/Institut für Europäische Politik, 2000).

Booth, Ken, 'Security and Self: Reflections of a Fallen Realist,' in *Critical Security Studies: Concepts and Cases* (London: UCL Press, 1997), pp. 83–119.

Bourdieu, Pierre, 'The Forms of Capital,' in John G. Richardson, ed., *Handbook of Theory and Research for the Sociology of Education* (New York: Greenwood Press, 1986), pp. 241–58.

Bourdieu, Pierre, 'The Cult of Unity and Cultivated Differences,' in Pierre Bourdieu with Luc Boltanski, Robert Castel, Jean-Claude Chamboredon and Dominique Schnapper, *Photography: A Middle-brow Art* (Cambridge: Polity Press, 1990), pp. 14–72.

Bourdieu, Pierre, 'The Social Definition of Photography,' in Pierre Bourdieu with Luc Boltanski, Robert Castel, Jean-Claude Chamboredon and Dominique Schnapper, *Photography: A Middle-brow Art* (Cambridge: Polity Press, 1990), pp. 73–98.

Brosseau, Marc, 'Geography's Literature,' *Progress in Human Geography*, Vol. 18, No. 3 (1994), pp. 333–53.

Brosseau, Marc, 'The City in Textual Form: Manhattan Transfer's New York,' *Ecumene*, Vol. 2, No. 1 (1995), pp. 89–114.

Browning, Christopher, *The Construction of Europe in the Northern Dimension* (Copenhagen: Copenhagen Peace Research Institute, 2001).

Browning, Christopher, *The Region-Building Approach Revisited: The Continued Othering of Russia in Discourses of Region-Building in the European North* (Copenhagen: Peace Research Institute, 2001).

Bukaev, Sergei, Viktor Samokhvalov, Vladimir Kulikov and Sergei Anufriev, 'Vorkuta,' in Marketta Seppälä, ed., *Strangers in the Arctic. 'Ultima Thule' and Modernity* (Pori: Pori Art Museum, 1996), pp. 125–30.

Bunkśe, Edmunds V., 'Saint-Exupéry's Geography Lesson: Art and Science in the Creation and Cultivation of Landscape Values,' *Annals of the Association of American Geographers*, Vol. 80, No. 1 (1990), pp. 96–108.

Burke, Peter, *Eyewitnessing: The Uses of Images as Historical Evidence* (London: Reaktion Books, 2001).

Buzan, Barry, *People, States & Fear: An Agenda for International Security Studies in the Post-Cold War Era*. Second Edition (Hemel Hempstead: Harvester Wheatsheaf, 1991).

Byles, Jeff, 'Maps and Chaps. The New Geography Reaches Critical Mass,' *The Village Voice*, Vol. XLVI, No. 31 (August 2001), electronic version at <http://www.villagevoice.com/issues/0131/edbyles.php>.

Catellani, Nicola, 'The Multilevel Implementation of the Northern Dimension,' in Hanna Ojanen, ed., *The Northern Dimension: Fuel for the EU?* (Helsinki/Bonn: The Finnish Institute of International Affairs/Institut für Europäische Politik, 2001), pp. 54–77.

Chatwin, Bruce, *Songlines* (Berkshire: Picador, 1988).

Checkel, Jeffrey T., 'The Constructivist Turn in International Relations Theory,' *World Politics*, Vol. 50, No. 2 (January 1998), pp. 324–48.

Childs, Marquis, *Sweden – The Middle Way* (New Haven: Yale University Press, 1936).

Chorlton, Windsor, *Ice Ages* (Alexandria: Time, 1983).

Christiansen, Thomas, Fabio Petito and Ben Tonra, 'Fuzzy Politics Around Fuzzy Borders: The European Union's Near Abroad,' *Cooperation and Conflict*, Vol. 35, No. 4 (December 2000), pp. 417–31.

Classen, Constance, *The Colour of Angels. Cosmology, Gender and the Aesthetic Imagination* (London and New York: Routledge, 1998).

Claval, Paul, *Introduction to Regional Geography* (Malden: Blackwell, 1998).

Cloke, Paul and Jo Little, 'Introduction: Other Countrysides,' in Paul Cloke and Jo Little, eds., *Contested Countryside Culture. Otherness, Marginalization and Rurality* (London: Routledge, 1997), pp. 1–18.

Coates, Kenneth, 'The Discovery of the North: Towards a Conceptual Framework for the Study of Northern/Remote Regions,' *The Northern Review*, No. 12/13 (1993/4), pp. 31–44.

Cocteau, Jean, *Die Ritter von der Tafelrunde* (Munich *et al.*: Kurt Desch, 1962).

Coetzee, J.M., 'The Marvels of Walter Benjamin,' *The New York Review of Books*, Vol. XLVIII, No. 1 (11 January 2001), pp. 28–33.

Conefrey, Mick and Tim Jordan, *Icemen. A History of the Arctic and its Explorers* (London: Boxtree, 1999).

Connerton, Paul, *How Societies Remember* (Cambridge: Cambridge University Press, 1989).

Cook, Ian, David Crouch, Simon Naylor and James R. Ryan, eds., *Cultural Turns/ Geographical Turns: Perspectives on Cultural Geography* (Harlow: Pearson Education, 2000).

Couldry, Nick, *Inside Culture: Re-imagining the Method of Cultural Studies* (London, Thousand Oaks, New Delhi: Sage, 2000).

Crang, Mike, *Cultural Geography* (London and New York: Routledge, 1998).

Croll, James, *Discussions on Climate and Cosmology* (Edinburg: Adam and Charles Black, 1885).

Cruikshank, Julie, 'Glaciers and Climate Change: Perspectives from Oral Tradition,' *Arctic*, Vol. 54, No. 4 (December 2001), pp. 377–93.

Crush, Jonathan, 'Post-colonialism, De-colonialism, and Geography,' in Anne Godlewska and Neil Smith, eds., *Geography and Empire* (Oxford: Blackwell, 1994), pp. 333–50.

Dahl, Hans Fredrik, 'Those Equal Folk,' in Stephen R. Graubard, ed., *Norden – The Passion for Equality* (Oslo: Norwegian University Press, 1986), pp. 97–111.

Dahlström, Margareta, Heikki Eskelinen and Ulf Wiberg, eds., *The East-West Interface in the European North* (Uppsala: Nordisk Samhällsgeografisk Tidskrift, 1995).

Daniels, Stephen, 'Arguments for Humanistic Geography,' in R.J. Johnston, ed., *The Future of Geography* (Cambridge: Cambridge University Press, 1985), pp. 143–57.

Daniels, Stephen and Denis Cosgrove, 'Introduction: Iconography and Landscape,' in Denis Cosgrove and Stephen Daniels, eds., *The Iconography of Landscape: Essays on the Symbolic Representation, Design and Use of Past Environments* (Cambridge: Cambridge University Press, 1988), pp. 1–10.

Daniels, Stephen and Roger Lee, 'Editor's Introduction,' in Stephen Daniels and Roger Lee, eds., *Exploring Human Geography* (London: Arnold, 1996), pp. 1–7.

Derrida, Jacques, 'Structure, Sign and Play in the Discourse of the Human Sciences,' in Robert Con Davis, ed., *Contemporary Literary Criticism* (New York: Longman, 1986), pp. 480–98.

Desch, Michael C., 'Culture Clash: Assessing the Importance of Ideas in Security Studies,' *International Security*, Vol. 23, No. 1 (Summer 1998), pp. 141–70.

Diamond, Jared, 'Living on the Moon,' *The New York Review of Books*, Vol. XLIX, No. 9 (23 May 2002), pp. 59–60.

Diez, Thomas, 'Europe as a Discursive Battleground: Discourse Analysis and European Integration Studies,' *Cooperation and Conflict*, Vol. 36, No. 1 (March 2001), pp. 5–38.

Dijkink, Gertjan, *National Identity and Geopolitical Visions: Maps of Pride and Pain* (London and New York: Routledge, 1996).

Domosh, Mona, 'Towards a Feminist Historiography of Geography,' *Transactions of the Institute of British Geographers*, NS Vol. 16, No. 1 (1991), pp. 95–104.

Drivenes, Einar Arne, 'Polarforskning – vitenskap eller politikk?' *Ottar*, No. 197 (1993), pp. 4–12.

Driver, Felix, 'Editorial: Field-work in Geography,' *Transactions of the Institute of British Geographers*, NS Vol. 25, No. 3 (2000), pp. 267–8.

Dufek, Antonin, 'Photography, Society and Time,' in The Municipal House and The Moravian Gallery, *Spolecnost pred objektivem (Society Through the Lens) 1918–1989* (Brno: Obecni dum, 2000), pp. 18–27.

Duncan, James, 'The Superorganic in American Cultural Geography,' *Annals of the Association of American Geographers*, Vol. 70, No. 2 (1980), pp. 181–98.

Duncan, James and Derek Gregory, 'Introduction,' in James Duncan and Derek Gregory, eds., *Writes of Passage: Reading Travel Writing* (London: Routledge, 1999), pp. 1–13.

Duncan, Nancy, ed., *BodySpace. Destabilizing Geographies of Gender and Sexuality* (London and New York: Routledge, 1996).

Eder, Klaus, *Kulturelle Identität zwischen Tradition und Utopie. Soziale Bewegungen als Ort gesellschaftlicher Lernprozesse* (Frankfurt/New York: Campus, 2000).

Edwards, Elizabeth, 'Essay (Part 3): Anthropological (de)constructions,' in Jorma Puranen, *Imaginary Homecoming* (Oulu: Pohjoinen, 1999), pp. 42–6.

Edwards, Elizabeth, 'Essay (Part 4): Dichotomies and Disjunctions,' in Jorma Puranen, *Imaginary Homecoming* (Oulu: Pohjoinen, 1999), pp. 60–64.

Eek, Ann Christine, 'The Roald Amundsen Photographs of the Netsilik People, 1903-5,' in J.C.H. King and Henrietta Lidchi, eds., *Imaging the Arctic* (London: The British Museum Press, 1996), pp. 117-24.

Eide, Espen Barth, 'Adjustment Strategy of a Non-Member: Norwegian Foreign and Security Policy in the Shadow of the European Union,' *Cooperation and Conflict*, Vol. 31, No. 1 (March 1996), pp. 69-104.

Eisenstadt, Shmuel N., 'Die Konstruktion kollektiver Identität im modernen Nationalstaat,' in Bernd Henningsen and Claudia Beindorf, eds., *Gemeinschaft. Eine zivile Imagination* (Baden-Baden: Nomos, 1999), pp. 197-211.

Elkins, David J. and Richard Simeon, 'A Cause in Search of Its Effect, or What Does Political Culture Explain,' *Comparative Politics*, Vol. 11, No. 2 (January 1979), pp. 127-45.

Enquist, Per Olov, 'On the Art of Flying Backward with Dignity,' in Stephen R. Graubard, ed., *Norden – The Passion for Equality* (Oslo: Norwegian University Press, 1986), pp. 65-78.

Entrikin, Nicholas, *The Betweenness of Place* (Baltimore: The Johns Hopkins University Press, 1991), pp. 66-78.

Eskilsson, Lena, *Masculinity and the Northern Space* (Umeå: The International Research Network on the History of Polar Science, 1996).

Esping-Andersen, Gøsta, *The Three Worlds of Welfare Capitalism* (Princeton: Princeton University Press, 1990).

Europeisk Ungdom, *Norge og Unionen Overnasjonalitet i det nye Europa* (Flekkefjord: Hegland trykkeri, 1992).

Evjen, Bjørg, 'Women in Polar Research – Exotic Elements, Intruders or Equals?' *Ottar*, No. 226 (1999), pp. 31-41.

Eyles, John, 'Interpreting the Geographical World,' in John Eyles and David M. Smith, eds., *Qualitative Methods in Human Geography* (Oxford: Blackwell, 1992), pp. 1-16.

Fabian, Rainer and Hans-Christian Adam, *Frühe Reisen mit der Kamera* (Hamburg: Stern, 1981).

Faltings, Volkert F., *Die Dingprotokolle der Westerharde Föhr und Amrum 1658-1671*, 2 vols. (Neumünster: Wachholtz, 1990).

Fischer, Joschka, 'Vom Staatenbund zur Föderation – Gedanken über die Finalität der europäischen Integration,' speech delivered at Humboldt University Berlin, 12 May 2000, at < http://www.auswaertiges-amt.de/6_archiv/2/r/r000512a.htm >.

Fleming, James Rodger, *Historical Perspectives on Climate Change* (Oxford: Oxford University Press, 1998).

Forbes, Bruce, 'Wounds on the Tundra,' *N66. Culture in the Barents Region*, No. 3 (1997), pp. 16-18.

Forgacs, David, 'National-popular: Genealogy of a Concept,' in Simon During, ed., *The Cultural Studies Reader. Second Edition* (London/New York: Routledge, 1999), pp. 210-19.

Foucault, Michel, *The Archaeology of Knowledge* (London: Tavistock, 1972).

Foucault, Michel, 'Questions on Geography,' in Colin Gordon, ed., *Power/Knowledge. Selected Interviews and Other Writings by Michel Foucault 1972-1977* (New York: Pantheon Books, 1980), pp. 63-77.

Frängsmyr, Tore, 'Nordenskiöld och polarforskningen – den idéhistoriska bakgrunden,' *Ymer*, Vol. 100 (1980), pp. 7–38.

Friis, Peter and Rasmus Ole Rasmussen, *The Development of Greenland's Main Industry – the Fishing Industry* (Roskilde: Roskilde University Center, North Atlantic Regional Studies, 1989).

Frome, Karen, 'A Forced Perspective: Aerial Photography and Fascist Propaganda,' *Aperture*, No. 132 (Summer 1993), pp. 76–7.

Galtung, Johan, 'A Structural Theory of Imperialism,' *Journal of Peace Research*, Vol. 13, No. 2 (1971) pp. 81–117.

Garthoff, Raymond L., *The Great Transition: American-Soviet Relations and the End of the Cold War* (Washington: The Brookings Institution, 1994).

Gebhardt, Jürgen, 'Politische Kulturforschung – ein Beitrag zur vergleichenden Analyse soziokultureller Ordnungszusammenhänge,' in Constantin von Barloeven and Kai Werhahn-Mees, eds., *Japan und der Westen, Band 3: Politik, Kultur, Gesellschaft* (Frankfurt: Fischer, 1986), pp. 60–77.

Gilbert, E.W., 'The Idea of Region,' *Geography*, Vol. 45, No. 3 (1960), pp. 157–75.

Goldmann, Stefan, 'Wilde in Europa: Aspekte und Orte ihrer Zurschaustellung,' in Thomas Theye, ed., *Wir und die Wilden. Einblicke in eine kannibalische Beziehung* (Reinbek: Rowohlt, 1985), pp. 243–69.

Goldstein, Walter, ed., *Clash in the North: Polar Summitry and NATO's Northern Flank* (Washington *et al.*: Pergamon-Brassey's, 1988).

Graburn, Nelson N.H., 'The Present as History: Photography and the Inuit, 1959–94,' in J.C.H. King and Henrietta Lidchi, eds., *Imaging the Arctic* (London: The British Museum Press, 1996), pp. 160–67.

Granberg, Leo, ed., *The Snowbelt. Studies on the European North in Transition* (Helsinki: Kikimora Publications, 1998).

Granö, Johannes Gabriel, *Pure Geography* (Baltimore and London: The Johns Hopkins University Press, 1997).

Gregory, Derek, *Geographical Imaginations* (Oxford: Blackwell, 1994).

Greiffenhagen, Martin, *Kulturen des Kompromisses* (Opladen: Leske + Budrich, 1999).

Gribbin, John and H.H. Lamb, 'Climatic Changes in Historical Times,' in John Gribbin and H.H. Lamb, eds., *Climatic Changes* (Cambridge: Cambridge University Press, 1978).

Grubb, Michael, *The Kyoto Protocol. A Guide and Assessment* (London: Earthscan and Royal Institute of International Affairs, 1999).

Haavio, Ari, 'Approaches to Man's Relation to Nature,' in Anna-Mari Konttinen, ed., *Green Moves, Political Stalemates. Sociological Perspectives on the Environment* (Turku: University of Turku, 1996), pp. 9–15.

Haila, Yrjö, 'The North as/and the Other,' in Marketta Seppälä, ed., *Strangers in the Arctic. 'Ultima Thule' and Modernity* (Pori: Pori Art Museum, 1996), pp. 112–9.

Haila, Yrjö, 'The North as/and the Other: Ecology, Domination, Solidarity,' in Frank Fischer and Maarten A. Hajer, eds., *Living with Nature* (Oxford: Oxford University Press, 1999), pp. 42–57.

Haila, Yrjö and Richard Levins, *Humanity and Nature* (London: Pluto Press, 1992).

Haldrup, Michael, 'Living on the Edge – Considerations on the Production of Space in the Faroe Islands,' *Nordisk Samhällsgeografisk Tidskrift*, No. 23 (October 1996), pp. 18–27.

Halliday, Fred, 'Culture and International Relations: A New Reductionism?' in Michi Ebata and Beverly Neufeld, eds., *Confronting the Political in International Relations* (London: Macmillan, 2000), pp. 47–71.

Ham, Peter van, 'Testing Cooperative Security in Europe's North: American Perspectives and Policies,' in Dmitri Trenin and Peter van Ham, *Russia and the United States in Northern European Security* (Helsinki/Bonn: The Finnish Institute of International Affairs/Institut für Europäische Politik, 2000), pp. 57–88.

Hansen, Guttorm, *Der er det godt å sitte: hverdag på Løvebakken gjennom hundre års parlamentarisme* (Oslo: Aschehoug, 1984).

Harley, J. Brian, 'Maps, Knowledge, and Power,' in Stephen Daniels and Roger Lee, eds., *Exploring Human Geography* (London: Arnold, 1996), pp. 377–94.

Harvey, David, *The Condition of Postmodernity* (Oxford: Blackwell, 1989).

Harvey, David, 'Cosmopolitanism and the Banality of Geographical Evils,' *Public Culture*, Vol. 12, No. 2 (2000), pp. 529–64.

Harvey, David, *Spaces of Hope* (Berkeley and Los Angeles: University of California Press, 2000).

Hassi, Satu, 'Opening Remarks,' in *10 Years of Arctic Environmental Cooperation* (Helsinki: Ministry for Foreign Affairs of Finland, 2001), pp. 13–14.

Hastadt, Elis Wilhelm, *The Parliament of Sweden* (London: Hansard Society, 1957).

Hastrup, Kirsten, 'Uchronia and the Two Histories of Iceland,' in Marketta Seppälä, ed., *Strangers in the Arctic. 'Ultima Thule' and Modernity* (Pori: Pori Art Museum, 1996), pp. 36–49.

Haukkala, Hiski, 'Introduction,' in Hiski Haukkala, ed., *Dynamic Aspects of the Northern Dimension* (Turku: University of Turku, Jean Monnet Unit, 1999), pp. 9–20.

Hautajärvi, Harri, *Lapin läänin matkailuarkkitehtuurin historia* (Oulu: University of Oulu, 1995).

Heble, Ajay, *Landing on the Wrong Note: Jazz, Dissonance and Critical Practice* (New York and London: Routledge, 2000).

Heininen, Lassi, 'The International Situation and Cooperation in Change,' in Leo Granberg, ed., *The Snowbelt. Studies on the European North in Transition* (Helsinki: Kikimora Publications, 1998), pp. 199–230.

Heininen, Lassi, *Euroopan pohjoinen 1990-luvulla. Moniulotteisten ja ristiriitaisten intressien alue* (Rovaniemi: University of Lapland, Arctic Centre, 1999).

Heininen, Lassi, 'Ideas and Outcomes: Finding a Concrete Form for the Northern Dimension Initiative,' in Hanna Ojanen, ed., *The Northern Dimension: Fuel for the EU?* (Helsinki/Bonn: The Finnish Institute of International Affairs/Institut für Europäische Politik, 2001), pp. 20–53.

Heiskala, Risto, *Toiminta, tapa ja rakenne: kohti konstruktionista synteesiä yhteis-kuntateoriassa* (Helsinki: Gaudeamus, 2000).

Helander, Elina, 'The Status of the Sámi People in the Inter-State Cooperation,' in Jyrki Käkönen, ed., *Dreaming of the Barents Region: Interpreting Cooperation in the Euro-Arctic Rim* (Tampere: Tampere Peace Research Institute, 1996), pp. 296–306.

Henson, Matthew A., *A Black Explorer at the North Pole* (Lincoln: University of Nebraska Press, 1989 [1912]).

Heurlin, Bertel, *Denmark and the Northern Dimension* (Copenhagen: Dansk Uden-rigspolitisk Institutt, 1999).

Hille, Jochen, *Das norwegische EU-Referendum von 1994: Nationale Identität versus europäische Integration* (Berlin: Institut für Internationale Politik und Regionalstudien e.V. in Zusammenarbeit mit dem Otto-Suhr-Institut für Politikwissenschaft der Freien Universität Berlin, 2000).

Hoeg, Peter, *Frøken Smillas fornemmelse for sne* (Copenhagen: Munksgaard/Rosinante, 1992).

Hoffmann, Stanley, 'An American Social Science: International Relations,' in James Der Derian, ed. *International Theory: Critical Investigations* (Houndmills and London: Macmillan, 1995), pp. 212–41.

Holenberg, Sören and Peter Esaiasson, *De folkvalda: en bok om riksdagsstedamöterna och den representativa demokratin i Sverige* (Stockholm: Bonnier, 1998).

Hollis, Martin and Steve Smith, *Explaining and Understanding International Relations* (Oxford: Clarendon Press, 1991).

Holm, Ulla and Pertti Joenniemi, *North, South and the Figure of Europe: Changing Relationships* (Copenhagen: Copenhagen Peace Research Institute, 2001).

Holmberg-Harva, Uno, 'Petsamonmaan kolttain pyhät paikat,' *Suomalais-ugrilaisen seuran toimituksia* LVIII (1928), pp. 15–25.

Holsti, Kalevi J., *International Politics: A Framework for Analysis. Seventh Edition* (Englewood Cliffs: Prentice Hall, 1995).

Hopf, Ted, 'The Promise of Constructivism in International Relations Theory,' *International Security*, Vol. 23, No. 1 (Summer 1998), pp. 171–200.

Hultblad, Filip, *Övergång från nomadism till agrar bosättning i Jokkmokks socken*, Acta Lapponica XIV (Stockholm: Nordiska Museet, 1968).

Huntford, Roland, *Roald Amundsens oppdagelsesreiser i bilder* (Oslo: Grøndahl & Søn Forlag, 1988).

Huntington, Samuel P., 'The Clash of Civilizations?' *Foreign Affairs*, Vol. 72, No. 3 (Summer 1993), pp. 22–49.

Huseby, Beate and Ola Listhaug, 'Identifications of Norwegians with Europe: The Impact of Values and Centre-Periphery Factors,' in Ruud de Moor, ed., *Values in Western Societies* (Tilburg: Tilburg University Press, 1995), pp. 137–62.

Häkli, Jouni, 'Manufacturing Provinces. Theorizing the Encounters between Governmental and Popular "Geographs" in Finland,' in Gearóid Ó Tuathail and Simon Dalby, eds., *Rethinking Geopolitics* (London: Routledge, 1998), pp. 131–51.

Häkli, Jouni, 'Cultures of Demarcation: Territory and National Identity in Finland,' in Guntram Herb and David Kaplan, eds., *Nested Identities: Identity, Territory, and Scale* (Lanham: Rowman & Littlefield, 1999), pp. 123–49.

Høydalsnes, Eli, 'Bilder i nordnorsk ramme: Kunsthistoriske betraktninger om essens og konstruksjon,' in Trond Thuen, ed., *Landskap, region og identitet: Debatter om det nordnorske* (Bergen: Norges forskningsråd, 1999), pp. 39–60.

Ingebritsen, Christine, 'Norwegian Political Economy and European Integration: Agricultural Power, Policy Legacies and EU Membership,' *Cooperation and Conflict*, Vol. 30, No. 4 (December 1995), pp. 349–63.

Ingebritsen, Christine, *The Nordic States and the European Unity* (Ithaca and London: Cornell University Press, 1998).

Ingold, Tim, *The Scolt Lapps Today* (Cambridge: Cambridge University Press, 1976).

Isaksson, Pekka, *Kumma kuvajainen* (Jyväskylä: Kustannus-Puntsi, 2001).

Jachtenfuchs, Markus und Beate Kohler-Koch, *The Transformation of Governance in the European Union* (Mannheim: Mannheimer Zentrum für Europäische Sozialforschung, 1995).

Jackson, Peter, *Maps of Meaning: An Introduction to Cultural Geography* (London/New York: Routledge, 1987).

Jackson, Peter, 'Berkeley and Beyond: Broadening the Horizons of Cultural Geography,' *Annals of the Association of American Geographers*, Vol. 83, No. 3 (1993), pp. 519-20.

Jackson, Peter, 'Rematerializing Social and Cultural Geography,' *Social & Cultural Geography*, Vol. 1, No. 1 (September 2000), pp. 9-14.

Jacobs, Jane M., 'Editorial: Difference and Its Other,' *Transactions of the Institute of British Geographers*, Vol. 25, No. 4 (2000), pp. 403-407.

Jahn, Detlef, Pertti Pesonen, Tore Slaata and Leif Åberg, 'The Acteurs and the Campaign,' in Anders T. Jenssen, Pertti Pesonen and Mikael Gilljam, eds., *To Join or Not to Join: Three Nordic Referendums on Membership of the European Union* (Oslo: Scandinavian University Press, 1998), pp. 61-81.

Jama, Olavi, 'Haaparannan lukiosta sipirjaan. Torniolaakson kirjallisuus kahden kansalliskulttuurin marginaalissa,' in Matti Savolainen, ed., *Marginalia ja kirjallisuus. Ääniä suomalaisen kirjallisuuden reunoilta* (Helsinki: Suomalaisen Kirjallisuuden Seura, 1995), pp. 93-143.

Jay, Martin, 'Scopic Regimes of Modernity,' in Scott Lash and Jonathan Friedman, eds., *Modernity and Identity* (Oxford: Blackwell, 1992), pp. 178-95.

Jensen, Torben K., *Politik i praxis: aspekter af danske folketingsmedlemmers politiske kultur og lisverden* (Frederiksberg: Samfundslitteratur, 1993).

Jenssen, Anders T., *Hva gjorde EU-striden med det ideologiske landskapet i Norge?* (Trondheim: Universitet i Trondheim, Institutt for sosiologi og statsvitenskap, 1995).

Jenssen, Anders T., 'Ouverturen,' in Anders T. Jenssen and Henry Valen, eds., *Brussel midt imot Folkeavstemningen om EU* (Oslo: Gyldendal, 1996), pp. 18-29.

Jenssen, Anders T., 'Personal Economies and Economic Expectations,' in Anders T. Jenssen, Pertti Pesonen and Mikael Gilljam, eds., *To Join or Not to Join: Three Nordic Referendums on Membership of the European Union* (Oslo: Scandinavian University Press, 1998), pp. 194-234.

Jenssen, Anders T., Kathrine L. Moen, Inge Ramberg, Kristen Ringdal and Arve Østgaard, *Medlemsundersøkelsen: holdninger og sosial bakgrunn blant medlemmer i Nei til EU: Europabevegelsen og Sosialdemokrater mot EU: dokumentasjon* (Trondheim: Universitet i Trondheim, Institutt for sosiologi og statsvitenskap, 1994).

Jenssen, Anders T., Pertti Pesonen and Mikael Gilljam, eds., *To Join or Not to Join: Three Nordic Referendums on Membership of the European Union* (Oslo: Scandinavian University Press, 1998).

Jenssen, Anders T. and Henry Valen, eds., *Brussel midt imot Folkeavstemningen om EU* (Oslo: Gyldendal, 1996).

Joenniemi, Pertti, 'Regionality? A Sovereign Principle in International Relations?' in Heikki Patomäki, ed., *Peaceful Changes in World Politics* (Tampere: Tampere Peace Research Institute, 1995), pp. 337–79.

Joenniemi, Pertti, 'Changing Politics along Finland's Borders. From Norden to the Northern Dimension,' in Pirkkoliisa Ahponen and Pirjo Jukarainen, eds., *Tearing Down the Curtain, Opening the Gates. Northern Boundaries in Change* (Jyväskylä: SoPhi, 2000), pp. 114–32.

Johansen, Øystein Kock, ed., *Norske maritime oppdagere og ekspedisjoner gjennom tusen år* (Oslo: Index, 1999).

Jones, Michael, 'The Elusive Reality of Landscape,' *Norsk Geografisk Tidsskrift*, Vol. 45, No. 4 (1991), pp. 229–44.

Jones, Michael, *Perspektiver på landskap og hvordan det kan anvendes i sørsamisk sammenheng* (Trondheim: Norwegian University of Science and Technology, Department of Geography, 1999), at < http://www.uit.no/ssweb/dok/seminar/sorsamisk/JONES.html >.

Jónsson, Finnur, *Det islandse Altings historie i omrids* (Copenhagen: Hest, 1922).

Järvikoski, Timo, *Kemihaaran allasalueen väestötutkimus* (Helsinki: Vesihallitus, 1975).

Järvikoski, Timo and Juha Kylämäki, *Isohaaran padosta Kemijoen karvalakkilähetystöön. Tutkimus Kemijoen kalakorvauskiistasta* (Turku: University of Turku, 1981).

Järvinen, Jukka, Arja-Liisa Räisänen, Pirkko Siitari and Jussi Vilkuna, eds., *Pohjoinen valokuva – dokumenttivalokuvaus Pohjois-Suomessa* (Oulu: Pohjoinen valokuvakeskus and Kustannus-Puntsi, 2000).

Kaartvedt, Alf, Rolf Danielsen and Tim Greve, *Det Norske storting gjennom 150 år*, 4 vols. (Oslo: Gyldendal, 1964).

Kabakov, Ilya and Pavel Pepperstein, 'Tennis Game,' in Marketta Seppälä, ed., *Strangers in the Arctic. 'Ultima Thule' and Modernity* (Pori: Pori Art Museum, 1996), pp. 148–66.

Kajala, Liisa, *Lappilaisten näkemyksiä metsien hoidosta ja käytöstä* (Helsinki: Metsäntutkimuslaitos, 1997).

Kajala, Liisa, ed., *Lapin metsästrategia* (Helsinki: Maa- ja metsätalousministeriö, 1996).

Karemaa, Outi, *Vihollisia, vainoojia, syöpäläisiä* (Helsinki: Suomen Historiallinen Seura, 1998).

Karjalainen, Pauli Tapani, *Geodiversity as a Lived World. On the Geography of Existence* (Joensuu: University of Joensuu, Faculty of Social Sciences, 1986).

Karjalainen, Pauli Tapani, 'Elämä tässä talossa: lähikartoitusta kirjallisuuden kautta,' *Alue ja Ympäristö*, Vol. 24, No. 2 (1995), pp. 15–24.

Karjalainen, Pauli Tapani, 'Mahdollisten maisemien semantiikkaa,' *Terra*, Vol. 107, No. 2 (1995), pp. 123–5.

Karjalainen, Pauli Tapani, 'The Significance of Place: An Introduction,' in Pauli Tapani Karjalainen and Pauline von Bonsdorff, eds., *Place and Embodiment* (Helsinki: Lahti Research and Training Centre, 1995), pp. 9–12.

Karlqvist, Anders and Olle Melander, 'Sverige åter i Antarktis,' *Ymer*, Vol. 110 (1990), pp. 11–21.

Karppi, Kristiina and Johan Eriksson, eds., *Conflict and Cooperation in the North* (Umeå: Kulturgräns Norr, 2002).

Katz, Cindi and Andrew Kirby, 'In the Nature of Things: the Environment and Everyday Life,' *Transactions of the Institute of British Geographers*, Vol. 16, No. 3 (1991), pp. 259–71.

Katzenstein, Peter J., 'Introduction: Alternative Perspectives on National Security,' in Peter J. Katzenstein, ed., *The Culture of National Security: Norms and Identity in World Politics* (New York: Columbia University Press, 1996), pp. 1–32.

Kekkonen, Urho, *Onko maallamme malttia vaurastua?* (Helsinki: Otava, 1952).

Keller, Evelyn Fox, *Reflections on Gender and Science* (New Haven: Yale University Press, 1985).

Kellogg, W.W., 'The Poles: A Key to Climate Change,' in Gunter Weller and Sue Ann Bowling, eds., *Climate of the Arctic* (Fairbanks: Geophysical Institute, 1975), pp. v–vi.

Kelly, Aileen, 'In the Promised Land,' The New York Review of Books, Vol. XLVIII, No. 19 (29 November 2001), pp. 45–8.

Kent, Neil, *The Soul of the North: A Social, Architectural and Cultural History of the Nordic Countries, 1700–1940* (London: Reaktion Books, 2000).

Khorkina, Svetlana A., *Russia and Norway in the Arctic 1890–1917. A Comparative Study of Russian and Norwegian Traditions of Polar Exploration and Research* (Tromsø: University of Tromsø, 1999).

King, A.D., 'Opening Up the Social Sciences to the Humanities: A Response to Peter Taylor,' *Environment and Planning A*, Vol. 28, No. 11 (November 1996), pp. 1954–9.

King, J.C.H. and Henrietta Lidchi, 'Introduction,' in J.C.H. King and Henrietta Lidchi, eds., *Imaging the Arctic* (London: The British Museum Press, 1996), pp. 11–18.

Kitchin, R.M. and P.J. Hubbard, 'Research, Action and "Critical" Geographies,' *Area*, Vol. 31, No. 3 (1999), pp. 195–8.

Kivikoski, Mikko, 'Onko Itä-Eurooppaa enää olemassa?' in Pauli Kettunen, Auli Kultanen and Timo Soikkanen, eds., *Jäljillä. Kirjoituksia historian ongelmista. Osa 2* (Turku: Kirja-Aurora 2000), pp. 163–79.

Kivimäki, Timo, 'Integration and Regionalization in the Northern Calotte,' in The Finnish Institute of International Affairs, *Northern Dimensions* (Helsinki: The Finnish Institute of International Affairs, 1998), pp. 73–85.

Knuuttila, Seppo, Ilkka Liikanen, Pertti Rannikko, Hannu Itkonen, Merja Koistinen, Jukka Oksa and Sinikka Vakimo, *Kyläläiset, kansalaiset* (Joensuu: University of Joensuu, Karelian Institute, 1996).

Kobayashi, Audrey, 'Coloring the Field: Gender, "Race," and the Politics of Fieldwork,' *Professional Geographer*, Vol. 46, No. 1 (1994), pp. 73–80.

Kokkonen, Pellervo, 'Kartan sosiaalinen todellisuus,' in Tuukka Haarni, Marko Karvinen, Hille Koskela and Sirpa Tani, eds., *Tila, paikka ja maisema* (Tampere: Vastapaino, 1997), pp. 53–72.

Konttinen, Esa, 'Uusien liikkeiden tuleminen subjektiviteetin puolustamisen kulttuuri-ilmastossa,' in Kaj Ilmonen and Martti Siisiäinen, eds., *Uudet ja vanhat liikkeet* (Tampere: Vastapaino, 1998), pp. 187–217.

Korpijaakko, Kaisa, 'Saamelaiset ja suomalainen historiankirjoitus,' *Historiallinen aikakauskirja*, Vol. 88, No. 3 (1990), pp. 241–6.

Kristeva, Julia, *Revolution in Poetic Language* (New York: Columbia University Press, 1984).

Kuhnle, Stein, 'Norwegen,' *Aus Politik und Zeitgeschichte*, B 43 (1992), pp. 12–21.

Kurth, James, 'America and the West: Global Triumph or Western Twilight?' *Orbis*, Vol. 45, No. 3 (Summer 2001), pp. 333–41.

Käkönen, Jyrki, *Perspectives on Environment, State and Civil Society. The Arctic in Transition* (Uppsala: University of Uppsala, Environmental Policy and Society, 1994).

Käkönen, Jyrki, 'North Calotte as a Political Actor,' in Jyrki Käkönen, ed., *Dreaming of the Barents Region: Interpreting Cooperation in the Euro-Arctic Rim* (Tampere: Tampere Peace Research Institute, 1996), pp. 55–88.

Lagerlöf, Selma, *Nils Holgerssons underbara resa genom Sverige*, 2 vols. (Stockholm: Albert Bonnier, 1906–1907).

Laïdi, Zaki, *A World Without Meaning: The Crisis of Meaning in International Politics* (London and New York: Routledge, 1998).

Lamers, Karl and Wolfgang Schäuble, 'Überlegungen zur europäischen Politik. Positionspapier der CDU/CSU-Bundestagsfraktion vom 1. September 1994,' *Blätter für deutsche und internationale Politik*, No. 10 (October), 1994, pp. 1271–80.

Lantto, Patrik, *Tiden börjar på nytt – en analys av samernas etnopolitiska mobilisering i Sverige 1900–1950* (Umeå: Kulturgräns Norr, 2000).

Lapid, Yosef, 'Culture's Ship: Returns and Departures in International Relations Theory,' in Yosef Lapid and Friedrich Kratochwil, eds., *The Return of Culture and Identity in IR Theory* (Boulder and London: Lynne Rienner, 1996), pp. 3–20.

Lapin lääninhallitus, *Lapin läänin kehityksestä 1970- ja 1980-lukujen vaihteessa* (Rovaniemi: Lapin lääninhallitus, 1982).

Lapin Seutukaavaliitto, *Väestö ja työvoima Lapissa 1960–2030* (Rovaniemi: Lapin Seutukaavaliitto, 1988).

Larramendi, Ramón Hernando de, 'Perilous Journey. Three Years Across the Arctic,' *National Geographic*, Vol. 187, No. 1 (January 1995), pp. 120–38.

Lash, Scott and John Urry, *Economies of Signs and Space* (London: Sage, 1994).

Lebow, Richard Ned, 'The Long Peace, the End of the Cold War, and the Failure of Realism,' in Richard Ned Lebow and Thomas Risse-Kappen, eds., *International Relations Theory and the End of the Cold War* (New York: Columbia University Press, 1995), pp. 23–56.

Lefebvre, Henri, *The Production of Space* (Oxford: Blackwell, 1991).

Lehti, Marko, 'Competing or Complementary Images: The North and the Baltic World from the Historical Perspective,' in Hiski Haukkala, ed., *Dynamic Aspects of the Northern Dimension* (Turku: Turku University, Jean Monnet Unit, 1999), pp. 21–45.

Lehtinen, Ari Aukusti, 'Kalottipolitiikka ja saamelainen regionalismi,' *Terra*, Vol. 99, No. 1 (1987), pp. 13–18.

Lehtinen, Ari Aukusti, 'The Northern Natures – A Study of the Forest Question Emerging Within the Timber-line Conflict in Finland,' *Fennia*, Vol. 169, No. 1 (1991), pp. 57–169.

Lehtinen, Ari Aukusti, 'Geography and Biopower: In Search of Authentic Land- and Lifescapes (A Nordic Interpretation),' Paper presented at the Nordic Symposium for Critical Human Geography, University of Oslo, 5–6 October, 2001.

Lehtinen, Ari Aukusti and Teijo Rytteri, 'Backwoods Provincialism: The Case of Kuusamo Forest Common,' *Nordia Geographical Publications*, Vol. 27, No. 1 (1998), pp. 27–37.

Lehtola, Veli-Pekka, *Rajamaan identiteetti. Lappilaisuuden rakentuminen 1920- ja 1930-luvun kirjallisuudessa* (Helsinki: Suomalaisen Kirjallisuuden Seura, 1997).

Lehtola, Veli-Pekka, *Nickul - rauhan mies, rauhan kansa* (Jyväskylä: Kustannus-Puntsi, 2000).

Lehtonen, Mikko, *Pikku jättiläisiä. Maskuliinisuuden kulttuurinen rakentuminen* (Tampere: Vastapaino, 1995).

Leighly, John, *Land and Life. A Selection from the Writings of Carl Ortwin Sauer* (Berkeley: University of California Press, 1993).

Leighton, Marian K., *The Soviet Threat to NATO's Northern Flank* (New York: National Strategy Information Center, 1979).

Lemberg, Hans, 'Zur Entstehung des Osteuropabegriffs im 19. Jahrhundert. Vom "Norden" zum "Osten" Europas,' *Jahrbuch für Geschichte Osteuropas*, Vol. 33, No. 1 (1985), pp. 48–91.

LeTourneau, Michael, 'Salmon in Sachs. Film Depicts Temperature Change,' *Northern News Service* (27 November 2000).

Ley, David and Marwyn S. Samuels, 'Introduction: Context of Modern Humanism in Geography,' in David Ley and Marwyn S. Samuels, eds., *Humanistic Geography: Prospects and Problems* (London: Groom Helm, 1978), pp. 1–17.

Liksom, Rosa, *Yhden yön pysäkki* (Espoo: Weilin + Göös, 1985).

Liksom, Rosa, *Unohdettu vartti* (Espoo: Weilin + Göös, 1986).

Liksom, Rosa, *Tyhjän tien paratiisit* (Porvoo: WSOY, 1989).

Liksom, Rosa, *One Night Stands* (New York: Serpent's Tail, 1993).

Liljeqvist, Gösta H., *High Latitudes. A History of Swedish Polar Travels and Research* (Stockholm: Swedish Polar Secretariat & Streiffert, 1993).

Limerov, Pavel, 'When the Northern Wind Begins to Blow,' *N66. Culture in the Barents Region*, No. 4 (1998), pp. 20–22.

Lindner, Wolf, Prisca Lanfranchi and Ewald R. Weibel, eds., *Schweizer Eigenart - eigenartige Schweiz: Der Kleinstaat im Kräftefeld der europäischen Integration* (Bern, Stuttgart, Vienna: Akademische Kommission der Universität Bern, 1996).

Lindström, Eric, *Riksdag och regering: ansvarsfördeling, arbetsformer, beslutsprocesser* (Stockholm: Liber, 1981).

Lipponen, Paavo, '10 Years of Arctic Environmental Cooperation,' in *10 Years of Arctic Environmental Cooperation* (Helsinki: Ministry for Foreign Affairs of Finland, 2001), pp. 15–18.

Listhaug, Ola, Anders T. Jenssen and Per A. Pettersen, 'The EU Referendum in Norway: Continuity and Change,' *Scandinavian Political Studies*, Vol. 19, No. 3 (September 1996), pp. 257–79.

Loderer, Benedikt, 'Zwei Schweizen,' *Ästhetik und Kommunikation*, Vol. 30, No. 107 (December 1999), pp. 45–50.

Lorenz, Kathrin, *Europäische Identität? Theoretische Konzepte im Vergleich* (Berlin: Freie Universität Berlin, Arbeitsschwerpunkt Hauptstadt Berlin, 1999).

Lund, Swein and Carlotte Persen, 'Norske samer og EU,' in Nei til EU, *Norge og EU: virkninger av medlemskap i Den europeiske union* (Trondheim: Aktietrykkeriet i Trondhjem, 1994), pp. 247–53.

Lundström, Sven, 'Andrée-expeditionen i vår tids ljus,' *Ymer*, Vol. 117 (1998), pp. 39–49.

Luzin, Gennady P., Michael Pretes and Vladimir V. Vasiliev, 'The Kola Peninsula: Geography, History and Resources,' *Arctic*, Vol. 47, No. 1 (March 1994), pp. 1-15.

Lyck, Lise, ed., *The Faroese Economy in a Strategic Perspective* (Stockholm: Nordic Institute for Regional Policy Research, 1997).

Lynge, Aqqaluk, 'From Environmental Protection to Sustainable Development: An Inuit Perspective,' in *10 Years of Arctic Environmental Cooperation* (Helsinki: Ministry for Foreign Affairs of Finland, 2001), pp. 51-3.

Lyotard, Jean-Francois, *La condition post-moderne* (Paris: Minut, 1979).

Makine, Andreï, *Requiem for the East* (London: Sceptre, 2001).

Malaurie, Jean (in conversation with Jan Borm), 'Walrossuppe, Seehundblut. Die Kultur der Inuit – ein Ethnographenleben im ewigen Eis,' *Lettre International*, No. 56 (Spring 2002), pp. 80-90.

Marmier, Xavier, *Pohjoinen maa: 1800-luvun Lappia ja Suomea ranskalaisen silmin* (Helsinki: Suomalaisen Kirjallisuuden Seura, 1999).

Massa, Ilmo, 'Problem of the Development of the North Between the Wars: Some Reflections of Väinö Tanner's Human Geography,' *Fennia*, Vol. 162, No. 2 (1984), pp. 201-15.

Massa, Ilmo, *Pohjoinen luonnonvalloitus: suunnistus ympäristöhistoriaan Lapissa ja Suomessa* (Helsinki: Gaudeamus, 1994).

Massachusetts Institute of Technology (MIT), *Inadvertent Climate Modification. Report of the Study of Man's Impact on Climate* (Cambridge: MIT, 1971).

Massachusetts Institute of Technology (MIT), *Man's Impact on the Global Environment. Assessment and Recommendations for Action. Report of the Study of Critical Environmental Problems* (Cambridge: MIT, 1970).

Mathewson, Kent, 'Cultural Landscape and Ecology II: Regions, Retrospects, Revivals,' *Progress in Human Geography*, Vol. 23, No. 2 (1999), pp. 267-81.

Mathiesen, Thomas, *Makt och motmakt* (Gothenburg: Korpen, 1982).

Maunuksela, Arja, 'Kuvia pohjoisesta identiteetistä,' *Helsingin Sanomat*, 3 August 2000, p. B8.

McCannon, John, *Red Arctic, Polar Exploration and the Myth of the North in the Soviet Union 1932-1939* (Oxford: Oxford University Press, 1998).

McGhee, Robert, *Ancient People of the Arctic* (Vancouver: UNB Press, 1996).

McRobbie, Angela, 'The Place of Walter Benjamin in Cultural Studies,' in Simon During, ed., *The Cultural Studies Reader. Second Edition* (London/New York: Routledge, 1999), pp. 78-96.

Medvedev, Sergei, 'The Blank Space: Glenn Gould, Russia, Finland and the North,' *International Politics*, Vol. 38, No. 1 (March 2001), pp. 91-102.

Miekkavaara, Leena, 'A.E. Nordenskiöld Jäämerellä,' in Markku Löytönen, ed., *Matka-arkku. Suomalaisia tutkimusmatkailijoita* (Helsinki: Suomalaisen Kirjallisuuden Seura, 1990), pp. 104-37.

Mikko, Kenneth, 'Dear Reader!' *N66. Culture in the Barents Region*, No. 1 (1997), p. 2.

Mikko, Kenneth, 'Hans Ragnar Mathisen. Samernas kartograf,' *N66. Culture in the Barents Region*, No. 4 (1998), pp. 2-4.

Mikko, Kenneth, 'Screaming all over the World,' *N66. Culture in the Barents Region*, No. 4 (1998), pp. 6-7.

Mikko, Kenneth, 'Barents från ovan,' *N66. Culture in the Barents Region*, No. 7 (1999), pp. 10–15.

Mikko, Kenneth, 'Just Kicking Around,' *N66. Culture in the Barents Region*, No. 10 (2000), p. 4.

Ministry for Foreign Affairs of Finland, 'Conclusions of Chair,' in *10 Years of Arctic Environmental Cooperation* (Helsinki: Ministry for Foreign Affairs of Finland, 2001), pp. 9–12.

Mitchell, Don, *Cultural Geography: A Critical Introduction* (Oxford: Blackwell, 2000).

Mitchell, W.J.T., *Iconology: Image, Text, Ideology* (Chicago and London: The University of Chicago Press, 1986).

Moisio, Sami, 'Pohjoisen ulottuvuuden geopolitiikka: pohjoinen periferia ja uuden Euroopan alueellinen rakentaminen,' *Terra*, Vol. 112, No. 3 (2000), pp. 117–28.

Moravcsik, Andrew, 'Preferences and Power in the European Community: A Liberal Intergovernmentalist Approach,' in Simon Bulmer and Andrew Scott, eds., *Economist and Political Integration in Europe: Internal Dynamics and Global Context* (Oxford: Blackwell, 1994), pp. 29–80.

Morin, Edgar, *Penser l'Europe* (Paris: Gallimard, 1990).

Morin, Karen M. and Lawrence D. Berg, 'Emplacing Current Trends in Feminist Historical Geography,' *Gender, Place & Culture*, Vol. 6, No. 4 (December 1999), pp. 311–30.

Mouritzen, Hans, 'Thule and Theory: Democracy vs. Elitism in Danish Foreign Policy,' in Bertel Heurlin and Hans Mouritzen, eds., *Danish Foreign Policy Yearbook 1998* (Copenhagen: Danish Institute of International Affairs, 1998), pp. 79–101.

Muiznieks, Nils, 'A European Northern Dimension – A Latvian and a Swedish Perspective,' Report from a seminar organized by the Olof Palme International Center and the Latvian Institute of International Affairs, Riga, Latvia, 10–11 December 1999.

Murden, Simon, 'Cultural Conflict in International Relations: The West and Islam,' in John Baylis and Steve Smith, eds., *The Globalization of World Politics: An Introduction to International Relations* (Oxford: Oxford University Press, 1997), pp. 374–90.

Mänty, Jorma and Neil Pressman, eds., *Cities Designed for Winter* (Tampere: Tampere University of Technology, 1988).

Möttölä, Kari, ed., *The Arctic Challenge: Nordic and Canadian Approaches to Security and Cooperation in an Emerging International Region* (Boulder and London: Westview, 1988).

Münch, Richard, *Globale Dynamik, lokale Lebenswelten. Der schwierige Weg in die Weltgesellschaft* (Frankfurt: Suhrkamp, 1998).

Nairn, Karen, 'Doing Feminist Fieldwork about Geography Fieldwork,' in Pamela Moss, ed., *Feminist Geography in Practice. Research and Methods* (Oxford: Blackwell, 2002), pp. 146–59.

Nansen, Fridtjof, *Farthest North*, with introductions from Jon Krakauer and Roland Huntford (New York: Modern Library, 1999 [1897]).

National Geographic, Introduction to 'The Millennium series,' *National Geographic*, Vol. 193, No. 2 (February 1998), pp. 4–5.

Nei til EU, *Basic Programme*, at < http://www.neitileu.no/man/organisasjon/serv/ bjelker.html > .

Nei til EU, *Norge og EU: virkninger av medlemskap i Den europeiske union* (Trondheim: Aktietrykkeriet i Trondhjem, 1994).

Nesanelis, Dmitri and Victor Semjonov, 'The Village Universe,' *N66. Culture in the Barents Region*, No. 10 (1999), pp. 13–16.

Neumann, Iver B., 'A Region-building Approach to Northern Europe,' *Review of International Studies*, Vol. 20, No. 1 (January 1994), pp. 53–74.

Neumann, Iver B., *Uses of the Other. The 'East' in European Identity Formation* (Minneapolis: University of Minnesota Press, 1999).

Nevalainen, Jaana, 'Ikuisesta kesästä talvikaupunkiin – vuodenajat suomalaisessa kaupunkiympäristössä,' *Alue ja Ympäristö*, Vol. 24, No. 2 (1995), pp. 6–14.

Newsome, Brian, '"Dead Lands" or "New Europe"? Reconstructing Europe, Reconfiguring Eastern Europe: "Westerners" and the Aftermath of the Cold War,' *East European Quaterly*, Vol. XXXVI, No. 1 (March 2002), pp. 39–62.

Nickul, Karl, 'Suenjel, kolttain maa,' *Terra*, Vol. 45, No. 2 (1933), pp. 68–86.

Nickul, Karl, 'Petsamon eteläosan koltankieliset paikannimet kartografiselta kannalta,' *Fennia*, Vol. 60, No. 1 (1934).

Nickul, Karl, 'Eräs Petsamokysymys. Suonikylän alueesta kolttakulttuurin suojelualueena,' *Terra*, Vol. 47 (1935), pp. 81–105.

Nickul, Karl, *Skolt Lapp Community Suenjelsijd During the Year 1938*, ed. Ernst Manker, Acta Lapponica V (Stockholm: Nordiska Museet, 1948).

Nickul, Karl, 'Place Names in Suenjel – a Mirror of Skolt History,' *Studia Ethnographica Uppsaliensia* V (Lund, 1964).

Noble, Allen G. and Ramesh Dhussa, 'Image and Substance: A Review of Literary Geography,' *Journal of Cultural Geography*, Vol. 10, No. 2 (1990), pp. 49–65.

Nordenskiöld, A.E., *Vegas färd kring Asien och Europa* (Stockholm: Biblioteksförlaget, 1960 [1880]).

Nordic Council, *Öppet för världens vindor. Norden* (Copenhagen: Nordic Council, 2000).

Norton, William, *Cultural Geography: Themes, Concepts, Analyses* (Don Mills: Oxford University Press, 2000).

Novack, Jennifer, 'The Northern Dimension in Sweden's EU Policies: From Baltic Supremacy to European Unity?' in Hanna Ojanen, ed., *The Northern Dimension: Fuel for the EU?* (Helsinki/Bonn: The Finnish Institute of International Affairs/Institut für Europäische Politik, 2001), pp. 78–106.

Nowotny, Helga, *Time. The Modern and Postmodern Experience* (Cambridge: Polity Press, 1994).

Numelin, Ragnar, 'Voittamaton Ruija: Pohjois-Norjan jälleenrakentaminen,' *Terra*, Vol. 73, No. 2 (1961), pp. 87–99.

Nuorgam, Anne, '10 Years of Arctic Environmental Cooperation,' in *10 Years of Arctic Environmental Cooperation* (Helsinki: Ministry for Foreign Affairs of Finland, 2001), pp. 39–42.

Official Statistics of Norway, *Statistical Yearbook of Norway 2000. 119th Issue* (Oslo/Kongsvinger: Gyldendal, 2000).

Ojanen, Hanna, 'How to Customize Your Union: Finland and the Northern Dimension of the EU,' in *Yearbook of the Finnish Institute of International Affairs* (Helsinki: Finnish Institute of International Affairs, 1999), pp. 13–26.

Ojanen, Hanna, 'Conclusions: Northern Dimension – Fuel for the EU's External Relations?' in Hanna Ojanen, ed., *The Northern Dimension: Fuel for the EU?* (Helsinki/Bonn: Finnish Institute of International Affairs/Institut für Europäische Politik, 2001), pp. 217–37.

Okey, Robin, 'Central Europe/Eastern Europe: Behind the Definitions,' *Past & Present*, No. 137 (1992), pp. 102–33.

Olsson, Gunnar, *Lines of Power/Limits of Language* (Minneapolis: University of Minnesota Press, 1991).

Olsson, Gunnar, 'Heretic Cartography,' *Ecumene*, Vol. 1, No. 3 (1994), pp. 215–34.

Olwig, Kenneth, 'Eurooppalaisen kansakunnan pohjoinen luonne,' in Svenolof Karlsson, ed., *Vapauden lähde. Pohjolan merkitys Euroopalle* (Helsinki: VAPK Publishing, 1992), pp. 158–82.

Olwig, Kenneth, 'Recovering the Substantive Nature of Landscape,' *Annals of the Association of American Geographers*, Vol. 86, No. 4 (1996), pp. 630–53.

Olwig, Kenneth, 'Landscape as a Contested Topos of Place, Community and Self,' in Paul Adams, Steven Hoelscher and Karen Till, eds., *Textures of Place: Geographies of Imagination, Experience, and Paradox* (Minneapolis: University of Minnesota Press, 2001), pp. 95–119.

Olwig, Kenneth, 'Landskabet som samfundsmæssig forførelse,' *Nordisk Samhällsgeografisk Tidskrift Debat*, at < http://www.geo.ruc.dk/NST/Debat/DebatOlwig.pdf >.

Osborne, Brian S., 'The Iconography of Nationhood in Canadian Art,' in Denis Cosgrove and Stephen Daniels, eds., *The Iconography of Landscape: Essays on the Symbolic Representation, Design and Use of Past Environments* (Cambridge: Cambridge University Press, 1988), pp. 162–78.

Oskarson, Maria and Kirsten Ringdal, 'The Arguments,' in Anders T. Jenssen, Pertti Pesonen and Mikael Gilljam, eds., *To Join or Not to Join: Three Nordic Referendums on Membership of the European Union* (Oslo: Scandinavian University Press, 1998), pp. 149–67.

Paasi, Anssi, *Neljä maakuntaa: maantieteellinen tutkimus aluetietoisuuden kehittymisestä* (Joensuu: University of Joensuu, 1986).

Paasi, Anssi, 'The Rise and the Fall of Finnish Geopolitics,' *Political Geography Quarterly*, Vol. 9, No. 1 (1990), pp. 53–65.

Paasi, Anssi, 'Deconstructing Regions: Notes on the Scales of Spatial Life,' *Environment and Planning A. Society and Place*, Vol. 23, No. 2 (1991), pp. 239–56.

Paasi, Anssi, 'Kulttuuri: maantieteellisiä näkökulmia,' *Alue ja Ympäristö*, Vol. 20, No. 2 (1991), pp. 2–19.

Paasi, Anssi, *Territories, Boundaries and Consciousness: The Changing Geographies of the Finnish-Russian Border* (Chichister: John Wiley & Sons, 1996).

Paasilinna, Erno, *Petsamo – historiaa ja muistoja* (Helsinki: Otava, 1992).

Pakkala, Teuvo, 'Kaksi viimeistä naparetkeä,' in Anneli Kajanto, ed., *Namsarai* (Helsinki: Suomalaisen Kirjallisuuden Seura, 1999), pp. 46–56.

Panofsky, Erwin, *Studien zur Ikonologie der Renaissance. Zweite Auflage* (Cologne: Dumont, 1997).

Parker, Noel, 'Integrated Europe and its "Margins": Action and Reaction,' in Noel Parker and Bill Armstrong, eds., *Margins in European Integration* (Wiltshire: Macmillan Press, 2000), pp. 3–27.

Paterson, J.H., 'The Novelist and His Region: Scotland Through the Eyes of Sir Walter Scott,' *Scottish Geographical Magazine*, Vol. 81, No. 3 (1965), pp. 146–52.

Patten, Chris, 'Statement on Transatlantic Relations. Speech to a Pleanary Session of the European Parliament,' Strasbourg, 16 May 2001, at <http://www.Europa.eu.int/comm/externalrelations/patten/speech01204.htm>.

Paulaharju, Samuli, *Kolttain mailta* (Helsinki: Kirja, 1921).

Pehkonen, Samu, *"Tänne, muttei pidemmälle!"* – *Alta-kamppailu ja Pohjois-Norjan monikulttuurinen maisema* (Tampere: Tampere Peace Research Institute, 1999).

Peters, Evelyn J., 'Aboriginal People and Canadian Geography: A Review of the Recent Literature,' *Canadian Geographer*, Vol. 44, No. 1 (Spring 2000), pp. 44–55.

Petersen, Nikolaj, 'Denmark, Greenland, and Arctic Security,' in Kari Möttölä, ed., *The Arctic Challenge: Nordic and Canadian Approaches to Security and Cooperation in an Emerging International Region* (Boulder and London: Westview, 1988), pp. 39–73.

Petersen, Nikolaj, 'The H.C. Hansen Paper and Nuclear Weapons in Greenland,' *Scandinavian Journal of History*, Vol. 23, Nos. 1–2 (June 1998), pp. 21–44.

Petrone, Penny, ed., *Northern Voices: Inuit Writing in English* (Toronto: University of Toronto Press, 1998).

Pile, Steve, *The Body and the City. Psychoanalysis, Space and Subjectivity* (London: Routledge, 1996).

Pocock, Douglas C.D., 'Geography and Literature,' *Progress in Human Geography*, Vol. 12, No. 1 (1988), pp. 87–102.

Poikela, Kirsi-Marja, *Leipää vai luonnonsuojelua: sanomalehtikirjoittelun näkökulma Pohjois-Suomen vanhojen metsien suojeluohjelmaan* (Rovaniemi: University of Lapland, 1998).

Professional Geographer, Special Issue on 'Women in the Field: Critical Feminist Methodologies and Theoretical Perspectives,' *Professional Geographer*, Vol. 46, No. 1 (February 1994), pp. 54–102.

Puranen, Jorma, 'Foreword,' in Jorma Puranen, *Imaginary Homecoming* (Oulu: Pohjoinen, 1999), pp. 11–12.

Puranen, Jorma, *Imaginary Homecoming* (Oulu: Pohjoinen, 1999).

Puranen, Jorma, *Language Is a Foreign Country. Photographs 1991–2000* (Helsinki: The Finnish Museum of Photography, 2000).

Raik, Kristi, 'Estonian Perspectives on the Northern Dimension,' in Hiski Haukkala, ed., *Dynamic Aspects of the Northern Dimension* (Turku: Turku University, Jean Monnet Unit, 1999), pp. 151–66.

Raivo, Petri J., 'Maiseman kulttuurinen transformaatio. Ortodoksinen kirkko suomalaisessa kulttuurimaisemassa,' *Nordia Geographical Publications*, Vol. 25, No. 1 (1996), pp. 1–370.

Raivo, Petri J., 'The Limits of Tolerance: The Orthodox Milieu as an Element in the Finnish Cultural Landscape,' *Journal of Historical Geography*, Vol. 23, No. 3 (1997), pp. 327–39.

Rannikko, Pertti, 'Ympäristökamppailujen aallot,' in Ari Lehtinen and Pertti Rannikko, eds., *Pasilasta Vuotokselle. Ympäristökamppailujen uusi aalto* (Helsinki: Gaudeamus, 1994), pp. 11-28.

Rannikko, Pertti, 'Ympäristötietoisuus ja ympäristöristiriidat,' in Timo Järvikoski, Pekka Jokinen and Pertti Rannikko, *Näkökulmia ympäristösosiologiaan* (Turku: Turun täydennyskoulutuskeskus, 1995), pp. 65-91.

Rannikko, Pertti, 'Local Environmental Conflicts and Change in Environmental Consciousness,' *Acta Sociologica*, Vol. 39, No. 1 (1996), pp. 57-72.

Rannikko, Pertti, 'Combining Social and Ecological Sustainability in the Nordic Forest Periphery,' *Sociologia Ruralis*, Vol. 39, No. 3 (1999), pp. 394-410.

Rehn, Gösta, 'The Wages of Success,' in Stephen R. Graubard, ed., *Norden – The Passion for Equality* (Oslo: Norwegian University Press, 1986), pp. 143-75.

Renbeteskommionens af 1913 handlingar I:1 (Stockholm, 1917).

Renbeteskommissionens af år 1909 handlingar I (Helsingfors, 1912).

Rhodes, Edward, 'The United States and the Northern Dimension: America's Northern European Initiative,' in *International Perspectives on the Future of the Barents Euro-Arctic Region and the Northern Dimension*. Report from a think-tank seminar, Björkliden, Sweden, June 2001, pp. 41-56.

Rickard, Jolene, 'Sovereignty: A Line in the Sand,' *Aperture*, No. 139 (Summer 1995), pp. 51-60.

Ridanpää, Juha, 'Postcolonialism in a Polar Region? Relativity Concerning a Postcolonialist Interpretation of Literature from Northern Finland,' *Nordia Geographical Publications*, Vol. 27, No. 1 (1998), pp. 67-77.

Ridanpää, Juha, 'The Discursive North in a Few Scientific Discoveries,' *Nordia Geographical Publications*, Vol. 29, No. 2 (2000), pp. 25-37.

Ridanpää, Juha, 'Pentti Haanpää's Kairanmaa in Cultural Transformation,' in Sirpa Leppänen and Joel Kuortti, eds., *Inescapable Horizon: Culture and Context* (Jyväskylä: University of Jyväskylä, 2000), pp. 133-54.

Rikkinen, Kalevi, *Suuri Kuolan retki* (Helsinki: Otava, 1980).

Ringdal, Kirsten, 'Velgernes argumenter,' in Anders T. Jenssen and Henry Valen, eds., *Brussel midt imot Folkeavstemningen om EU* (Oslo: Gyldendal, 1996), pp. 45-66.

Ripley, Brian, 'Cognition, Culture, and Bureaucratic Politics,' in Laura Neack, Jeanne A.K. Hey and Patrick J. Haney, eds., *Foreign Policy Analysis: Continuity and Change in Its Second Generation* (Englewood Cliffs: Prentice Hall, 1995), pp. 85-97.

Rokkan, Stein and Derek W. Urwin, *Economy, Territory, Identity: Politics of West European Peripheries* (London, Beverly Hills, New Delhi: Sage, 1983).

Rolfsen, Nordahl, *Læsebog for folkeskolen*, 5 vols. (Kristiania: Jacob Dybwad, 1892-1895).

Rosberg, J.E., 'Napaseututtutkimus Lapinmeren rannoille, lähinnä Frans Josefin maassa ja Novaja Zemljassa,' *Terra*, No. 37 (1925), pp. 57-74.

Rose, Gillian, *Feminism and Geography. The Limits of Geographical Knowledge* (London: Polity Press, 1993).

Routledge, Marie, 'The North, Inuit Art Today,' in Pekka Lehmuskallio, Markku Lehmuskallio and Kati Kivimäki, eds., *Arctic Inua: Contemporary Eskimo Art from the Lehmuskallio Collection* (Rauma: Lönnström Art Museum, 2000), pp. 30-53.

Routledge, Paul, 'The Third Space as Critical Engagement,' *Antipode*, Vol. 28, No. 4 (1996), pp. 399–419.

Rowntree, Lester, 'Cultural/humanistic Geography,' *Progress in Human Geography*, Vol. 11, No. 4 (1987), pp. 558–64.

Ruggie, John Gerard, 'Territoriality and Beyond: Problematizing Modernity in International Relations,' *International Organization*, Vol. 47, No. 1 (Winter 1993), pp. 139–74.

Rundstrom, Robert A. and Martin S. Kenzer, 'The Decline of Fieldwork in Human Geography,' *Professional Geographer*, Vol. 41, No. 3 (1989), pp. 294–303.

Russo, Antonella, with research by Diego Mormorio, 'The Invention of Southernness: Photographic Travels and the Discovery of the Other Half of Italy,' *Aperture*, No. 132 (Summer 1993), pp. 58–69.

Ruth, Arne, 'The Second New Nation: The Mythology of Modern Sweden,' in Stephen R. Graubard, ed., *Norden – The Passion for Equality* (Oslo: Norwegian University Press, 1986), pp. 240–82.

Ryall, Anka, 'Antarctica and the Question of Women,' Paper presented at Writing the Journey: a Conference on American, British and Anglophone Travel Writers and Writing, University of Pennsylvania, 10–13 June 1999.

Rytchëu, Juri, *Die Suche nach der letzten Zahl* (Zurich: Unionsverlag, 2001).

Sack, Robert David, *Homo Geographicus* (Baltimore and London: The Johns Hopkins University Press, 1997).

Sagan, Scott D., *The Limits of Safety: Organizations, Accidents and Nuclear Weapons* (Princeton: Princeton University Press, 1993).

Said, Edward W., *Orientalism* (London: Routledge, 1978).

Said, Edward W., *Orientalism: Western Conceptions of the Orient. With a New Afterword* (London: Penguin, 1995).

Sairinen, Rauno, *Suomalaiset ja ympäristöpolitiikka* (Helsinki: Tilastokeskus, 1996).

Salovaara, Hannes, *Naparetkeilijöitä. Kuvauksia uljaista miehistä ja heidän seikkailurikkaista matkoistaan maapallon autioimmille ja vaikeapääsyisimmille seuduille* (Helsinki: Otava, 1929).

Saressalo, Lassi, 'History of the Sámi Area and People,' in Juha Pentikäinen and Marja Hiltunen, eds., *Cultural Minorities in Finland* (Helsinki: Finnish National Commission for Unesco, 1995), pp. 109–14.

Sauer, Carl, 'The Education of a Geographer,' *Annals of the Association of American Geographers*, No. 46 (1956), pp. 287–99.

Saussure, Ferdinand de, 'The Nature of the Linguistic Sign [1916],' in Lucy Burke, Tony Crowley and Alan Girvin, eds., *The Routledge Language and Cultural Theory Reader* (London: Routledge, 2000), pp. 21–32.

Savolainen, Matti, 'Keskusta, marginalia, kirjallisuus,' in Matti Savolainen, ed., *Marginalia ja kirjallisuus: Ääniä suomalaisen kirjallisuuden reunoilta* (Helsinki: Suomalaisen Kirjallisuuden Seura, 1995), pp. 7–35.

Sawhill, Steven G., 'Cleaning-up the Arctic's Cold War Legacy: Nuclear Waste and Arctic Military Environmental Cooperation,' *Cooperation and Conflict*, Vol. 35, No. 1 (March 2000), pp. 5–36.

Schartzbach, Martin, *Alfred Wegener. The Father of Continental Drift* (Berlin: Springer-Verlag, 1986).

Schlözer, August Ludwig, *Allgemeine Nordische Geschichte. Aus den neuesten und besten Nordischen Schriftstellern und nach eigenen Untersuchungen beschrieben und als eine Geographische und Historische Einleitung zur richtigern Kenntniss aller Skandinavistischen, Finnischen, Slavischen, Lettischen und Sibirischen Völker, besonders in alten und mittleren Zeiten* (Halle, 1771).

Semple, Ellen C., *Influences of Geographic Environment on the Basis of Ratzel's System of Anthropogeographie* (New York: Henry Holt, 1911).

Seppänen, Janne, 'Who Stole the Landscape? Some Remarks on Critical Practices in Landscape Photography,' in Leila Laukkanen, Janne Rissanen and Pirkko Siitari, eds., *Pohjoinen valokuva 91. Rajoilla – valokuva ja kulttuuri-identiteetti* (Oulu: Pohjoinen valokuvakeskus, 1991), pp. 13–23.

Seppänen, Janne, *Valokuvaa ei ole* (Helsinki: Musta Taide and The Finnish Museum of Photography, 2001).

Seurujärvi-Kari, Irja, 'Legal Position of the Sámi Today,' in Juha Pentikäinen and Marja Hiltunen, eds., *Cultural Minorities in Finland* (Helsinki: Finnish National Commission for Unesco, 1995), pp. 134–6.

Seurujärvi-Kari, Irja, 'Sámi Area and Populations in Finland,' in Juha Pentikäinen and Marja Hiltunen, eds., *Cultural Minorities in Finland* (Helsinki: Finnish National Commission for Unesco, 1995), pp. 101–105.

Sharapov, V.E. and D.A. Nesanelis, 'The Theme of Swinging in Shamanism and Folk Culture,' *N66. Culture in the Barents Region*, No. 2 (1997), pp. 2–3.

Sharp, Joanne P., 'Locating Imaginary Homelands: Literature, Geography and Salman Rushdie,' *GeoJournal*, Vol. 38, No. 1 (1996), pp. 119–27.

Shields, Rob, *Places on the Margin. Alternative Geographies of Modernity* (London: Routledge, 1991).

Sibley, David, *Geographies of Exclusion* (London: Routledge, 1995).

Silk, John A., 'Beyond Geography and Literature,' *Environment and Planning D: Society and Space*, Vol. 2, No. 2 (1984), pp. 151–78.

Smith, Neil, 'Socializing Culture, Radicalizing the Social,' *Social & Cultural Geography*, Vol. 1, No. 1 (September 2000), pp. 25–8.

Smith, Neil, *American Century: Roosevelt's Geographer and the Prelude to Globalization* (Berkeley and Los Angeles: University of California Press, 2003 forthcoming).

Smith, Susan J., 'Constructing Local Knowledge,' in John Eyles and David M. Smith, eds., *Qualitative Methods in Human Geography* (Oxford: Blackwell, 1988/1992), pp. 17–38.

Soja, Edward W., *Thirdspaces. Journeys to Los Angeles and Other Real-and-Imagined Places* (Oxford: Blackwell, 1996).

Sontag, Susan, *On Photography* (London: Penguin, 1978).

Stefánsson, Vilhjálmur, *Unsolved Mysteries of the Arctic* (Freepoint: Books for Libraries Press, 1972).

Steger, Will, 'Dispatches From the Arctic Ocean,' *National Geographic*, Vol. 189, No. 1 (January 1996), pp. 78–89.

Stenbaek, Marianne, 'Human Dimensions of Arctic Global Change and Trends in the Human Dimension of the Arctic,' in Peter A. Friis, ed., *The Internationalization Process and the Arctic. Proceedings from Arctic Research Forum Symposium* (Roskilde: Roskilde University, 1994), pp. 21–5.

Stenelo, Lars-Göran and Magnus Jerneck, eds., *The Bargaining Democracy* (Lund: Lund University Press, 1996).

Stern, Pamela, 'The History of Canadian Arctic Photography: Issues of Territorial and Cultural Sovereignty,' in J.C.H. King and Henrietta Lidchi, eds., *Imaging the Arctic* (London: The British Museum Press, 1996), pp. 47–52.

Stoddard, D.R., *On Geography and Its History* (Oxford: Basil Blackwell, 1986).

Stokke, Olav Schram and Ola Tunander, eds., *The Barents Region. Cooperation in Arctic Europe* (London, Thousand Oaks, New Delhi: Sage, 1994).

Stråth, Bo, *Folkhemmet mot Europa: Ett historisk perspektiv på 90-talet* (Stockholm: Tiden, 1992).

Suopajärvi, Leena, *Vuotos ja Ounasjoki-kamppailujen kentät ja merkitykset Lapissa* (Rovaniemi: University of Lapland, 2001).

Susiluoto, Paulo, 'Suomen ajan ihmismaantiedettä Petsamosta,' in Väinö Tanner, *Ihmismaantieteellisiä tutkimuksia Petsamon seudulta. 1. Kolttalappalaiset*, ed. Paulo Susiluoto (Helsinki: Suomalaisen Kirjallisuuden Seura, 2000), pp. 9–31.

Sylvester, Christine, 'Picturing the Cold War: An Art Graft/Eye Graft,' *Alternatives*, Vol. 21, No. 4 (October–December 1996), pp. 393–418.

Sæter, Martin, 'Norway and the European Union: Domestic Debate versus External Reality,' in Lee Miles, ed., *The European Union and the Nordic Countries* (London and New York: Routledge, 1996), pp. 133–49.

Sørensen, Øystein, ed., *Jakten på det norske* (Oslo: Gyldendal, 1998).

Sörlin, Sverker, *Framtidslandet. Debatten om Norrland och naturresurserna under det industriella genombrottet* (Stockholm: Carlsson Bokförlag, 1988).

Sörlin, Sverker, *Hans W:sson Ahlman. Arctic Research and Polar Warming. From a National to an International Scientific Agenda 1929–1952* (Umeå: Umeå University, Centre for Regional Studies, 1997).

Sörlin, Sverker, 'Hemkomsten: De dödas färd från Vitön,' *Ymer*, Vol. 117 (1998), pp. 50–59.

Sörlin, Sverker, Robert Marc Friedman, Michael Harbsmeier and Urban Wråkberg, *Det nordliga rummet: Polarforskingen och de nordiska länderna* (Umeå: The International Research Network on the History of Polar Science, 1995).

Tanner, Väinö, 'Voidaanko Petsamon aluetta käyttää maan hyödyksi?' *Fennia*, Vol. 49, No. 3 (1927).

Tanner, Väinö, 'Petsamon alueen paikannimiä,' *Fennia*, Vol. 49, No. 2 (1928).

Tanner, Väinö, 'Antropogeografiska studier inom Petsamo-området. 1. Skoltlapparna,' *Fennia*, Vol. 49, No. 4 (1929).

Tanner, Väinö, 'Om Petsamo-kustlapparnas sägner om forntida underjordiska boningar, s. k. jennam'vuölas'kuatt,' *Finskt Museum* XXXV (1929), pp. 1–24.

Tanner, Väinö, *Ihmismaantieteellisiä tutkimuksia Petsamon seudulta. 1. Kolttalappalaiset*, ed. Paulo Susiluoto (Helsinki: Suomalaisen Kirjallisuuden Seura, 2000).

Telò, Mario, 'Introduction: Globalization, New Regionalism and the Role of the European Union,' in Mario Telò, ed., *European Union and New Regionalism: Regional Actors and Global Governance in a Post-Hegemonic Era* (Aldershot, Burlington, Singapore: Ashgate, 2001), pp. 1–21.

Terebikhin, Nikolaj M., 'Bjarmaland. Spiritual Culture in the Barents Region,' *N66. Culture in the Barents Region*, No. 1 (1997), pp. 21-3.

Thaysen, Uwe, ed., *Der Zentrale Runde Tisch der DDR. Wortprotokoll und Dokumente* (Opladen: Westdeutscher Verlag, 1999).

Thordarson, Mathias, *The Althing. Iceland's Thousand Year Parliament, 930-1930* (Reykjavik: Jonsson, 1930).

Thorsen, Svend, *Danmarks Folketing: om dets hus og historie* (Copenhagen: Schultz, 1961).

Thorsteinsson, Björn and Thorstein Josepsson, *Thingvellir: Birthplace of a Nation* (Reykjavik: Heimskringla, 1961).

Thrift, Nigel, 'Introduction: Dead or Alive?' in Ian Cook, David Crouch, Simon Naylor and James R. Ryan, eds., *Cultural Turns/Geographical Turns: Perspectives on Cultural Geography* (Harlow: Pearson Education, 2000), pp. 1-6.

Tiilikainen, Teija, *Europe and Finland: Definining the Political Identity of Finland in Western Europe* (Aldershot, Brookfield, Singapore, Sydney: Ashgate, 1998).

Tikkanen, Matti and Pentti Viitala, 'Pieni Kuolan retki,' *Terra*, Vol. 99, No. 1 (1987), pp. 3-12.

Todorov, Tzvetan, *The Conquest of America. The Question of the Other* (New York: Harper Perennial, 1992).

Tuan, Yi-Fu, 'Humanistic Geography,' *Annals of the Association of American Geographers*, Vol. 66, No. 2 (1976), pp. 266-76.

Tuan, Yi-Fu, 'Literature and Geography: Implications for Geographical Research,' in David Ley and Marwyn S. Samuels, eds., *Humanistic Geography: Prospects and Problems* (London: Groom Helm, 1978), pp. 194-206.

Tuan, Yi-Fu, 'Language and the Making of Place: A Narrative-Descriptive Approach,' *Annals of the Association of American Geographers*, Vol. 81, No. 4 (December 1991), pp. 684-96.

Tuan, Yi-Fu, 'Desert and Ice: Ambivalent Aesthetics,' in Thomas Trummer, ed., *The Waste Land: Desert and Ice. Barren Landscapes in Photography* (Vienna: edition selene, 2001), pp. 68-99.

Tunander, Ola, *Cold Water Politics: The Maritime Strategy and Geopolitics of the Northern Front* (London, Newbury Park, New Delhi: Sage, 1989).

Tuohimaa, Sinikka, 'Pohjois-Suomen unohdetut naiskirjailijat,' in Katja Majasaari and Marja Rytkönen, eds., *Silmukoita verkossa. Sukupuoli, kirjallisuus ja identiteetti* (Oulu: University of Oulu, 1997), pp. 41-55.

Ursin, Martti, *Pohjois-Suomen tuhot ja jälleenrakennus saksalaissodan 1944-1945 jälkeen* (Oulu: Pohjoinen, 1980).

Vahtola, Jouko, 'Petsamo Suomen tieteen tutkimuskohteena,' in Jouko Vahtola and Samuli Onnela, eds., *Turjanmeren maa. Petsamonhistoria 1920-1944* (Jyväskylä: Petsamoseura, 1999), pp. 485-507.

Valentine, Gill, 'Whatever Happened to the Social? Reflections on the "Cultural Turn" in British Human Geography,' *Norsk Geografisk Tidsskrift*, Vol. 55, No. 3 (2001), pp. 166-72.

Valkonen, Tapani, 'Alueelliset erot,' in Tapani Valkonen, Risto Alapuro, Matti Alestalo, Riitta Jallinoja and Tom Sandlund, *Suomalaiset. Yhteiskunnan rakenne teollistumisen aikana* (Juva: WSOY, 1985), pp. 201-42.

Varanka, Piia, *Lappi matkailun näyttämöllä. Saamelaiskulttuuri ja luonto matkailun kulisseina* (Rovaniemi: University of Lapland, 2001).

Varjo, Uuno, *Lapin maatalous 1950-1959* (Rovaniemi: Lapin seutusuunnittelun kuntainliitto, 1967).

Varjo, Uuno,'Agriculture in North Lapland, Finland: Profitableness and Trends since World War II,' *Fennia*, Vol. 132 (1974), pp. 1–73.

Vaughan, Richard, *The Arctic. A History* (Phoenix Mill *et al.*: Alan Sutton, 1994).

Vilkuna, Kustaa, *Lohi. Kemijoen ja sen lähialueen lohenkalastuksen historia* (Helsinki: Otava, 1975).

Wagner, Peter, *Reading Iconotexts: From Swift to the French Revolution* (London: Reaktion Books, 1995).

Walker, R.B.J., 'The Subject of Security,' in Keith Krause and Michael C. Williams, eds., *Critical Security Studies: Concepts and Cases* (London: UCL Press, 1997), pp. 61–81.

Wallenius, Kurt Martti, *Ihmismetsästäjiä ja erämiehiä* (Helsinki: Otava, 1962 [1933]).

Wallenius, Kurt Martti, *Petsamo - mittaamattomien mahdollisuuksien maa* (Keuruu: Otava, 1994).

Waltz, Kenneth N., *Theory of International Politics* (Reading: Addison-Wesley, 1979).

Wamsley, Douglas and William Barr, 'Early Photographers of the Canadian Arctic and Greenland,' in J.C.H. King and Henrietta Lidchi, eds., *Imaging the Arctic* (London: The British Museum Press, 1996), pp. 36–45.

Warwick, Jack, *The Long Journey: Literary Themes of French Canada* (Toronto: University of Toronto Press, 1968).

Weatherford, Elizabeth, 'Native Visions: The Growth of Indigenous Media,' *Aperture*, No. 119 (Early Summer 1990), pp. 58–61.

Wendt, Alexander, 'Anarchy Is What States Make of It: The Social Construction of Power Politics,' *International Organization*, Vol. 46, No. 2 (Spring 1992), pp. 391–425.

Wendt, Alexander, 'Constructing International Politics,' *International Security*, Vol. 20, No. 1 (Summer 1995), pp. 71–81.

Wendt, Alexander, *Social Theory of International Politics* (Cambridge: Cambridge University Press, 1999).

Whitman, Nicholas, 'Technology and Visions: Factors Shaping Nineteenth-Century Arctic Photography,' in J.C.H. King and Henrietta Lidchi, eds., *Imaging the Arctic* (London: The British Museum Press, 1996), pp. 29–35.

Wiberg, Ulf, 'Regionformationer och nordeuropeiska integrationsperspektiv,' *Nordisk Samhällsgeografisk Tidskrift*, No. 21 (1995), pp. 55–67.

Willbanks, Thomas J. and Robert W. Kates, 'Global Change in Local Places: How Scale Matters,' *Climatic Change*, Vol. 43, No. 3 (1999), pp. 601–29.

Williams, Raymond, *Keywords: A Vocabulary of Culture and Society* (London: Fontana Press, 1988).

Wilson, J. Tyzo, *I.G.Y. The Year of the Moon* (London: Michael Joseph, 1961).

Wolff, Larry, *Inventing Eastern Europe* (Stanford: Stanford University Press), 1994.

Wråkberg, Urban, 'Minnets land: om den geografiske namngivningens historia i Arktis,' in Roger Sørheim and Leif Jonny Johannessen, eds., *Svalbard - fra ingenmannsland til del av Norge* (Trondheim: University of Trondheim, Centre for Environment and Development, 1995), pp. 121–42.

Wæver, Ole, 'Territory, Authority and Identity. The Late 20th Century Emergence of Neo-Medieval Political Structures in Europe,' paper presented at the first European Peace Research Association (EUPRA) conference in Florence, November 1991.

Wæver, Ole, 'Nordic Nostalgia: Northern Europe after the Cold War,' *International Affairs*, Vol. 68, No. 1 (January 1992), pp. 77–102.

Wæver, Ole, 'Regionalization in Europe – and in the Baltic Sea Area,' in *Cooperation in the Baltic Sea Area. Report from the Second Parliamentary Conference at the Storting, Oslo 22–24 April 1992* (Stockholm: Nordic Council 1992), pp. 16–21.

Wæver, Ole, 'Culture and Identity in the Baltic Sea Region,' in Pertti Joenniemi, ed., *Cooperation in the Baltic Sea Region* (New York: Taylor & Francis, 1993), pp. 23–48.

Wæver, Ole, 'The Baltic Sea: A Region after Post-Modernity?' in Pertti Joenniemi, ed., *Neo-Nationalism or Regionality: The Restructuring of Political Space around the Baltic Rim* (Stockholm: NordREFO, 1997), pp. 293–342.

Yearley, Steven, *Sociology, Environmentalism, Globalization* (London: Sage, 1996).

Young, Oran, 'The Age of the Arctic,' *Foreign Policy*, No. 61 (Winter 1985–1986), pp. 160–79.

Young, Robert, *White Mythologies: Writing History and the West* (London: Routledge, 1990).

Zetterberg, Hans L., 'The Rational Humanitarians,' in Stephen R. Graubard, ed., *Norden – The Passion for Equality* (Oslo: Norwegian University Press, 1986), pp. 79–96.

Žižek, Slavoj, *Tarrying with the Negative: Kant, Hegel and the Critique of Ideology* (Durham: Duke University Press, 1993.

Åquist, Ann-Cathrine, 'Idéhistorisk översikt,' in Bjørn T. Asheim, Sune Berger, Frank Hansen, Perttu Vartiainen and Ann-Cathrine Åquist, eds., *Traditioner i Nordisk kulturgeografi* (Uppsala: Nordisk Samhällsgeografisk Tidskrift, 1994), pp. 1–13.

Index

T - #0495 - 101024 - C0 - 219/149/17 - PB - 9781138722460 - Gloss Lamination